# FUNDAMENTALS OF CONSTRUCTION ESTIMATING

# FUNDAMENTALS OF CONSTRUCTION ESTIMATING

## Second Edition

## David J. Pratt

**THOMSON**

**DELMAR LEARNING**

Australia  Canada  Mexico  Singapore  Spain  United Kingdom  United States

**THOMSON**

**DELMAR LEARNING**

Fundamentals of Construction Estimating, Second Edition
David J. Pratt

**Vice President, Technology and Trades SBU:**
Alar Elken

**Editorial Director:**
Sandy Clark

**Acquisitions Editor:**
Allison Weintraub

**Development Editor:**
Jennifer A. Thompson

**Marketing Director:**
Cyndi Eichelman

**Channel Manager:**
Fair Huntoon

**Marketing Coordinator:**
Brian McGrath

**Production Director:**
Mary Ellen Black

**Production Manager:**
Larry Main

**Production Editor:**
Ruth Fisher

**Editorial Assistant:**
Jennifer Luck

**Cover Design:**
Julie Moscheo

**Background Photo:**
©Don Bishop/Gettyimages

**Inset Photo:**
©Corbis Images/PictureQuest

Library of Congress Cataloging-in-Publication Data:

ISBN-13: 978-1-4018-0959-1
ISBN-10: 1-4018-0959-6

# CONTENTS

**Preface    xi**

**Chapter 1    INTRODUCTION    1**

Objectives    1
The Role of Estimating in the Construction Industry    2
Conceptual Estimates    3
Preliminary Estimates and Cost Planning    4
Project Delivery Systems and Estimating    5
Estimates for Different Types of Contracts    8
Methods of Estimating    9
Estimating and Construction Safety    14
Summary    15
Review Questions    16

**Chapter 2    THE ESTIMATING PROCESS AND PRELIMINARY PROCEDURES    17**

Objectives    17
The Estimating Process    17
Projects Out for Bid    18
Open Bidding    20
The Decision to Bid    23
Scheduling the Estimating Process    24
Bid Record and Bid Documents    25
Obtaining Bid Documents    25
Review Bid Documents    27
The Query List    27
The Team Approach    28
Site Visit    29
Computer Estimating Systems    29
Other Technology    32
Summary    32
Review Questions    33

**Chapter 3    MEASURING QUANTITIES GENERALLY    35**

Objectives    35
The Quantity Takeoff    35

What Is Measured?   36
Units of Measurements   36
Measuring "Net in Place"   37
Takeoff Rules   39
Accuracy of Measurement   41
Organization of the Takeoff   42
Estimating Stationery   43
Takeoff by Computer   44
Formulas and Perimeter Centerline Calculations   45
Summary   53
Review Questions   54


**Chapter 4    MEASURING SITEWORK, EXCAVATION AND PILING    55**

Objectives   55
Generally   55
Soils Report   56
Bank Measure, Swell and Compaction Factors   57
Excavation Safety Considerations   57
Use of Digitizers   58
Measuring Notes—Excavation and Backfill   59
Calculation of Cut and Fill Using the "Grid Method"   59
Calculation of Cut and Fill Using the Section Method   62
Measuring Notes—Piling   62
Drawings   66
Sitework Takeoff   80
Summary   95
Review Questions   96


**Chapter 5    MEASURING CONCRETE WORK    99**

Objectives   99
Concrete Work Generally   99
Measuring Notes—Concrete   100
Measuring Formwork Generally   102
Measuring Notes—Formwork   102
Measuring Notes—Finishes and Miscellaneous Work   104
Example 1—House   105
Example 2—Office/Warehouse Building   108
Summary   120
Review Questions   122


**Chapter 6    MEASURING MASONRY WORK    125**

Objectives   125
Masonry Work Generally   125
Measuring Masonry Work   126
Brick Masonry   127
Concrete Blocks   127
Conversion Factors   129
Measuring Notes—Masonry   130

Masonry Work—House (Brick Facings Alternative)   131
Masonry Work—Office/Warehouse Building   134
Summary   139
Review Questions   139


**Chapter 7**    **MEASURING CARPENTRY AND MISCELLANEOUS ITEMS    141**

Objectives   141
Measuring Rough Carpentry   141
Board Measure   142
Metric Units   142
Measuring Notes—Rough Carpentry   143
Measuring Finish Carpentry and Millwork   145
Measuring Notes—Finish Carpentry   145
Doors and Frames   146
Windows   146
Miscellaneous Metals   147
Specialties   148
Finish Hardware   148
Measuring Exterior and Interior Finishes   149
Carpentry and Miscellaneous Work Takeoff—House Example   149
Exterior and Interior Finishes Takeoff—House Example   166
Summary   173
Review Questions   174


**Chapter 8**    **PRICING GENERALLY    175**

Objectives   175
Introduction   175
Contractor's Risk   176
Pricing Labor and Equipment   177
Pricing Materials   184
Pricing Subcontractors' Work   186
Summary   187
Review Questions   188


**Chapter 9**    **PRICING CONSTRUCTION EQUIPMENT    191**

Objectives   191
Introduction   191
Renting versus Purchasing Equipment   192
Depreciation   193
Maintenance and Repair Costs   196
Financing Expenses   197
Taxes, Insurance and Storage Costs   197
Fuel and Lubrication Costs   198
Equipment Operator Costs   199
Company Overhead Costs   199
Use of Spreadsheets   203
Summary   203
Review Questions   207

**Chapter 10    PRICING EXCAVATION AND BACKFILL    209**

Objectives    209
Excavation Equipment and Methods    209
Excavation Productivity    212
Excavation Work Crews    216
Excavation Materials    220
Excavation and Backfill Recap and Pricing Notes Example 1—House    223
Excavation and Backfill Pricing Notes Example 2—
   Office/Warehouse Building    227
Summary    232
Review Questions    233

**Chapter 11    PRICING CONCRETE WORK    235**

Objectives    235
Cast-in-Place Concrete Work Generally    235
Supply and Placing Concrete    236
Concrete Materials    241
Formwork    245
Reinforcing Steel    253
Miscellaneous Concrete Work Items    254
Wage Rates    256
Concrete Work Recap and Pricing Notes Example 1—House    257
Concrete Work Pricing Notes Example 2—Office/Warehouse Building    265
Summary    269
Review Questions    271

**Chapter 12    PRICING MASONRY, CARPENTRY AND FINISHES WORK    273**

Objectives    273
Introduction    273
Masonry    274
Rough Carpentry    276
Finish Carpentry and Millwork    281
Exterior and Interior Finishes    281
Wage Rates    281
Masonry, Rough Carpentry and Finish Carpentry Recap and Pricing Notes
   Example 1—House    284
Masonry Work Pricing Notes Example 2—Office/Warehouse Building    298
Summary    300
Review Questions    301

**Chapter 13    PRICING SUBCONTRACTORS' WORK    303**

Objectives    303
Introduction    303
List of Subtrades    305
Unknown Subcontractors    306
Evaluating Subcontractors    308
Bonding of Subtrades    309

Pre-Bid Subtrade Proposals    310
Analyzing Subtrade Bids    311
Bid Depositories    313
Scope of Work    313
Masonry Scope of Work    314
Windows Scope of Work    315
Glass and Glazing Scope of Work    317
Summary    318
Review Questions    318

**Chapter 14**    **PRICING GENERAL EXPENSES    321**

Objectives    321
Introduction    321
Project Schedule    323
Site Personnel    325
Safety and First Aid    326
Travel and Accommodation    327
Temporary Site Offices    328
Temporary Site Services    330
Hoardings and Temporary Enclosures    331
Temporary Heating    331
Site Access and Storage Space    333
Site Security    334
Site Equipment    334
Trucking    337
Dewatering    337
Site Cleanup    337
Miscellaneous Expenses    337
Labor Add-Ons    339
Bid Total Add-Ons    341
General Expenses Pricing Notes Example 1—House    342
General Expenses Pricing Notes Example 2—Office/Warehouse Building    344
Summary    348
Review Questions    348

**Chapter 15**    **CLOSING THE BID    349**

Objectives    349
Introduction    349
In Advance of the Bid-Closing    350
The Pre-Bid Review    358
The Closing    361
Unit-Price Bids    365
After the Bid    371
The Estimate–Cost Control Cycle    373
Summary Example 1—House    373
Summary Example 2—Office/Warehouse Building    375
Summary    389
Review Questions    390

**Appendix A**     **EXTRACT FROM A TYPICAL SOILS REPORT**    **391**

**Appendix B**     **LIST OF ITEMS AND CSI-BASED CODES**    **395**

**Appendix C**     **GLOSSARY**    **399**

**Appendix D**     **THE METRIC SYSTEM AND CONVERSIONS**    **403**

**INDEX**    **407**

# PREFACE

## Intended Use and Level

*Fundamentals of Construction Estimating* is written to provide the reader with the resources necessary to learn how to estimate the construction costs of building projects making use of modern technology and following the methods employed by successful general contractors. The book is primarily intended for the person who is beginning to learn the process of construction cost estimating whether they are a student taking a course in estimating at college, or someone who has recently assumed estimating responsibilities in a construction organization. The text will also be of interest to construction managers, supervisors and practicing estimators who, from time to time, may wish to refer to a source of estimating data or simply investigate how other estimators approach this subject.

## Approach

The goal of the text is to describe an easy to follow step-by-step method of estimating and preparing construction bids that can be used with the latest technology to produce an accurate construction cost estimate in the minimum of time; a cost estimate that is easy to review and one that is recognized by experienced estimators as the product of a professional. The procedures demonstrated are intended to reflect the basic core of the estimating discipline that is found in many different types of construction estimates. While the techniques developed here can be applied, with little modification, to the wide variety of construction work, the worked examples offered relate to a housing project and a commercial project selected for their relative simplicity so that the reader can concentrate on the estimating technique rather than having to spend time unravelling detail.

The estimating process presented is not intended to be some radical new method of estimating, instead, it is estimating as it is currently pursued by professional estimators doing essentially what estimators have always done but with far more speed, accuracy and thoroughness using all the modern innovations that advance the efficiency and effectiveness of the process.

Note that all prices used in the text are for illustrative purposes only, actual prices of construction work vary considerably from place to place and from time to time and should be carefully considered before using in actual estimates.

## Text Layout

Chapters 1 and 2 introduce readers to the estimating process and Chapter 3 describes the general principles of measuring work and preparing quantity takeoffs, which is one of the main components of the estimating process. In Chapter 4 the quantity takeoff of excavation work is discussed in detail and the two examples of a full estimate begin in this chapter, then continue throughout the rest of the book. One example shows a cost estimate of a house and the second is an estimate of an office-warehouse building. These examples of cost estimates are based on drawings that are included with the text. The drawings of a house are referred to as Figure 4.4 and the drawings of the office-warehouse building are Figure 4.5.

Chapters 5 through 7 consider the take-off process for concrete work, masonry work and a number of other miscellaneous trades that a general contractor may need to measure. Chapters 8 through 12 concern the pricing of general contractor's work including the pricing of equipment involved in the work.

In Chapter 13 the pricing of subcontracted work is examined and Chapter 14 concerns the pricing of general expenses or site overheads for a construction project. In the final chapter, Chapter 15, the process of closing a bid is explored in some detail and bid summaries for the two example estimates are discussed.

Appendix A furnishes a sample of a soils report, which is discussed in a number of chapters.

Appendix B is new for second edition and provides a list of common takeoff items with CSI based code numbers suitable for use in an estimating database.

Appendix C, a glossary of terms, is also new for this edition and consists of a collection of construction and estimating terms found in the text with explanations and definitions of these terms.

Appendix D describes the metric units used in estimating and provides unit conversion factors for English to metric and metric to English conversions.

## New to This Editiion

This second edition contains a number of features to assist students of estimating in their studies. Each chapter now begins with a list of learning objectives identifying the learning outcomes that will be achieved by studying the chapter material. Readers will also find a summary at the end of each chapter outlining the main concepts addressed in the chapter. Furthermore, a glossary of common terms used in estimating is included in Appendix C.

Many examples of computer estimating have been added to the text for the second edition. In this edition, there is a CD containing estimating software designed by Timberline Software Corporation. The CD is titled "Precision Collection Basic Edition." The text explains the use of this software for estimating and includes a demonstration of how to compile the complete estimate of an office/warehouse project using the Precision Estimating program.

The second edition introduces the use of the SI (system international) metric system of measurement in estimating. There are a number of newly worked examples and end-of-chapter questions that involve SI units in takeoff and pricing operations. In addition, Appendix D contains a description of SI units used in estimating processes with a number of conversion factors for converting English units to metric units and vice versa.

## Supplement

*Instructor's Guide* provides solutions to all end-of-chapter review questions. ISBN: 140180960X

## About the Author

**David J. Pratt** is a Professor of Civil Engineering Technology at Southern Alberta Institute of Technology. Professor Pratt has had a long career as a construction consultant and before that as senior estimator with the Commonwealth Construction Division of Guy F. Atkinson Inc of San Francisco. He is a member of the Canadian Institute of Quantity Surveyors and holds a degree in Quantity Surveying from Liverpool College of Building and a degree in Economics from the University of Calgary.

## Acknowledgements

I would like to thank the large number of people who contributed to this text. Particular thanks go to the following:

- Jennifer Thompson of Delmar for her help with this second edition
- Bill White and Ron Leach for providing drawings
- Gail Farrar and the copy editors at Publishers' Design and Production Services, Inc.
- The group of professional estimators at Stuart Olson Construction, PCL Constructors Inc. and Ellis-Don Construction Ltd. who continue to provide an insight into the estimator's task.

I would also like to extend my gratitude to the following organizations:

- The American Institute of Architects
- Alberta Construction Tendering System Ltd. (ACTS)
- The Canadian Institute of Quantity Surveyors
- Construction Specifications Institute
- Timberline Software Corporation

In addition, many thanks go to the reviewers, who provided guidance throughout the development of the text:

William Maloney
University of Kentucky
Lexington, KY

Daniel Farhey
University of Dayton
Dayton, OH

Dr. Alan Atalah
Bowling Green State University
Bowling Green, OH

**How to install this software**

You must install the software using the Setup command. Please note: These instructions are based on the use of Microsoft® Windows XP Professional. Your installation may vary slightly depending on the system you are using.

1. If you use Microsoft Windows NT Workstation 4.0, Windows 2000 Professional, or Windows XP Professional, log on as the administrator (not as an operator with administration rights).
2. If you use anti-virus software, turn it off for the duration of the installation. Shut down other unnecessary programs or services for the duration of the installation.
3. Insert the CD-ROM into your CD drive.
4. From the Windows Start menu, select Run. In the Open box, type d:\Setup.exe (if d: is the letter assigned to your CD drive.)
5. In the Welcome window, click Next.
6. After you read the end user license agreement, click Accept.
7. To install the Pervasive Work Group Engine, click Yes. After installation, click Continue to restart your computer. Log in again, if necessary. Estimating Basic installation will resume.
8. Click Next to accept the default destination folder, or click Browse to select a different folder.
9. In the Select Components window, select Estimating--Basic and Sample Database--Standard. Click Next. In the Select Program Folder window, accept the default choices and click Next.
10. In the Start Copying Files window, review the list of components and destination folders and click Next to begin the installation.
11. Click yes if you would like to view the release notice now. Click No to continue.
12. Restart your anti-virus software and any other programs that you closed in step 1.

For technical and installation support, call Thomson/Delmar Learning, 1-800-477-3692, Monday through Friday, 8:30 a.m. to 5:30 p.m. EST. You may also e-mail a request to help@delmar.com.

# 1

# INTRODUCTION

## OBJECTIVES

*After reading this chapter and completing the review questions, you should be able to do the following:*

- List and briefly describe the different functions served by estimating in the construction industry.
- Outline the role of the estimator in each stage of the construction process.
- Identify conceptual estimates and explain how they are used.
- Identify preliminary estimates and explain how they are used.
- Identify detailed estimates and explain how they are used.
- Briefly describe the cost planning process.
- Describe the three main project delivery systems—traditional design-bid-build, construction management and design-build—and explain how conceptual, preliminary and detailed estimates are used with each.
- Identify what type of estimating is required for lump-sum, unit-price and cost-plus contracts.
- Describe preliminary estimating methods including price per unit, price per unit area, price per unit volume approximate quantities and use these techniques to prepare estimates.
- Outline the detailed estimating procedure.
- Recognize the need for knowledge of Occupational Safety and Health Administration safety standards and the impact that compliance with the standards has on construction costs.

The goal of this book is to present a method of compiling consistently accurate construction cost estimates in a minimum of time. The method can easily be integrated with the latest technology available to obtain soaring productivity; it is a method of estimating that offers extensive review and control capabilities because it is consistent with the basic procedures followed by professional estimators and quantity surveyors in the construction industry.

The method presented is intended to represent a standard or basic core that can be adopted in the many types of construction estimating used across the wide variety of construction work. Worked examples and explanations that are offered, however, will come from small building projects of minimal complexity so that the reader can concentrate on the technique involved rather than spend time unraveling detail.

The book is intended primarily for the person who is beginning to learn the process of construction cost estimating. This person may be employed in a contractor's office taking on estimating responsibilities for the first time, or he or she may be a student starting a course in estimating at college. The text will also be of interest to many supervisors, construction managers and practicing estimators who, from time to time, may need to refer to an estimating standard or simply investigate how other estimators approach this subject.

## The Role of Estimating in the Construction Industry

Estimates serve a number of different functions in the construction industry (see Figure 1.1). In the early stages of a construction program, the owner needs an estimate of the probable cost of construction to assess the financial feasibility of the project. This **conceptual estimate** has to be prepared from a minimum amount of information because it is required at a time when the project is often little more than

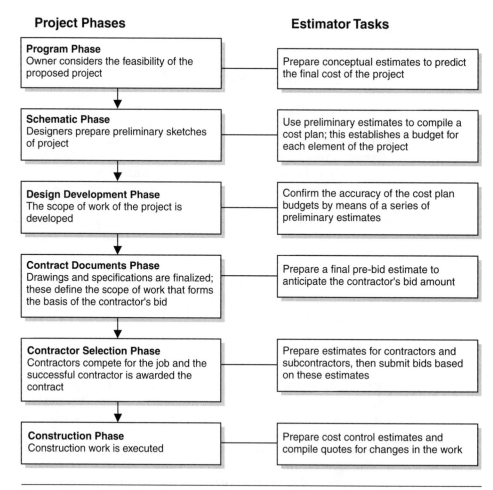

**Project Phases**

**Estimator Tasks**

**Program Phase**
Owner considers the feasibility of the proposed project

Prepare conceptual estimates to predict the final cost of the project

**Schematic Phase**
Designers prepare preliminary sketches of project

Use preliminary estimates to compile a cost plan; this establishes a budget for each element of the project

**Design Development Phase**
The scope of work of the project is developed

Confirm the accuracy of the cost plan budgets by means of a series of preliminary estimates

**Contract Documents Phase**
Drawings and specifications are finalized; these define the scope of work that forms the basis of the contractor's bid

Prepare a final pre-bid estimate to anticipate the contractor's bid amount

**Contractor Selection Phase**
Contractors compete for the job and the successful contractor is awarded the contract

Prepare estimates for contractors and subcontractors, then submit bids based on these estimates

**Construction Phase**
Construction work is executed

Prepare cost control estimates and compile quotes for changes in the work

**Figure 1.1**    Role of the Estimator in the Construction Process

a vague idea in the mind of the owner. There will be few if any design details at this stage because the design process will not begin until the owner is satisfied that the cost of proceeding with it is justified.

Once the design of the project is underway, budget amounts can be established for the various elements of the project using procedures for a **preliminary estimate**. These cost budgets are compiled in a **cost plan** that is a summary of all anticipated project expenditures. The budget amounts contained in the cost plan are verified from time to time during the design phase using more accurate estimating methods based on the specific design details that emerge in this phase of the project. This cost management process also includes estimating the cost of alternative designs so that informed decisions can be made on what to include in the design. When the design is completed, a final pre-bid estimate can be compiled to anticipate the contractor's bid price for the work. If this estimate is accurate, the bid prices obtained will be within the owner's budget for the project.

Most contracts that transpire in the construction industry result from competing bids from contractors to supply goods and services to meet certain specifications for a stipulated sum of money. The sum of money specified in such a bid represents the total amount that the contractor will receive for performing the work described in the contract; clearly, an accurate forcast of the cost of the work is necessary if the contractor is to profit from his or her endeavors and also be competitive. Providing this cost forecast is the prime function of the contractor's bid estimate prepared from the drawings and specifications supplied by the owner to define the scope of the contract work.

Estimates are also required after work starts on the project. In cost control programs, estimating is required to facilitate the control of expenditure of funds on a project. Contractors set cost targets based on their estimates of the cost of each component of the work, and then they compare the actual cost of work against these target amounts to discover where corrective action is needed to bring productivity up to required levels. Often during construction operations the owner or the designer asks the contractor to quote prices for proposed changes in the scope of work. Each of these quotes amounts to a *mini-bid* that involves an estimate of the full cost of the change followed by an offer to make the change for the price quoted to the owner.

In this Chapter we briefly discuss conceptual and preliminary estimates, but it is the *bid estimate*—the estimate of a general contractor, a subcontractor or supplier used to determine the total payment they are willing to accept in exchange for the performance of work and supply of materials or equipment to meet the requirements of their contract—that is the focus of this text.

## Conceptual Estimates

Even though there may still be owners who proceed with a project on the basis of no more than the feeling that it will succeed, most of the people and organizations that decide to build come to this decision after careful analysis of two primary factors: the value of the development and the cost of the development.

The value of a proposed facility can be appraised from the profits that are expected to flow after its construction, or, if the concept of profits is not applicable to the venture, it can be based on an assessment of the benefits that are expected to materialize from the completion of the project. In either case an attempt will be made to quantify the utility of the proposed development in terms of a monetary value. As costs and benefits are usually extended over a number of years, monetary value will normally be determined by means of "present worth" analysis or other "time value of money" concepts.

Using this analysis, a feasible project can be defined as one in which the anticipated value of all benefits exceeds the estimated total cost of putting the project in place. The cost profile of any project embraces many constituents including the cost of the land required, the cost of financing the project, the legal and general administrative costs, the cost of designing and administrating the work and, of course, the construction cost of the work. Further costs may also need to be considered such as commissioning costs, operating costs and, possibly, marketing costs. All of these amounts must be determined by estimate.

Some costs are relatively easy to establish. The costs of land and financing, for instance, are not difficult to determine as current market prices and rates are normally accessible, but the amount of what is most often the major cost component, the construction cost, is far more difficult to ascertain with any certainty. We will see that the most accurate way to predetermine the cost of construction work is by means of a detailed estimate using the methods employed by contractors. However, a detailed estimate requires a defined scope of work, and as we have said, this is generally not available at this stage in the project program.

So a conceptual estimate is normally produced from merely the notion the owner has of what he or she would like to see constructed. If the owner's analysis has begun with the assessed value of the project, he or she should be in a position to say that the project is viable if it can be built for a certain price, where this price is the maximum amount the owner is willing to pay for obtaining the benefits that are anticipated from the project. Alternatively, analysis may lead an owner to conclude that a certain structure of definite size and scope is necessary to generate the specific benefits that are sought. Typically the owner's financial situation reveals that there is only a certain amount of funds available to spend on this structure. In this case the obvious question will be "Can I build the structure for this amount?"

In either case the owner needs an estimate of the cost of the work, which, at this stage, is referred to as the conceptual estimate. This estimate, because of the lack of design details, must be prepared using one of the approximate estimating techniques considered in the following, but this is not to say a crude approximation of costs will suffice. The feasibility decision, which may involve thousands, if not millions, of dollars, is of major importance to most owners, so the accuracy of the predicted costs used in the calculations is crucial if the decisions that follow are to be sound.

## Preliminary Estimates and Cost Planning

As we have previously suggested, a decision to proceed with the venture signifies that the perceived benefits justify the project cost. The largest component of this cost is often the construction cost established by the conceptual estimate. This estimated sum becomes the construction budget and, if the project is to remain feasible, it is clear that the actual construction cost must not exceed this budget. Cost planning and subsequent cost control are pursued with the objective of meeting the budget.

After the decision to continue the project has been made, the design team will form and begin to prepare a first schematic design of the work. This design consists of preliminary drawings and specifications that depict the general scope of the project including the shape, size and layout of the design but with little detail at this stage.

A number of preliminary estimates can be prepared as the design develops, informing the designers that their proposed design does or does not meet the project budget. The preliminary estimates can also assist the designers by providing cost information about alternative design details so that they can make more informed design decisions. By evaluating the benefits and costs of a proposed design improvement, the designers can use principles of **value analysis** to determine if the improvements are justified.

In order to facilitate a more detailed evaluation of the benefits and costs of a project and its constituent parts, preliminary estimates can be subdivided into prices for groupings of building components that are common to most buildings. These groupings are referred to as *assemblies* or *elements* and include substructure, superstructure, exterior cladding, interior partitions and doors, vertical movement, and so on. The set of prices for these elements is referred to as the cost plan.

In the process of value analysis, the estimated cost of each element in the cost plan is compared with the perceived value of that element to consider if the sum allocated to that component part of the building is justified by the value provided by the component. At the time of the conceptual estimate, the estimator will have made numerous assumptions about these elements based on discussions with the owner and perception of the owner's needs. For instance, the exterior cladding of the building may have been assumed to be concrete block masonry. During the design stage it might be suggested that the cladding be changed to brick masonry. Cost estimates of the alternatives and the relative benefits of the two systems will be evaluated to determine if the extra cost of the more expensive brick cladding is justified by the increased value of a brick masonry over a concrete block exterior. Then, if a decision to spend the extra amount on exterior cladding is made and the overall budget is still to be maintained, a saving in another element must be found to balance the additional cost of the cladding.

In this fashion the design and accompanying estimates proceed until the design is complete and we have an budget in place that reflects all the key design decisions. There are many possible reasons why this final budget may differ from that prepared at the conceptual stage, but if cost planning has been properly applied, each step in reaching this point will have been made in awareness of its cost implications, and when contractor's bids are received, there will be no surprises for the owner.

## Project Delivery Systems and Estimating

Different project delivery systems make use of different types of estimates—conceptual, preliminary and detailed—at various stages in the development of the project depending on how well the scope of work is defined at that particular stage. While the traditional design-bid-build delivery system is, perhaps, still the most common way to organize the delivery of construction work, there has been a growing number of innovative delivery systems adopted, which, in some cases, has called for some innovative estimating methods.

### Traditional (Design-Bid-Build) Delivery

The traditional delivery system (see Figure 1.2) requires the project to be fully designed before work begins, and then makes use of a single **general contractor** who assumes responsibility for constructing the entire project on the terms of a lump-sum contract with the owner. With this system, a series of estimates is used as outlined in Figure 1.1. The role of conceptual estimates and preliminary estimates that accompany the development of the design is to manage the cost of the project so that the bid price obtained when the design is complete is within the budget of the owner. These early estimates generally do not relate directly to any construction contracts but they may be used as a basis of the amount of fee charged by consultants involved in the design process. For instance, the conceptual estimate of a proposed development may

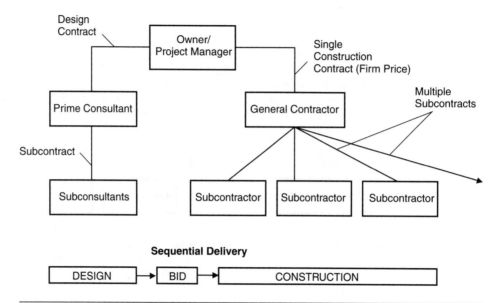

**Figure 1.2**   Traditional (Design-Bid-Build) Delivery System

indicate that the construction cost will be $5 million. The prime consultant could use this sum to calculate the fee of $300,000 quoted to the owner as the price to design the project. The design contract between the owner and the prime consultant could also refer to the $5 million estimate as the owner's budget and the terms of the contract may require the consultant to design a project that meets this budget. From this you can appreciate the need for accuracy in the conceptual estimate that is developed from so little information.

### Construction Management Delivery

One alternative to the traditional system of project delivery is the use of a construction management organization. This form of organization (see Figure 1.3) seeks to facilitate overlap between project stages, often referred to as **fast tracking**, to allow earlier completion of the project. Fast tracking is achieved by dividing the project into a number of phases. This approach allows construction work to begin after the design and bid for phase one only are complete. Phase one may consist of nothing more than demolition and site preparation that require very little time for design. While the work on this first phase proceeds, the design continues for phase two, which may be the building foundation system. Bids for this phase are then obtained so that a second contractor can be hired to start work on phase two as the first phase nears completion. The project continues in this fashion, phase after phase, until the entire project is in place.

With this organization structure, the **construction manager** assumes the responsibility for overall control and coordination of the construction work that was provided by the general contractor under the traditional organization. One of the main duties of the construction manager is cost control. This function is accomplished using conceptual and preliminary estimates to produce a cost plan as previously described. The cost plan, however, now has to be broken down into a number of subplans corresponding to the number of phases involved.

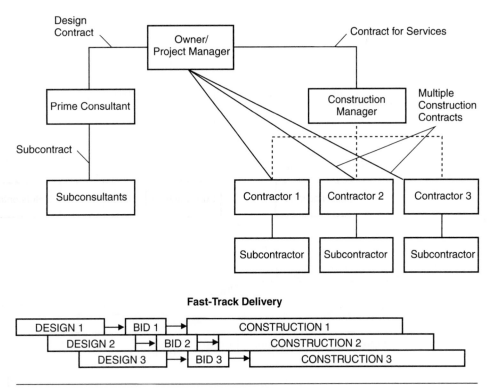

**Figure 1.3**   Construction Management Delivery System

With the traditional organization, the general contractor's bid will indicate, before work begins, the total money required for construction, but the owner will have to wait until the bid on the final phase of work is received before the total price is known when using the construction management approach. If cost planning has not been effective and the contractor's bid is unacceptable, there is an opportunity to make changes or even cancel the project before incurring construction expenses with the traditional organization. These opportunities are not possible with a fast-tracked project as construction work will be well under way and much of it paid for by the time the bid for the final phase is received.

Because, on a fast-tracked project, the owner has to commit to the project long before the total price is certain, the owner has to make serious decisions about the continuing viability of the project based solely on the information provided by preliminary estimates. The need for accurate estimates on this type of project is paramount—estimates need to be continually reviewed and updated as phases are completed and the design proceeds on subsequent phases.

Consequently, to achieve any success with a fast-tracked project, the estimating abilities of the construction manager and his or her team have to be of the highest standard; if they are not, cost and schedule objectives will not be met.

### Design-Build Delivery

Another alternative to traditional project delivery is the **design-build** concept (see Figure 1.4). In this approach, the owner deals with a single organization that assumes the responsibility for both the design and the construction of the project.

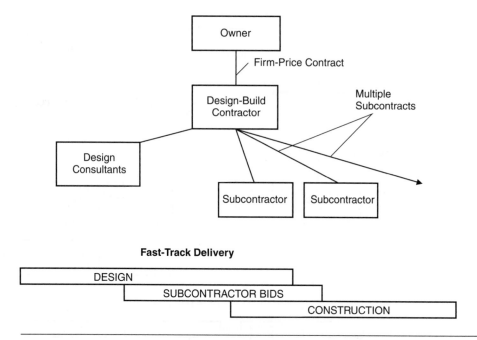

**Figure 1.4**    Design-Build Delivery System

Design-build is not a new approach to construction; it has been around in turnkey and package deal forms for many years, but recently there has been such a resurgence of interest in this format that design-build has become the delivery system of choice for many types of projects.

With the design-build method, the owner indicates that he or she requires a certain facility and usually outlines some parameters relating to location, size, design and the desired time of occupancy. The owner may have prepared some preliminary design sketches but, more often, design-build contractors competing for the work will each submit outlines of designs that they believe will satisfy the owner's needs together with a price for the project. On the face of it, this concept may resemble a general contractor's firm-price bid but, in fact, it is very different from the contractor's bid associated with traditional project delivery.

Unlike a general contractor's bid that is obtained by means of a detailed estimate of a precisely defined scope of work, the price included in the design-build contractor's proposal will have to come from a conceptual estimate. This estimate will be based on a design that, at this time, will probably consist of little more than a few rough sketches, some outline specifications and, possibly, an architect's three-dimensional model of the concept. Here again the need for care and accuracy in this conceptual estimate is quite obvious when you consider that the design-build contractor is assuming the responsibility for both the design and the construction of a project for the price quoted.

## Estimates for Different Types of Contracts

As mentioned, most general contractors and subtrades enter into **lump-sum contracts** on the terms that they will perform a defined scope of work for the sum that they stipulated when they submitted their bids. But there are a number of different types of contracts, other than lump-sum, that call for different types of estimates.

### Cost-Plus Contracts

**Cost-plus contracts** are usually the owner's choice of last resort because they can place the owner in a vulnerable financial position. Not only is the total cost of construction uncertain when the work begins but also, under this type of contract, the owner may not be able to avoid paying for the contractor's mistakes and inefficiencies

There are, however, a number of reasons why an owner may make this selection. A cost-plus contract may be used if it is imperative that the work get underway quickly, or if the work is very difficult to define, such as extensive renovations to old buildings. In both situations the fact that work can begin without the need for detail drawings can make a cost-plus contract a viable proposition. Also, if an owner has worked with a contractor in the past with good results, the owner may wish to avoid the time and expense of the bid process by awarding a cost-plus contract.

As cost-plus contracts are followed by rather than preceded by the costing of the operations, there is little need for estimating with these types of contracts. However, under the terms of some cost-plus contracts, the contractor offers to perform the work for a guaranteed maximum price. In this case careful estimating is required to determine the maximum price offered by the contractor, but much of the estimate will be based on "allowances" as the lack of specific design details will prevent the establishment of firm prices.

### Unit-Price Contracts

**Unit-price**, or measurement and payment, **contracts** are based on a pricing schedule that consists of a breakdown of the work and estimated quantities for each of the items on the breakdown as shown in Figure 1.5. This payment schedule, together with outline drawings, is provided by the designers for the use of prospective contractors in preparing their bids.

The bidding contractor enters a unit-price against each of the items and the total bid is determined from the aggregate of these prices multiplied by the estimated quantity of each item (see Figure 1.6).

The contract subsequently awarded to the successful bidder provides that this contractor is paid for the actual quantity of each work item executed, multiplied by the bid unit-price for that item. This approach enables the owner to obtain competitive prices and exert some control over the total amount paid for a project in which the final quantity of work is difficult to predetermine. Examples of where this type of contract is used include civil engineering works, especially those that contain much underground construction in which the nature of the ground encountered is not accurately discernible before the work begins.

The contractor's estimating procedure for unit-price contracts is similar to the detailed method used for lump-sum contracts but the bid price must be dissected into unit prices each of which have to include an overhead and profit component.

See Chapter 15 for the spreadsheet estimating procedure for unit-price contracts.

## Methods of Estimating

Methods of estimating can be divided into two main categories: preliminary estimating methods and detailed estimating methods. The price per unit, price per unit area, price per unit volume and approximate quantities methods examined here are all preliminary methods and would be used mostly by designers in the cost planning processes just described. The detailed estimating method, because it offers so much greater accuracy, is the method of choice for most construction contractors; this

| | 66 AVENUE UNDERPASS — SCHEDULE OF PRICES | | | | |
|---|---|---|---|---|---|
| | **ITEM** | **QUANTITIY** | **UNIT** | **UNIT PRICE** | **AMOUNT $** |
| | | | | | |
| 02100 | SITE CLEARANCE | | | | |
| 02101 | On force account | P.C. | SUM | — | 5,000.00 |
| | | | | | |
| 02200 | EXCAVATION | | | | |
| 02201 | Common dry | 6100 | CY | | |
| | | | | | |
| 02250 | BACKFILL | | | | |
| 02251 | Common | 1500 | CY | | |
| 02252 | Pit-run | 4100 | CY | | |
| | | | | | |
| 03200 | REINFORCING STEEL | 195 | Tons | | |
| | | | | | |
| 03000 | CONCRETE | | | | |
| 03301 | CLASS "A" | 37 | CY | | |
| 03302 | CLASS "B" | 1030 | CY | | |
| 03303 | CLASS "D" | 17 | CY | | |
| 03304 | Slope protection | 250 | CY | | |
| 03305 | Wearing surface | 27 | CY | | |
| 03601 | Nonshrink grout | 1 | CY | | |
| | | | | | |
| 03400 | PRECAST STRINGERS | 26 | No. | | |
| | | | | | |
| 05010 | GUARDRAIL | 60 | LF | | |
| | | | | | |
| 05800 | EXPANSION JOINTS | 67 | LF | | |
| | | | | | |
| 99000 | MISC. FORCE ACCOUNT | P.C. | SUM | — | 7,500.00 |
| | | | SUBTOTAL | | |
| | | 10% Allowance for extra work | | | |
| | | | BID TOTAL | | |

**Figure 1.5**   Blank Schedule of Prices for a Unit-Price Contract

| 66 AVENUE UNDERPASS — SCHEDULE OF PRICES | | | | |
|---|---|---|---|---|
| **ITEM** | **QUANTITIY** | **UNIT** | **UNIT PRICE** | **AMOUNT $** |
| 02100 SITE CLEARANCE | | | | |
| 02101 On force account | P.C. | SUM | — | 5,000.00 |
| | | | | |
| 02200 EXCAVATION | | | | |
| 02201 Common dry | 6100 | CY | $ 15.00 | 91,500.00 |
| | | | | |
| 02250 BACKFILL | | | | |
| 02251 Common | 1500 | CY | $ 33.50 | 50,250.00 |
| 02252 Pit-run | 4100 | CY | $ 51.00 | 209,100.00 |
| | | | | |
| 03200 REINFORCING STEEL | 195 | Tons | $1,925.00 | 375,375.00 |
| | | | | |
| 03000 CONCRETE | | | | |
| 03301 CLASS "A" | 37 | CY | $ 296.50 | 10,970.50 |
| 03302 CLASS "B" | 1030 | CY | $ 347.00 | 357,410.00 |
| 03303 CLASS "D" | 17 | CY | $2,467.50 | 41,947.50 |
| 03304 Slope protection | 250 | CY | $ 675.00 | 168,750.00 |
| 03305 Wearing surface | 27 | CY | $ 408.00 | 11,016.00 |
| 03601 Nonshrink grout | 1 | CY | $7,680.00 | 7,680.00 |
| | | | | |
| 03400 PRECAST STRINGERS | 26 | No. | $9,948.00 | 258,648.00 |
| | | | | |
| 05010 GUARDRAIL | 60 | LF | $ 19.57 | 1,174.20 |
| | | | | |
| 05800 EXPANSION JOINTS | 67 | LF | $ 262.50 | 17,587.50 |
| | | | | |
| 99000 MISC. FORCE ACCOUNT | P.C. | SUM | — | 7,500.00 |
| SUBTOTAL | | | | $1,613,908.70 |
| 10% Allowance for extra work | | | | $ 161,390.87 |
| BID TOTAL | | | | $1,775,299.57 |

**Figure 1.6**   Price Schedule of Prices for a Unit-Price Contract

method can also be employed by the designer, when drawings and specifications are sufficiently advanced, to obtain a more dependable forecast of cost than that achieved by preliminary estimating.

### Price per Unit

With the price per unit method of estimating, a certain unit that is common to all projects of this type is selected and the number of these units is calculated for each project, for example, beds in a hospital. Using this method, the construction cost of a newly constructed hospital is divided by the number of beds provided in the new building. This calculation provides the unit price per bed, which can then be used to price future hospital projects by multiplying the number of beds in the proposed facility by this unit price per bed.

The units adopted to facilitate this analysis depend on the type of project under consideration. Cost per stall may be used with a parking garage, cost per suite in an apartment building and cost per student if the construction of a college was under consideration. In each case the costs of completed projects are analyzed to determine unit prices by dividing the total construction cost by the number of units in the project. The unit price obtained is then applied to future projects to arrive at their anticipated total prices.

An accurate forecast of cost will be obtained only if the future project is very similar to those previously analyzed with regard to such items as the price of resources, the construction design, the total project size, the quality of finish, the geographical location and the time of year the work was undertaken. Clearly, differences between the previous projects and the new one can be expected in most of these areas. Such difference seriously reduce the accuracy of price per unit estimates and making the proper price adjustments for these project differences is most difficult.

Using this estimating method can generate a rough estimate quickly, but the lack of accuracy will render it of little use in the cost planning procedure outlined earlier. However, this method is often used to determine the very first notion of a price in early discussions of a project and as a crude means of comparing the known costs of different buildings.

### Price per Unit Area

Here the unit of analysis is the square foot or square meter of gross floor area of the project. The *gross floor area* is defined as the area of all floors measured to the outside of containing walls. As with the previous method, the analysis begins with the construction cost of a completed project. The known project construction cost is divided by the gross floor area of the project to obtain a cost per unit area, which can then be applied to future projects to estimate their construction cost.

This is the most common method of preliminary estimating as it is easy to understand and a large amount of data on prices per square foot is published that may supplement data derived from designers' analysis of their own projects. Because one of the first details of design to be defined is always the floor area of a proposed building, this method of estimating can be applied very early in the development of a project. However, virtually all the same variable factors that make accuracy difficult to attain with the price per unit method of estimating also apply to this price per square foot analysis.

### Price per Unit Volume

One variable that can be a significant influence on the cost of some projects is the story height. The height of warehouses, for instance, can vary considerably thus making the use of price per unit floor area inappropriate. A better method employs the price per unit of volume as the means of determining the preliminary estimate. The cost per unit volume is obtained by dividing the total construction cost by the volume of the building. The volume of the building is calculated by multiplying the area of the building by the height from the bottom of the gravel under the basement slab-on-grade to the top of the roof surface. The area of the building is again measured to the outside of containing walls.

## Assembly Estimates

To achieve greater accuracy in preliminary estimating, a price forecasting method is required that accounts for not just the overall size of the project but also for the prices of specific resources used in construction, the design details, the quality of finish, the geographical location and the time of year when the work was undertaken. All these variables are considered in a detailed estimate, but at the time of the preliminary estimate, many factors are not yet defined, so time spent measuring and pricing each factor in detail would be wasted as many of the specific details will change before the design is finalized. The assembly method attempts to give the estimator the opportunity to adjust for the previously listed variables without examining the fine detail of the design. This objective is achieved by measuring and pricing assemblies of work that comprise a group of related trade activities.

Continuous concrete footings would be an example of a possible assembly. When this assembly is estimated in detail, the estimator would measure and price the amount of each of the following items:

- Concrete to continuous footings in cubic yards
- Formwork to the sides of footings in square feet
- Forming a keyway in linear feet
- Excavating a trench by machine in cubic yards
- Hand excavation in cubic yards
- Backfill in cubic yards

Estimated as an assembly, continuous footings would be taken off as a single quantity measured in linear feet, and against this quantity a single price would be applied that is sufficient to include the cost of all the items listed.

Using this approach, a preliminary estimate can be assembled with relatively few items of takeoff, but each of the assemblies taken off can be carefully considered so that its price reflects the specific details anticipated by the estimator. Thus a more accurate estimate is obtained in far less time than that required to measure and price a detailed estimate.

## Detailed Estimates

The process of detailed estimating is the subject matter of the major portion of this book, but before we begin to examine all the particular aspects of this topic, it may be useful to consider what is the essence of this subject. Whether prepared by hand, by computer spreadsheet or by means of a totally computerized system, a detailed estimate can be analyzed in terms of the six principal stages of the process:

Stage 1. Quantity Takeoff—The work to be performed by the contractor is measured in accordance with standard rules of measurement.

Stage 2. Recap Quantities—The quantities of work taken off are sorted and listed to comply with the **CSI MasterFormat** or other standard to facilitate the process of pricing.

Stage 3. Pricing the Recap—Prices for the required labor, equipment and materials are entered against the quantities to determine the estimated cost of the contractor's work.

Stage 4. Pricing Subcontractor's Work—Prices are obtained from competing subtrades who quote to perform the work of their trade then, usually, the lowest bid from each trade is entered into the estimate.

Stage 5. Pricing General Expenses—The costs of the anticipated project overheads are calculated and added to the estimate.

Stage 6. Summary and Bid—All the estimated prices are summarized, the contractor's mark-up is added and the tender documents are completed. The bid can then be submitted and, finally, the bid results recorded and analyzed.

The detailed estimating method, rather than any of the other estimating methods considered, is far more likely to produce a price that is an accurate forecast of the actual costs of building a construction project. Because the very survival of contractors and subtrades who obtain work by offering firm price bids often rests on the accuracy of their estimating, it will come as no surprise that detailed estimating is the method of choice for bid preparation in the highly competitive construction industry.

The basis of a detailed estimate is the accurate assessment of the work in the form of a quantity takeoff that can only be obtained from the full design of the project. Because of this requirement and also because detailed estimating is such a time consuming process, preliminary estimates are usually prepared by the other quicker, but less accurate methods previously considered. However, on lump-sum contracts in which cost analysts wish to accurately anticipate the value of bids in an attempt to avoid surprises to the owner, they will prepare a detailed estimate in the same fashion as bidders, using the design drawings before they are made available to bidding contractors.

## Estimating and Construction Safety

The Occupational Safety and Health Administration safety standards (OSHA) are the law, and contractors who fail to comply with the requirements of this legislation face possible fines or even imprisonment. Most contractors take this law very seriously, so the estimator has to ensure that the estimate is based on methods and materials that comply with its provisions, and that the pricing of estimates reflects these requirements. However, safety in the construction industry is not just compliance with OSHA.

There is a growing awareness among construction companies of the cost of poor safety on the work site. Apart from the direct cost of construction accidents, which is reflected in higher worker's compensation insurance and property insurance premiums, there is a large number of what are often hidden costs including:

1. Tool and equipment repair and maintenance costs
2. Production interruptions and delay costs
3. Legal expenses
4. Expenditure on emergency supplies and equipment
5. Replacement equipment rentals
6. Investigative and administrative expenses
7. Cost of hiring and training replacement personnel
8. Overtime payments and other costs incurred trying to catch up
9. Decreased output of injured workers on return
10. Damage to company reputation and subsequent loss of business

Because these costs can be substantial and may have a significant impact on the competitiveness of an organization, many construction companies have been encouraged to introduce vigorous safety programs at their job sites. Such project safety programs do, admittedly, have a cost, which is accounted for in the general expenses section of an estimate. (See Chapter 14 for a discussion of general expenses.) However, with a successful company-wide safety program in place, estimators are able to use far more competitive prices in the estimates they prepare which not only improves the bid success rate, but helps to achieve the financial success of the project and, consequently, the improved profitability of the company.

## SUMMARY

- The intention of this book is to present a standard method of estimating that can be adopted for use with many different types of construction projects.
- Estimating serves a number of purposes in the construction process including preparation of bids and cost control.
- The role of the estimator includes preparing conceptual estimates, preliminary estimates and pre-bid and post-bid estimates over the course of all the phases of a construction project.
- Conceptual estimates are used to assess the feasibility of a project by comparing the anticipated cost with the value of a proposed project.
- Preliminary estimates are used throughout the design stage of a project, primarily to ensure that the project budget is not exceeded.
- A detailed estimate is the most accurate forecast of construction costs but can only be prepared given a defined scope of work in the form of detailed drawings and specifications.
- Project delilvery systems include traditional design-bid-build, construction management and design-build. They make use of conceptual, preliminary and detailed estimates at various stages during the course of a project.
- Although firm price lump-sum contracts are used on most construction projects, owners may adopt a cost-plus alternative when the scope of work is difficult to define or when there is insufficient time to prepare complete design documents before work begins.
- Detailed estimates may still be required with a cost-plus contract in which a guaranteed maximum price is involved.
- Unit-price contracts call for a breakdown of the owrk and a separate unit price for each item on the breakdown. This method requires the estimator to prepare a series of unit prices, each of which have to account for the labor, materials and equipment involved, and also to include overhead and profit components.
- Conceptual estimates and preliminary estimates are formulated using one or more of the following methods: price per unit, price per unit area, price per unit volume, assembly estimates.
- A detailed estimate comprises six elements:
  1. Takeoff quantities
  2. Recap quantities
  3. Recap pricing
  4. Subcontractors' work pricing
  5. General expenses pricing
  6. Estimate summary
- The estimator should be aware of Occupational Safety and Health Adminstration safety requirements and account for the cost of safety programs in construction estimates

## REVIEW QUESTIONS

1. What is a construction cost estimate?
2. List and describe the three main functions of estimates in the construction industry.
3. Which two factors are assessed to determine if a project is feasible?
4. What is a conceptual estimate?
5. Why are preliminary estimates prepared?
6. If the construction cost of a bungalow of size 34'6" (10.5 m) × 40'9" (12.4 m) was $120,790.00, what is the cost per square foot (square meter)?
7. If the average height of the building described in question 6 is 12'4" (3.8 m) to the roof surface, what is the cost per unit volume?
8. Based on your analysis in question 6, what would be the estimated price of a similar bungalow of this size: 30'0" (9.1 m) × 45'0" (13.7 m)?
9. List and describe the six elements of a detailed estimate.
10. What are the hidden costs of poor construction safety?
11. What effect does the introduction of construction company safety programs have on the process of estimating?

# 2

# THE ESTIMATING PROCESS AND PRELIMINARY PROCEDURES

## OBJECTIVES

*After reading this chapter and completing the review questions, you should be able to do the following:*

- Compile the goals and objectives of a contractor's estimating department.
- Describe a systematic approach to the estimating process.
- List the sources of information about projects out for bid.
- Identify and distinguish open bidding from closed bidding.
- Describe the process of prequalifying bidders and explain why it is done.
- Describe contractor's marketing strategies and explain why these strategies are pursued.
- List and explain the factors considered when a contractor is deciding whether to submit a bid on a project.
- Describe how contractors manage their estimating operations, including the process of bid record keeping.
- Explain what is involved in the preliminary review of bid documents and the use of a query list.
- Describe the team approach to estimating and the benefits this approach provides.
- Describe the purpose of the estimator's site visit and list the items that should be considered on a local project and on a remote project.
- Identify the role of computers in the estimating process and the benefits of also learning how to compile an estimate manually.
- Explain how modern communication technology, if properly managed, can enhance the estimating process.

## The Estimating Process

In this Chapter we will consider the activities and procedures that are undertaken in a construction company to facilitate the preparation of a cost estimate. Developing an estimate requires the estimator to interact with a large number of people and a mass of data that can be overwhelming if the process is not conducted in an orderly fashion. Lack of a systematic approach to estimating is not only inefficient, but it may also cause the estimator to lose control of the process, which, ultimately, can lead to a mountain of paper or computer material that is completely incomprehensible. Whether a single estimator or a whole team of estimators, the estimating department will be able to produce estimates more efficiently by following well-organized estimating procedures. Estimates will also be more accurate and the department resources will be employed more effectively.

To develop a well organized systematic approach to estimating, the objectives of the estimating function in the construction company should first be clearly understood. These objectives may include:

1. Preparing construction cost estimates that will enable the contractor to obtain profitable work by the competitive bidding process.
2. Maximizing the accuracy of the estimating process by including procedures for checking and verifying the precision of the work.
3. Maximizing the productivity of the estimating department in terms of generating the highest volume of estimating product with the resources available.
4. Using the estimating department in an effective manner that commits these resources to projects with the most likelihood of success.
5. Fostering a company-wide cooperative approach to estimating and bidding that, in recognition of the value of good estimating, commits all company personnel to improving the quality of estimating.

### Steps in the Estimating Process

The process of preparing a detailed estimate by a contractor is depicted in the form of a flow chart as shown in Figure 2.1. It will be noted that some of the steps involved must follow the completion of other activities (prerequisites), while other steps can be completed concurrently. The logic of this network must be carefully considered, especially with estimates of larger projects, when planning the schedule of estimating activities to be completed by the estimating team.

The key decision to bid or not to bid on the project is positioned on the flow chart immediately after bid information is obtained. However, the decision made at this stage is not final. A "no" decision may be reversed if the project suddenly looks more attractive and there is still time for estimating. A "yes" decision is often reviewed, perhaps several times during the course of the estimating process, as more information is gathered; nevertheless, if the decision reverts to "no," terminating early is clearly preferred to reduce the waste of estimating resources.

### Projects Out for Bid

A crucial factor in the success of an estimating department and, indeed, in the success of a construction company, is its ability to obtain information about projects that are out for bid or projects that are going to be out for bid in the near future. This intelligence gathering task is typically pursued by senior estimators and also by other company management personnel as part of the marketing effort of the construction company discussed in the following.

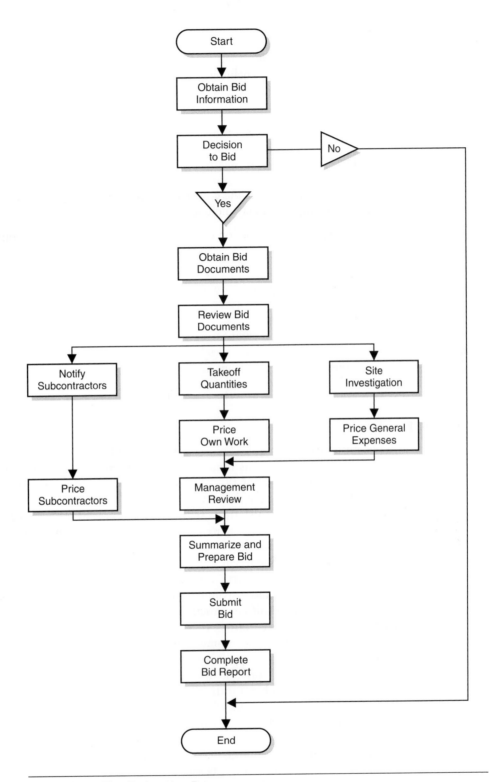

**Figure 2.1** The Estimating Process

### Sources of Bid Information

The primary sources of information regarding bid calls for public works and for projects in the private sector include:

*Public Bids*

- Advertisements in newspapers and trade journals
- Web pages of government agencies
- Government notices such as Pre-Solicitation Notice (Construction Contract)
- Construction associations and plan service centers
- Construction news services (McGraw-Hill, Southam, etc.)
- Bid information services that are now often Web-based, such as BIDS Inc. and Canada's MERX System

*Private Bids*

- Invitations from owners or consultants
- Business contacts
- Business news items
- Architectural and engineering consultants
- Construction news services (McGraw-Hill, Southam, etc.)
- Construction associations and plan service centers
- Bid information services

## Open Bidding

Public authorities are required to comply with regulations that control the award of construction contracts. This requirement usually means that they have to use an open bid system, that is, one that allows any and all qualified contractors to submit bids and be considered for the award of the work. Because open bidding involves publicly advertised bid calls, these are the major source of information about public projects that are out for bid. Figure 2.2 is a facsimile of the type of bid advertisement commonly posted for public projects.

Agencies of the U.S. government use a standard form called the "Pre-Solicitation Notice (Construction Contract)" to provide bid information. These notices are displayed in public places and distributed to contractors who have indicated an interest in bidding on government construction work in the area. Notices are also sent to local construction associations.

Both state and federal government agencies have now embraced the use of the Internet to publish bid notices so contractors are able to use the Web to find information about government work and upcoming bids. Private organizations have emerged on the Web that collect bid information from a wide range of sources and then offer this information to subscribers who are looking for projects of a particular type in a certain geographical area. One such organization is Business Information & Development Services Inc. (BIDS Inc), which provides leads on government projects out to bid and also helps small contractors market their services to government. In Canada, MERX connects suppliers to purchasers in the federal, provincial and MASH (municipal, academic, school and hospital) sectors providing contractors with information about projects and bid calls in these sectors.

**NYHC
1111 1st AVENUE
NEW YORK
NY 22222**

STANDARD SENIOR CITIZENS APARTMENT—40 UNITS
PROJECT NUMBER: HC789-1011

Sealed BIDS for the construction of the Standard Senior Citizens Apartment will be received until 2:00 p.m., January 19, 2006, in the main boardroom at the offices of NYHC.

This Federally funded project will consist of the construction of a two-story wood framed building. Bidding documents may be viewed at the above offices and purchased for $200.00 per set, payable to the State of New York.

There will be a MANDATORY pre-bid conference at the above offices on January 5th at 3:30 p.m.

Bids must be enclosed in sealed envelopes bearing the name and address of the bidder; designating the name of the project on the outside, addressed to NYHC. Bids must be accompanied by a Certified Check, Cashier's Check or Bid Bond, in the amount of ten (10%) percent of the bid amount furnished by a Surety Company authorized to do business in the State of New York.

Bids for this project must also be accompanied by:
Non-Collusion Affidavit and,
An Agreement of Surety (Consent of Surety) in which the surety company agrees to post a Performance Bond and a Labor and Materials Payment Bond if the contract is awarded to its principals.

Bidders are required to comply with the requirements of the State of New York relating to affirmative action requirements.

The successful bidder shall also be required to comply with the U.S. Federal Davis Bacon Wage Rates.

**NYHC reserves the right to reject any and all bids and waive immaterial informalities that will enable NYHC to more effectively procure performance of the work contemplated by the plans and specifications.**

**Figure 2.2**    Bid Advertisement

## Construction Associations

Construction groups in major cities have formed associations or plan service centers where both public and private bodies may deposit plans and specifications of projects calling for bids. At these centers, contractors, and particularly subcontractors, can view the documents, and facilities can be provided for estimators to work on their take-offs. Use of the facilities is usually restricted to members of the contractor's association. These associations often publish weekly bulletins to inform members about forthcoming bid calls and about bid documents that are available in their plan service center.

### Closed Bidding

Private owners, not being so restricted as public bodies, often directly invite contractors to bid on their construction projects. This is known as a closed bid system, and the number of companies invited to bid is typically limited to no more than six contractors who are known to the owner or designers and who meet the qualification criteria set by the owner. Thus, bids may be solicited only from contractors with whom the owner has previously worked, or contractors who have a proven ability and track record on the type of project to be built. Adopting a closed bid system also avoids the main problem of open bids: the large number of bids received, many of which are from contractors lacking the financial or technical capacity to successfully complete the project.

### Prequalified Bids

Another method of addressing the problem of bids from contractors of questionable capability is to call for bids from prequalified contractors only. To prequalify, contractors are required to complete questionnaires that probe such things as the contractor's financial situation, the experience of the company and its key personnel and the type and quality of company-owned construction equipment available for use on the project. From this information the owner and consultants try to judge the ability of the prospective bidder. The prequalification documents are usually required to be submitted by a date some time before the bid is to be received.

In some cases the owner requests a "qualification statement" to accompany all bids. Contractors complete these documents with reluctance when they include questions that call for disclosure of what the contractor considers to be confidential information. Some contractors are willing to risk having their bids rejected for being incomplete rather than complying with bid requirements for onerous "qualification statements."

### Construction News Services

Construction news services—McGraw-Hill's Dodge Reports and Southam Building Reports being prominent examples—offer bid reporting services that convey the status of construction projects from the feasibility stage to the completion of the project. Early knowledge of bids that are coming up can improve the effectiveness of planning in the estimating department. A contractor who is aware that a project is presently in the design stage can arrange the estimating schedule to ensure a bid is prepared on this project if it looks attractive.

### Marketing

Because private owners and their consultants do contact contractors directly for bids, some construction companies recognize a need for a marketing strategy to ensure that they are, at least, on bid lists. Many private projects are undertaken, if not under a veil of secrecy, then with very little publicity, so that the uninformed contractor first learns of the project when the site is fenced in and the equipment of one of his or her rivals moves onto the site. The key to avoiding this situation is the development of an information gathering system as part of the contractor's marketing effort. Aided by a network of business contacts and constant monitoring of the business activities of corporations that are most in need of construction services, some contractors aggressively pursue a marketing policy that goes beyond merely obtaining a place on bid lists.

Such a marketing policy is often pursued by contractors who offer construction management services. Many of these companies assert that their services will be of most benefit to the owner if they are incorporated into the construction program at an early stage. Accordingly, the marketing departments of these contractors are continually promoting the expertise of their companies and their services to prospective clients at the stage when projects are, ideally, no more than vague possibilities in the client's mind.

Even if construction management services are not offered by a contractor, a good marketing program can persuade an owner of the advantages of employing this contractor to the extent that the owner may forego the bid process and negotiate a deal directly with the contractor for the construction of the project. Some construction corporations recognize the advantages of this strategy to the extent that they now obtain most of their work by negotiation rather than expose themselves to the crushing competition for bid work.

## The Decision to Bid

When the economy is weak, the bid decision becomes easy for contractors who obtain their work by competitive bidding; they submit bids on every project that comes up in their area and even consider bidding on projects located outside of their usual area of operations. It is not difficult to find reasons for adopting these tactics because, in recessionary periods, there will be fewer projects to keep the contractor's equipment and crews fully employed, and, at the same time, there will be only a small number of projects out for bid.

When construction work is more abundant, some contractors still attempt to submit bids on all the projects that are out for bid. This strategy places a strain on the company's estimating department as it tries to expand the number of estimates produced per month while staff shortages in these boom times prevent an expansion of the department's human resources. This condition can lead to calamity as the estimating process becomes rushed and error prone.

Faced with limited resources in the estimating department, the decision to enter a bid on a project becomes more complex. Undoubtedly, estimating resources will be used more effectively if bids are submitted on selected projects only. The valuable time of the estimator will be wasted preparing a bid if there is little chance of obtaining a contract from that bid. By analyzing the company's bidding record, a contractor can identify the type and size of project that offers the highest probability of successfully securing a contract. Even when confronted with a project of the type and size preferred, there can be many reasons for the contractor to decide to avoid submitting a bid this time.

*Factors to Consider in the Bid Decision*

- Type of project—Most contractors develop a preference for certain types of work that they have been successful with and believe they can repeat their success.
- Size and (rough) estimate of contract value—Contractors often also have a preference for a specific size of project. This may be in terms of physical size (for example, volume of concrete involved), or it may be size in terms of the dollar value of the project.
- Location of the project—One contractor may choose to restrict operations to an area near his or her home city while another contractor may specialize in remote jobs involving camp accommodation.

- Quality of drawings and specifications—Many contractors consider poor quality drawings and specifications an indication that the work of the project is going to be subject to costly disruptions, as many design problems will have to be sorted out as the work proceeds. While many contractors view this as an unfavorable aspect of a project, some contractors view it as an opportunity to make a larger profit by charging high prices for the changes that will inevitably be made.

- Reputation of owners and designers—Suffice it to say that contractors find some owners and some designers easier to work with than others.

- Specialized work—A project may call for special expertise or specialized equipment that a contractor may find difficult to assemble if the contractor does not already have what is required.

- Anticipated construction problems—Many contractors will shy away from a project that they can see will expose them to difficult problems, but there are always some contractors who see an opportunity in situations in which other contractors are frightened away from a difficult project. These bidders may offer a high price, but they frequently encounter very little competition for the work.

- Safety considerations—If a contractor perceives that a project will involve a higher safety risk than is usual, the contractor may wish to avoid the job. This is especially true if the contractor has experienced safety problems and is concerned about the company's safety record.

- Need for the work—We have discussed the effect of economic conditions on the contractor's need for work, but at any time a contractor may have a supervisor or a crew that is going to be freed up, so there will be need for a new project.

- Bonding capacity—If bonds are required for the project, a major concern of the contractor is the question of whether or not the company's bonding capacity is sufficient to cover the bid price. Bonding companies place a limit on the value of the contractor's work they will bond. This limit is mainly a function of the financial position of the contractor. Bonding companies generally will not provide bonding when the value of the project added to the value of the contractor's work outstanding exceeds the limit they have set for this contractor.

These are some of the items that may be considered in the bid decision; concern about any one of them can cause the contractor to think twice before preparing a bid for a project. Note that, should the contractor decide to proceed with a bid on the project in question, all of the preceding factors will be reevaluated when the amount to be added to the estimate for profit is considered—see Chapter 15 for a discussion of the project fee.

## Scheduling the Estimating Process

Given limited time and other resources, and the goal of producing the required number of high quality estimates, the contractor may benefit from the use of some project management techniques in the organization of the estimating department. With clear goals and priorities in mind, the chief estimator sets about planning and scheduling the work of the department to attain the objectives of that department. An effective way to proceed with the planning of estimating operations is achieved by means of a simple bar chart form of time schedule. Figure 2.3 illustrates the use of a

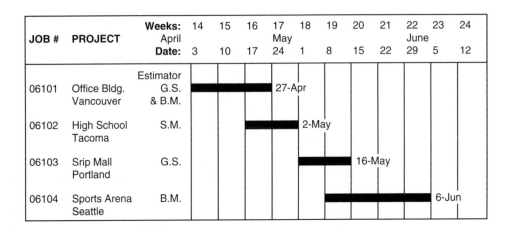

| JOB # | PROJECT | Weeks:<br>April<br>Date: | 14<br><br>3 | 15<br><br>10 | 16<br><br>17 | 17<br>May<br>24 | 18<br><br>1 | 19<br><br>8 | 20<br><br>15 | 21<br><br>22 | 22<br>June<br>29 | 23<br><br>5 | 24<br><br>12 |
|---|---|---|---|---|---|---|---|---|---|---|---|---|---|
| 06101 | Office Bldg.<br>Vancouver | Estimator<br>G.S.<br>& B.M. | ▬ | ▬ | ▬ | 27-Apr | | | | | | | |
| 06102 | High School<br>Tacoma | S.M. | | | ▬ | ▬ 2-May | | | | | | | |
| 06103 | Srip Mall<br>Portland | G.S. | | | | | | ▬ | ▬ 16-May | | | | |
| 06104 | Sports Arena<br>Seattle | B.M. | | | | | | ▬ | ▬ | ▬ | ▬ 6-Jun | | |

**Figure 2.3**   Project Estimator's Schedule

bar chart produced on a computer spreadsheet for scheduling the work of an estimating department. Normally one estimator is assigned to each project as it comes out for bid. Time for estimating varies with the amount and complexity of the work but, on average, an estimator spends two to three weeks on an estimate. Larger projects may call for the combined effort of a team of estimators, and sometimes with small projects an estimator may work on two or more estimates at the same time.

## Bid Record and Bid Documents

Once the decision to bid on a project has been made, a record of the bid begins. Some contractors in the past have maintained a bound Bid Record document, but now it is more common to record this information on a computer file with all the requirements preset so that the estimator merely types in each item of data in the correct places. Figure 2.4 represents the kind of preset items and responses that might be found on a computerized bid report. The collection of these bid report files that the contractor builds up over time forms a database that can be analyzed in a number of ways to obtain historic bid information. For instance, bid data can be deciphered to determine the type and size of project the contractor is most successful with in bids, as discussed previously.

## Obtaining Bid Documents

On most projects general contractors request two complete sets of bid documents for estimating a project. One set becomes the "estimate documents" that are used by the estimator to calculate takeoff quantities and are the source of all the information required to develop the bid price. The other set of documents is made available to subcontractors to enable them to do takeoffs and prepare estimates for their own particular trades. Many contractors set up a subtrade estimating area in their office where trade estimators can use the drawings and specifications to prepare their own takeoffs. Having a facility like this provides more control over the documents, which tend to get mislaid when they are lent out to trades. Nevertheless, the pressure from subtrades to access these documents often persuades contractors to lend out bid documents, at least over weekends.

| BID REPORT | | |
|---|---|---|
| Estimator: R. Jones | | Date: 9-Feb-06 |
| Job Number: CA6105 | | Location: Seattle |
| Project Name: XYZ High School | | |
| Description: | 3-Story Steel frame structure on conc. Spread footings —Brick Facings —— Flat roof — Landscaping and paving included. | |
| Owner: | XYZ School Board | |
| Consultant: | A.B. Cee | |
| Contact: | Joe | |
| Phone No. | 412 901 2345 | Fax No. 412 901 6789 |
| Documents Deposit: | $500.00 | |
| Bid Documents: | Specs + 25 Struct Dwgs; 13 Mech Dwgs; 9 Elect Dwgs. | |
| Addenda: | Add#1 issued 3rd. Feb. | |
| Closing Time/Date: | 2:00pm - February 24, 2006 | |
| Closing Location: | School Board Office 99 - 2nd. Street Room 209 | |
| Gross Floor Area: | 125,000 S.F. | |
| Prelim. Estimate: | $13,281,250 | |
| Owner's Budget: | $14,000,000 | |
| Bid Depository: | No | |
| Taxes Included: | Yes | |
| Bid Bond: | Yes | Percent: 10 |
| Consent of Surety: | Yes | |
| Performance Bond: | Yes | Percent: 50 |
| Labor & Materials Bond: | Yes | Percent: 50 |
| Award Period: | 30 days | |
| Start date: | Approx. April 1, 2006 | |
| Completion Date: | Not specified | |
| Notes: | | |

**Figure 2.4** Example of a Bid Report

Designers usually require a cash or certified check deposit from contractors who obtain bid documents, the sum amounting to as much as several hundred dollars on larger projects. The deposit is reimbursed when the drawings and specifications are returned to the designer's office in "good order." Contractors who take back documents that are badly damaged or incomplete will generally have to forfeit their deposit. In some cases consultants require bidders to pay a nonrefundable sum for bid documents to cover the cost of printing—the amount varies with the size of the project and the number of documents involved.

## Review Bid Documents

After bid documents have been listed on the bid report (see Figure 2.4), the documents are reviewed to obtain the information necessary to complete the rest of the bid report and also to highlight the data the estimator will need to refer to later in the estimating process. Anything unusual found in the specifications or on the drawings should also be highlighted in this review process. Items that are easy to miss, particularly, are searched out so that, in a later review that occurs when the estimate nears completion, the estimator can check off these items to confirm that they have been dealt with in the estimate.

Ideally every one of the bid documents should be fully examined in this review process; however, chiefly because word processors are now used to generate mammoth specifications, the number of documents in a bid package can be so enormous that some selection is unavoidable. The general contractor's estimator, therefore, will concentrate the review on the parts of the specifications that deal with the work of the general contractor; this is usually restricted to the earth work, concrete work, and carpentry trades. The "front end" of the specifications should also be considered, particularly the clauses dealing with the contract that might impact on the cost of the work or payment for the work done. If the contract designates a 60-day payment period, for instance, extra financing charges may be needed to be added in the estimate.

After the specifications have been reviewed, the estimator should turn to the drawings. Carefully studying the drawings and making detailed notes on what is observed help the estimator become familiar with the design and enables him or her to plan the takeoff procedure to take account of the particular nature of the project being estimated. The review of drawings can also allow estimators to recognize possible problems at an early stage, which provides them time to obtain any information necessary to complete the subsequent estimating tasks. For instance, the need for specialized scaffolding information may be apparent from reading the drawings; or there may appear to be a requirement for details about a new product shown on the drawings; or it may be evident that a special dewatering system will have to be used, which will involve consulting a specialist. All of these requirements can be dealt with more effectively if they are identified early in the estimating process and the last minute rush to check them off is avoided.

## The Query List

During the review and the quantity takeoff process that follows it, the estimator will uncover details that require clarification by the designer. The estimator may be tempted to call the designer each time this happens but doing so, on most bids, will lead to a great number of phone calls. A better method of handling this problem is to compile a list of all the questions that arise, the **query list**, then later make a single call to the designer to cover all the items that have accumulated. Often an item on the list will be clarified by further reading of the documents so the designer's assistance will not be required. And, while designers encourage contractors to ask questions during the bid period, they also wish to avoid a continuous stream of telephone calls from contractors. Sometimes design consultants adopt strategies to avoid telephone calls from bidders; one alternative is to ask contractors to use the fax to communicate all inquiries to the designer; another option is to indicate to all bidders that questions will be dealt with in a public setting at a pre-bid meeting or series of meetings.

Even though transmitting questions by fax may not be called for by the designer, some contractors still use this medium as it provides written evidence of the questions that were asked and, therefore, may help avoid arguments in the future about what was discussed. Other contractors are now using e-mail to communicate with design consultants. This can be not only a fast and effective way to get your questions to the designers, but it also has the same advantage as using the fax in that it provides a record of what was communicated and when the message was sent. Designers will usually respond orally to contractors regarding minor questions, but significant details will be incorporated into written documents that are distributed to all bidders. These publications that are known as **addenda** or bid bulletins become part of the bid documentation and will be included in the contract that the owner will establish with the successful bidder.

## The Team Approach

The nature of the estimating process means the estimator usually works alone on quantity takeoffs and cost estimates, but contractors agree that the quality of an estimate improves when the input of managers and experienced field personnel also are included in its preparation. The estimator should be encouraged to seek the assistance of these other company members as much as possible when problems and decisions need to be dealt with in the estimating process. Management and field supervisors together with the estimating staff should see the estimating function as a team effort to achieve accurate high-quality estimates.

Most companies hold an estimate review usually a day or two before the bid is due to be submitted. This meeting normally brings together the estimator, the chief estimator, one or two senior company managers, and, if possible, the person who is likely to be the superintendent of the project. The purpose of the meeting is to review the estimator's work in terms of the prices used and any underlying decisions or assumptions made in preparing the estimate. The estimator who has prepared the takeoff should have a comprehensive knowledge of the project work, and he or she should be encouraged to share this knowledge with the other participants at the bid meeting so that they can use their experience to consider the estimator's assessment of how the work will be pursued.

Alternative materials, methods, equipment, manpower and schedules may all be studied at this time with a view to achieving a more competitive bid price. The merits of proposed subcontractors may also be discussed at the meeting, but this may not be possible as the identity of subtrades is often unknown until shortly before the bid is submitted. Perhaps the most important item considered at the estimate review is the amount of fee to include in the bid price. There are many factors to account for when assessing a project fee; this is discussed in some detail in Chapter 15 under bid markup. Experience has shown that participation in the fee decision can be a valuable motivator of estimators and can avoid the finger pointing and blame that results when a single person decides on the fee and the decision turns out to be faulty. For these reasons, making the determination of fee a team decision by the participants at the bid review meeting can be a good idea.

## Site Visit

After the bid documents have been reviewed and the estimator has obtained a good general idea of the nature of the project, a visit to the site of the project should be arranged. Some estimators like to complete most, if not all, of the quantity takeoff before the site visit because, by working through takeoff process, the estimator can really get to know the nature of the job and may also be able to foresee possible construction problems the project holds for the builder. Then the estimator, as a result of the detailed knowledge of the work he or she has gained, will come to the site with a number of specific questions in mind, which may be answered by the site information gathered. While an experienced estimator will certainly know what to look for and assess when investigating the site, the presence of the chief estimator or a project superintendent on the visit will always be valuable if only to offer alternative viewpoints in the evaluation of the data obtained.

Figures 2.5 and 2.6 list the items that are normally considered on local and remote site visits. The condition of the site and any soil information that can be observed will impact the price of the excavation work estimated. Many of the other items examined are included in the general expenses of the project; pricing these items is dealt with in Chapter 14. It is recommended to take photographs on the site visit as the condition of the site may change in the time between the bid and the start of work. Also, the condition of structures adjacent to the site should be carefully documented and photographed at this time, especially if these structures are in poor condition. This information could be important in determining if damage to these occurred before or after the project work began.

Environmental concerns are another important area for the estimator to consider on the site visit. Complying with exacting environmental requirements can be a very costly proposition for a contractor, so the estimator must review the contract documents thoroughly to assess the extent of liability for possible environmental problems the contractor is assuming under the terms of the contract. Every effort should then be made to investigate the history of the site from local information and by reviewing soils reports and any environmental assessments that might be available. Special attention should be paid to the possibility of soil contamination in the location of the site. Only by assembling accurate data can the contractor quantify the financial risk imposed by the contract regarding environmental issues. Then an appropriate sum can be added to the bid to cover this risk.

## Computer Estimating Systems

Computers have become an indispensable tool of the estimator, but for the estimator to use computers effectively he or she must first have a thorough understanding of the basic concepts of estimating. Computer estimating systems perform many operations automatically and often invisibly, and the user may not completely understand what the system is doing. If an estimator does not know exactly what the computer is accomplishing in the preparation of an estimate, the estimator will not be in control of the process. Computer systems are not foolproof and errors do occur, but the experienced estimator can often see from the output that there is a problem. Without a thorough grounding in the nature of estimating, an estimator using a computer system can be at the mercy of the computer, which could result in seriously flawed estimates being produced.

One way to gain knowledge of the basic principles of estimating is to work through an estimate manually using nothing more than a pencil and paper together with a calculator. With this in mind, in this text we have provided a complete manual

| SITE VISIT CHECKLIST |
| Local Projects |

Estimator: _____    Date: _____

Job Number: _____    Location: _____

Project Name: _____    _____

Distance from Office: _____

Weather Conditions: _____

Access and Roads: _____

Sidewalk Crossing: _____

Site Conditions: _____

_____

_____

_____

Adjacent Structures: _____

_____

_____

_____

Obstructions: _____

Shoring or Underpinning: _____

Depth of Topsoil: _____

Soil Data: _____

Ground Water: _____

Soil Disposal Location: _____

Distance to Borrow Pit: _____

Local Sand & Gravel: _____

Electrical Service: _____

Telephone: _____

Sewer & Water Services: _____

Parking and Storage: _____

Security Needs: _____

Temporary Fences Required: _____

Garbage Disposal: _____

Toilets: _____

Site History: _____

Possible Contamination: _____

Other Comments: _____

_____

_____

_____

_____

**Figure 2.5**    Local Site Visits

| SITE VISIT CHECKLIST |
| :--- |
| Remote Projects |

| | |
| :--- | :--- |
| Estimator: _____ | Date: _____ |
| Job Number: _____ | Location: _____ |
| Project Name: _____ | _____ |

| | |
| ---: | :--- |
| Distance from Office: | _____ |
| Weather Conditions: | _____ |
| Access and Roads: | _____ |
| Sidewalk Crossing: | _____ |
| Site Conditions: | _____ |
| | _____ |
| | _____ |
| | _____ |
| Adjacent Structures: | _____ |
| | _____ |
| | _____ |
| | _____ |
| Obstructions: | _____ |
| Shoring or Underpinning: | _____ |
| Depth of Topsoil: | _____ |
| Soil Data: | _____ |
| Ground Water: | _____ |
| Soil Disposal Location: | _____ |
| Distance to Borrow Pit: | _____ |
| Local Sand & Gravel: | _____ |
| Electrical Service: | _____ |
| Telephone: | _____ |
| Sewer & Water Services: | _____ |
| Parking and Storage: | _____ |
| Security Needs: | _____ |
| Temporary Fences Required: | _____ |
| Garbage Disposal: | _____ |
| Toilets: | _____ |
| Site History: | _____ |
| Possible Contamination: | _____ |
| Local Weather Pattern: | _____ |
| Availability of Local Labor: | _____ |
| Local Accommodation: | _____ |
| Availability of Local Materials: | _____ |
| Local Subtrades: | _____ |
| Rates for Business Permit: | _____ |
| Rates for Building Permit: | _____ |
| | _____ |
| Other Comments: | _____ |

_____

_____

_____

**Figure 2.6**   Remote Site Visits

estimate of a small project so that the student can follow each step of the process. This approach is taken, not because manual methods are preferred, but, rather, because the process of working through these procedures by hand better develops the beginning estimator's comprehension of the whole estimating process.

This is not to say computers are ignored in this book. On the contrary, there are discussion on the use of computers for the work under consideration in several of the following chapters, and computer spreadsheets are used extensively in the pricing process we will describe. Also, in order to demonstrate how the use of a computer system can enhance the efficiency of the estimating process, we have provided a second worked example utilizing Timberline Precision Estimating software to produce an estimate of an office/warehouse building. A copy of Precision Estimating—Basic Edition that was used in this example is included in the inside back cover of this text. Students are directed to Appendix B for a primer on the use of Precision Estimating software.

## Other Technology

In a number of places in the text, we will refer to technological advances that are being, or could be, employed by estimators to improve the productivity of the estimating process. The development of fax machines and cellular telephones, for instance, has had a great impact on the bidding process, which is discussed in Chapter 15. Use of the Internet in the estimating process may be still in its infancy, but e-mail, direct access to Web pages and Web-based price lists are already having a significant impact on the efficiency of communication in the estimating process. Advances in computer-assisted design (CAD) and its integration into computer estimating systems are also boosting the productivity and accuracy of estimating. While the estimator needs to try to keep abreast of all these developments and remain open to the improvements they offer, he or she is warned of the need for careful management of the technology. A real danger exists in losing control of the process, resulting in a potential for errors to develop simply because the estimator does not fully understand what is going on when the technology button is pressed.

## SUMMARY

- A systematic approach to estimating is required because of the number of people and amount of data involved in an estimate.
- A contractor's estimating department should identify the goals and objectives of the department.
- Information about bid calls for projects are obtainable from a large number of sources for both public and private bids.
- With open bidding there is little restriction on who can submit a bid on the work, whereas in a closed-bid situation, only invited contractors are allowed to submit bids.
- Adopting a closed bid addresses the problem of receiving bids from undesirable contractors.
- Another way to deal with the problem of undesirable contractors is to require bidders to be prequalified; that is, contractors must complete questionnaires that enable owners to assess the capability of prospective contractors.
- Many government agencies are now publishing bid notices on the Internet; this is quickly becoming a very useful source of information about projects that are out for bid.

- Many contractors make an effort to constantly gather information about planned projects. This is part of their marketing strategy to get involved early in a project, which may allow them to obtain construction work by negotiation rather than by the highly competitive bid process.
- Contractors consider many factors when they decide to submit a bid on a project; the chief factors would be the size and type of project it is, the location of the project and the contractor's need for new work at that time.
- Contractors need to closely manage the estimating process; this involves keeping bid records, scheduling the estimator's time, and carefully reviewing bid documents when they are obtained.
- Careful attention should be paid to the "front-end" of specifications in the review process because it is here that general clauses that have a significant impact on the cost of the work are often found.
- During the review and the estimating process that follows it, the estimator should prepare a query list that collects all questions and items that need clarification from the consultants who produced the bid documents.
- Estimators should be encouraged to take a team approach to the process of compiling an estimate; this involves working closely with other members of their organization and not being afraid to seek help and advice when the estimator feels it is needed.
- The site visit is an important task in estimating a project. The estimator should use an extensive checklist to ensure that all the major items are properly considered on the visit to the site of the proposed work.
- Computers have become an indispensable tool of the estimator, but for the estimator to use computers effectively, he or she must first have a thorough understanding of the basic concepts of estimating. Learning how to compile an estimate manually can help the novice estimator gain the insight necessary to appreciate the role of computers in the process.
- New communication technology, especially the use of the Internet, if managed properly, can have a positive impact on the estimating efficiency.

## REVIEW QUESTIONS

1. What are the main objectives of the estimating department in a construction company?
2. Where can a contractor obtain information about which projects are out for bid for:
   a. public projects?
   b. private projects?
3. Describe the main features of an open bid and how problems associated with open bidding may be avoided.
4. Why are some construction companies aggressively pursuing marketing?
5. List and describe the major factors a contractor will consider when deciding whether to submit a bid on a project.
6. Why does the estimating process need to be scheduled?
7. Describe what estimators look for in their review of the bid documents.
8. What is a query list and how is it used by an estimator?
9. What is meant by a team approach to estimating, and how can it help an estimator in his or her work?
10. Why is it necessary for the estimator to visit the site even when extensive site information is provided in the bid documents?
11. Suggest how the Internet can be used in the estimating process.

# 3

# MEASURING QUANTITIES GENERALLY

## OBJECTIVES

*After reading this chapter and completing the review questions, you should be able to do the following:*

■ Define the quantity takeoff process.
■ Explain what a "Method of Measurement" for takeoffs means.
■ Describe the process of measuring "net in place" and explain why this process is adopted.
■ Describe the traditional English units and metric units of measurement used in estimating.
■ Describe how takeoff items are composed.
■ Discuss the level of accuracy required for a takeoff.
■ Describe a strategy for completing a takeoff of a large project.
■ Discuss the use of assemblies in the takeoff process.
■ Describe the specialized stationary used in estimating processes.
■ Discuss the role of computers in the takeoff process.
■ Calculate the area of regular and irregular plane shapes.
■ Calculate the volume of a variety of solid figures.
■ Given the dimensions of a building plan, calculate the centerline length of the perimeter trench/footings/wall.

## The Quantity Takeoff

An estimate begins with a quantity takeoff. A quantity takeoff is a process of measuring the work of the project in the form of a series of quantified work items. To prepare the takeoff, an estimator has to break down the design that is shown on the drawings and described in the specifications into predefined activities (work items) that correspond to the operations the contractor will perform to complete the work of the project. Many estimators maintain a catalog of standard items that represent activities encountered on a large variety of projects, but it is not unusual to have to deal with a new category of work unique to the project under consideration. The catalog of standard work items is often used as a checklist during the takeoff process, especially by junior estimators, to help ensure that all the categories of work in the project are accounted for.

Each item considered in the takeoff is measured according to a uniform set of rules with the object of producing a list of work items and their associated quantities in a format familiar to estimators. Because a standard format is used, estimators are able to more easily evaluate the work of each item and proceed to put a price on this work. In each subsequent chapter in which the measurement of the work of a particular trade is studied, we have included "measuring notes" that outline detailed rules regarding how the items of work of that trade are to be measured. These rules of measurement collectively amount to a **"Method of Measurement"** that can be shared by a group of estimators. It is strongly recommended that estimators, at least those in the same company, follow such a standard method of measuring work because price information that is gathered can only be shared if it relates to work that has been measured in the same fashion. There is further discussion about company estimating departments following standard rules of measurement in the section on "net in place" on page 37.

## What Is Measured?

The estimator should note that quantities of work are measured in the takeoff process, not quantities of materials. The difference between work quantities and materials quantities is a subtle one and can be difficult to grasp. It can be especially confusing for those who have come to estimating from a trade background, having been taught a subject called "estimating" in trade school that usually has to do with calculating quantities of project materials rather than cost estimating. The objective of a materials takeoff is to compute the quantity of materials that need to be purchased in order to construct a particular item of the project. A quantity takeoff for a cost estimate is different in three major respects:

1. Measurements are made **"net in place"** (see page 37) with cost estimate takeoffs. Materials takeoffs measure gross quantities.
2. A materials takeoff often does not provide sufficient information for pricing. For example "100 cubic yards of 3000 psi concrete" may be adequate for a material takeoff, but information about what the concrete is to be used for is required in a takeoff for a cost estimate.
3. There are a number of work items measured in a cost estimate takeoff that do not involve materials. For example, the item "Hand Troweling" has only a labor price associated with it; what is measured for this work item is the plan area of concrete to be troweled. There is no material to consider.

Note that whenever the term "estimate" appears in this text it refers to a cost estimate and not an estimate of quantities for purchasing materials.

## Units of Measurement

There are two systems of measurement in use in the North American construction industry: the English system and the metric system, each with its own set of units.* This text will utilize English units—a later metric edition is planned using interna-

---

*Information about metric units is available from the Institute of Science and Technology, which has a very informative Web site at www.NIST.gov.

tional metric units; "Système international d'unités" (SI units), on drawings and in all aspects of estimating. Estimators are advised to work entirely in one system or the other in accordance with the units used on the drawings of the project; mixing different units of measurement in a single estimate increases the probability of serious errors and should be avoided if at all possible.

Items of work are measured in the units most appropriate to the type of work involved, and this is reflected in the rules of measurement adopted by the estimating department. There are five basic categories of units used in estimating:

1.  Number—the quantity of these items is determined by counting the number of items, and the value obtained is designated number (No., Each or Ea.) to indicate that it is an enumerated item.

    Examples:      2' × 3' Mirrors                           —      6 No.

    Install ¾" × 12" anchor bolts     —      240 Each

    Cast iron manhole covers           —      60 Ea.

2.  Length—the quantity measured is the length of the item in feet and the value obtained is designated linear feet (lin. ft.). The equivalent metric units would be meters (m).

    Examples:      Form 2 × 4 keyway                    —      120 lin. ft. (37 m)

    6" Dia. ABS drain pipe               —      300 lin. ft. (91 m)

3.  Area—the quantity measured is the superficial area of the item in square feet and the value obtained is designated square feet (sq. ft.). The equivalent metric units would be square meters ($m^2$).

    Examples:      Form footings                           —      325 sq. ft. (30 $m^2$)

    ½" Gypsum wallboard                —      12,000 sq. ft. (1,115 $m^2$)

4.  Volume—the quantity measured is the volume of the item in cubic yards and the value obtained is designated cubic yards (cu. yd.). The equivalent metric units would be cubic meters ($m^3$).

    Examples:      Excavate trenches                    —      125 cu. yd. (96 $m^3$)

    3000 psi Concrete footings         —      89 cu. yd. (96 $m^3$)

5.  Weight—the quantity measured is the weight or mass of the item in pounds or tons and the value obtained is designated pounds or tons (lbs. or Ton). The equivalent metric units would be kilogram (kg) or metric ton (t). Metric tons are sometimes referred to as tonnes.

    Examples:      #3 Rebar in footings                 —      524 lb. (238 kg)

    W8 × 20 Structural beams          —      12 Ton (11 t)

See Appendix D for a full list of units used in estimating and conversion tables.

## Measuring "Net in Place"

Quantities of work are measured "net in place" in a takeoff for a cost estimate. This means that quantities are calculated using the sizes and dimensions indicated on the drawings with no adjustments to the values obtained for waste factors and suchlike.

To illustrate this principle consider the item ¾" Plywood floor sheathing that, let us say, is shown on the drawings applied to a floor 40 feet long and 34 feet wide. Multiplying these dimensions gives a net quantity of 1360 square feet. The takeoff quantity will remain as 1360 square feet, and prices for the labor and any equipment associated with this item of work will be applied to the quantity at a later stage of the estimating process to establish the estimated cost of this item. When the material component of this item is priced, however, a waste factor will be added to allow for cutting and waste on the material. It does not matter if the estimator increases the quantity of material or the price of material by the waste factor, the result will be the same in each case. In contrast, if this item were measured so that plywood materials can be ordered for the floor under construction, the person ordering the material will include in the calculations a waste factor that will result in a "gross" quantity of, say, 1408 square feet. This inflated amount takes into account the cutting and wastage of the material involved. Because field personnel are used to measuring "gross" quantities in this way, they often find it difficult to adjust to measuring "net" for an estimate, but there are a number of reasons why an effort should be made to takeoff "net in place" quantities for cost estimates:

1.  Consistency—Without a fixed ruling some estimators would measure "net," others measure "gross" and yet others may combine both "gross" and "net" quantities in a single estimate. As a result, it becomes far more difficult to make a meaningful comparison or assessment of different estimates because the basis of the estimates, the quantity of work involved, will not have been determined in the same way for each estimate. Also, the lack of a consistent standard method of measurement would require each estimator to maintain a personal database of prices to reflect the specific set of **"add-on"** factors used by that estimator. This is unacceptable if estimators in the same company are to share data, and it is a barrier to effective communications between all estimators.

2.  Objectivity—Measuring "net" quantities results in an objective appraisal of the design. **"Add-ons"** for waste factors, swell factors, compaction factors and the other adjustments applied to quantities taken off are all subjective assessments. One estimator may allow an additional 5% to the "net" quantity for wastage, while another estimator may consider 8% to be a more appropriate rate. When such disagreements are applied to a large number of items, clearly a different schedule of quantities will be produced by every different estimator. This adds a complication to the pricing of the work items, which is already a difficult process, and provides further reason why estimate comparisons are difficult when measuring "gross." It is much easier to review item unit prices when the quantities have been measured in an objective standard fashion than when they have been exposed to the subjective evaluation of the person who measured them.

3.  Unit Price Contracts—Measurement of work done on a unit price contract is invariably required to be calculated on a "net" quantity basis. The reason for this is quite obvious if we consider the consequences of allowing contractors to increase quantities by waste factors of their own choice when the work is to be paid for on the basis of the quantity of work done.

4.  Comparisons of Operation Efficiencies—The efficiency of construction operations is often analyzed in terms of cost or hours spent per unit of work done. Here again, it is clear that if the amounts of work done are measured on a subjective basis with a variety of "add-on" factors applied to the quan-

tities obtained, comparing the efficiencies determined in one analysis with those obtained in another is meaningless.

To illustrate this last point, consider the results of the analysis of Joe's project and Fred's project. These supervisors spent $1190 and $1200, respectively, placing concrete in footings, but Joe measured a "net" quantity of 100 cubic yards, while Fred measured a "gross" quantity of 105 cubic yards, which included a 5% waste factor. Using these values gives the following costs per cubic yard:

Joe's Project:     $$\frac{\$1190}{100 \text{ cu. yd.}} = \$11.90 \text{ per cu. yd.}$$

Fred's Project:     $$\frac{\$1200}{105 \text{ cu. yd.}} = \$11.41 \text{ per cu. yd.}$$

From this analysis it would appear that Fred's operation is more efficient because of the lower cost per cubic yards obtained. However, if the concrete amounts are calculated *on the same basis*, which is 100 cubic yards net in both cases, the result is reversed—Joe's cost per cubic yard ($11.90) is lower than Fred's ($12.00).

## Takeoff Rules

1.  Takeoff items comprise two components:
    a. Dimensions that define the size or quantity of the item in accordance with the required units of measurement for that item; and
    b. A description that classifies the item in terms of the requirements of the standard method of measurement.
2.  Dimensions are entered onto the takeoff in this order: length, width and depth (or height).
3.  Dimensions are written in feet to two decimal places, thus, 5'10" would be written into a dimension column as 5.83. See Figure 3.1 for decimal equivalents of inches in feet.

    When using the metric system, dimensions are recorded in meters to two decimal places. Thus, 3489 mm would be written as 3.49 m and 1200 meters would be 1200.00 m.

    (Some estimators write dimensions in duo decimals using a decimal point between feet, which are indicated to the left of the decimal, and inches to the right. So, using duo decimals, 5'10" is written 5.10 on the takeoff. The number is "decimalized" when it is later entered into a calculator or computer. This practice has not been followed in this text.)

| | | | |
|---|---|---|---|
| 1 inch | = 0.08 feet | 7 inches | = 0.58 feet |
| 2 inches | = 0.17 feet | 8 inches | = 0.67 feet |
| 3 inches | = 0.25 feet | 9 inches | = 0.75 feet |
| 4 inches | = 0.33 feet | 10 inches | = 0.83 feet |
| 5 inches | = 0.42 feet | 11 inches | = 0.92 feet |
| 6 inches | = 0.50 feet | 12 inches | = 1.00 feet |

**Figure 3.1**   Decimal Equivalents of Inches in Feet

4. If a dimension does not come directly from a bid drawing, evidence, in the form of side calculations showing how it was determined, should be included on the takeoff even for the simplest calculation.

5. Dimensions are rounded to the nearest whole inch. However, to avoid compounding rounding errors, fractions of an inch are not rounded in side calculations until the end result of the calculation is obtained.

6. Dimensions figured on the bid drawings, or calculated from figured dimensions, are used in preference to scaled dimensions. The estimator should scale drawings to obtain dimensions only as a last resort because drawings are not always accurately drawn to scale. (See discussion of the use of digitizers in Chapter 4.)

7. Deductions listed with the dimensions are written in red or enclosed in "brackets < >" and noted as deductions.

8. Throughout the takeoff, headings are inserted to indicate such things as the trade being taken off, the location of the work under consideration, and the phase in which the work is classified. Side notes are also recommended to explain what is being measured, especially when the work is complex or unusual. All of these headings and notations help provide an audit trail so that the estimator or any other interested party can more easily review the takeoff.

    The insertion of headings and side notes into the takeoff is not to be overlooked when using a computer estimating system, especially if dimensions are entered into the computer directly from the drawings without making handwritten notes. The computer may be able to print a report that shows the takeoff items in the order in which they were measured, but on a large job without headings and notes in the takeoff, it will be almost impossible to review this data to determine which part of the project the items relate to and whether all the work of the project has been taken into account.

9. It is recommended that the estimator make use of highlighter pens to check off items of work as the takeoff progresses. This allows him or her to identify what has been measured and distinguish what remains to be considered. This recommendation is made even though many designers specify that deposits for drawings will only be refunded when "unmarked" drawings are returned. The loss of the occasional deposit is a small price to pay for the advantage of using highlighters, and a loss may be avoided altogether as it can be argued that a good highlighting job improves the appearance of drawings.

10. Takeoff descriptions contain sufficient information for the estimator to later price the work involved. Estimators use abbreviations extensively in takeoffs to increase the speed of the process, and detail that can be easily added at the pricing stage may be omitted from descriptions at the time of the takeoff.

    To illustrate this point, consider a project that requires three mixes of concrete: Mix A for foundations, mix B for columns and mix C for all other concrete. Mix A consists of 3500 psi concrete with type V cement and 4% air entrainment. The estimator will describe the use of the concrete in the takeoff in terms of footings, foundation walls, columns up to the second floor, slabs-on-grade and so on. It is time wasting to include a full mix description with each concrete takeoff item. The types of mix and their prices have to be considered at the pricing stage, so why not leave the mix categorization until that time? See the recap examples in Chapter 11 for a description of how this mix classifying can easily be performed.

11. Generally, when describing a takeoff item, the estimator does not mention in the description or measure separately any of the following items because they will be dealt with in the process of pricing the takeoff items:
    a. Transportation or any other costs associated with the delivery of the materials involved
    b. Unloading materials
    c. Hoisting requirements
    d. Labor setting, fitting or fixing in position
    e. Lapping, cutting or waste of materials
    f. Stripping formwork
    g. Form oil
    h. Rough hardware
    i. Scaffold* and falsework

The estimator should be aware that while all the above rules are in place to obtain the consistency and objectivity required, there is always an occasional situation in which the work is not routine and a better result may be obtained if the rules are relaxed. So estimators, to continue to be effective, have to preserve a certain flexibility of approach keeping in mind their major goal of pricing the work of the project as efficiently as possible.

## Accuracy of Measurement

The quantity takeoff should accurately reflect the amount of work involved in a project, but how accurately should the work be measured? There is no clear answer to this question as the level of accuracy pursued by the estimator depends on the costs and benefits of attaining high accuracy. Devoting extra time to improving the accuracy of the measurement of certain items of work may not be justified. All we can say is that the takeoff has to be as accurate as possible given the nature of the work being measured and the cost of attaining high accuracy.

To demonstrate, consider the measurement of concrete work items. It is not difficult to calculate the quantities of concrete with a high degree of accuracy as concrete items are usually well-detailed on contract drawings. Also, the time spent carefully measuring concrete work can be justified as items of concrete are relatively expensive. In contrast, excavation items are not detailed well, if at all, on drawings, and the unit prices of these items are usually quite low. With the excavation trade, the estimator usually has to ascertain the dimensions of the work by applying judgment based on experience; this results in an assessment that may be quite different from the volumes of work actually excavated. Clearly, there is little benefit in spending much time developing and carefully measuring what is only a theoretical impression of the excavation requirements of a project. However, the estimator still needs to make a reasonable evaluation of excavations because, even though the price per cubic yard may be as low as $5.00, when there are in the region of 50,000 cubic yards to consider, the total price of the work is not insubstantial.

Therefore, an estimator constantly has to balance the cost of achieving high accuracy against the value obtained from the increase in accuracy. This is not an easy task, particularly for those new to estimating, but our advice to the estimator who is

---

*On certain projects that require extensive or complicated scaffold systems, it may be necessary to take-off the quantities of scaffold required. Otherwise scaffold requirements, and all the other items listed, are accounted for later in the process when the takeoff items are priced.

in some doubt about whether the time spent improving the accuracy of a measurement is justified, is to err on the side of caution and spend the extra time. A more accurate takeoff is always a better takeoff.

## Organization of the Takeoff

The order of the takeoff will generally follow the sequence of the work activities of the project and, conveniently for the estimator, the contract drawings are usually presented in this sequence. However, some estimators find it preferable to measure concrete work before excavation work even though excavation activities usually have to precede concrete work on the job. The reason for considering concrete work first is that the sizes and details of concrete items are clearly defined on the drawings, whereas excavation requirements have to be assessed. So measuring concrete first allows the estimator to become familiar with the project and, therefore, more efficient later at assessing the excavation requirements.

Other estimators begin with the measurement of excavation because they choose to proceed with their takeoff in the order in which the drawings are provided. Because site topography is usually shown on the first drawing, these estimators begin on this drawing with the excavation activities and proceed with the rest of the work of the project in the order of construction, which is generally the order in which the drawings depict the work. As an estimator you must decide which alternative to adopt; in this text we follow the second option.

An estimator needs a strategy to deal with large projects. Without a systematic approach, the estimator can easily get lost in the takeoff process and reach a point at which it is not clear what exactly has been measured and what is left to be considered. This ambiguity can cause extreme frustration to the estimator, which can lead to high stress and, in some cases, absolute panic. A simple yet effective strategy that can be used with any estimate consists of first dividing the project into manageable-size parts, proceeding through the takeoff one part at a time and, within each of these parts, measuring the work as a sequence of assemblies. How the project is divided up is entirely dependent on the nature of the project; high rise jobs, for instance, are most easily divided floor by floor or by groups of floors. Some projects can be divided into zones, phases or even separate buildings. The estimator will often find that the contract drawings reflect these divisions as it is common for a large project to be designed in parts rather than as one large whole.

Once the project is divided into manageable-size parts, assemblies are distinguished within. An **assembly** is a component of the work that can be considered separately from the other parts of the work. The notion of an assembly will develop as the reader progresses through the takeoffs that follow. For now, consider the example of a perimeter foundation wall as an assembly in the estimate of a house. The basic idea is that the estimator measures all the work involved in an identified assembly, then moves on to consider the next assembly. In this example, the estimator may takeoff all of the following items associated with this one assembly before passing on to the next assembly:

a. The concrete in the wall
b. The forms to the sides of the wall
c. The forms to openings and blockouts in the wall
d. The rubbed finish on the exposed concrete of the wall
e. The reinforcing steel in the wall

The major advantages of taking off by assemblies are that the assembly is evaluated only once and the same dimensions are shared by a number of items within the assembly, thus avoiding repetition. Alternatively, if the estimator was to measure all the concrete on the project, then return to consider the forms, and return a third and fourth time to this assembly to measure the cement finishes and embedded miscellaneous items, there would be a great deal of wasted time reassessing the component, repeating the same dimensions and, possibly, recalculating these same dimensions several times.

Another advantage of this approach is that takeoff by assemblies or "work packages" is utilized in many of the better computer estimating programs, which produces a very powerful and speedy estimating tool. A student who becomes familiar with the concept and use of assemblies when learning estimating will be more adept when these features are encountered in computer programs and better able to develop and customize them to obtain maximum benefits from their use.

Also, the process of scheduling the project is made easier when assemblies have been used in the takeoff because the activity breakdown used by schedulers corresponds quite closely to the assemblies measured by the estimator. Some computer estimating packages can integrate with scheduling packages to allow the scheduler to directly transfer the estimate assemblies to the schedule.

### Estimating Stationery

Specially printed forms designed for each of the various estimating procedures are used to increase productivity and contribute to the accuracy of an estimate. There are a number of different stationery formats available to the estimator; the decision on which to adopt depends on the method of estimating used. Figure 3.2 shows the form of quantity sheet used for quantity takeoffs in the following chapters, and Figure 3.3 shows the form of pricing sheets used.

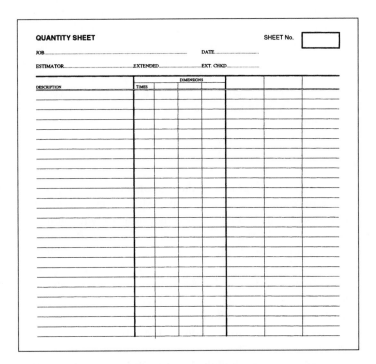

**Figure 3.2**   Sample Quantity Sheet

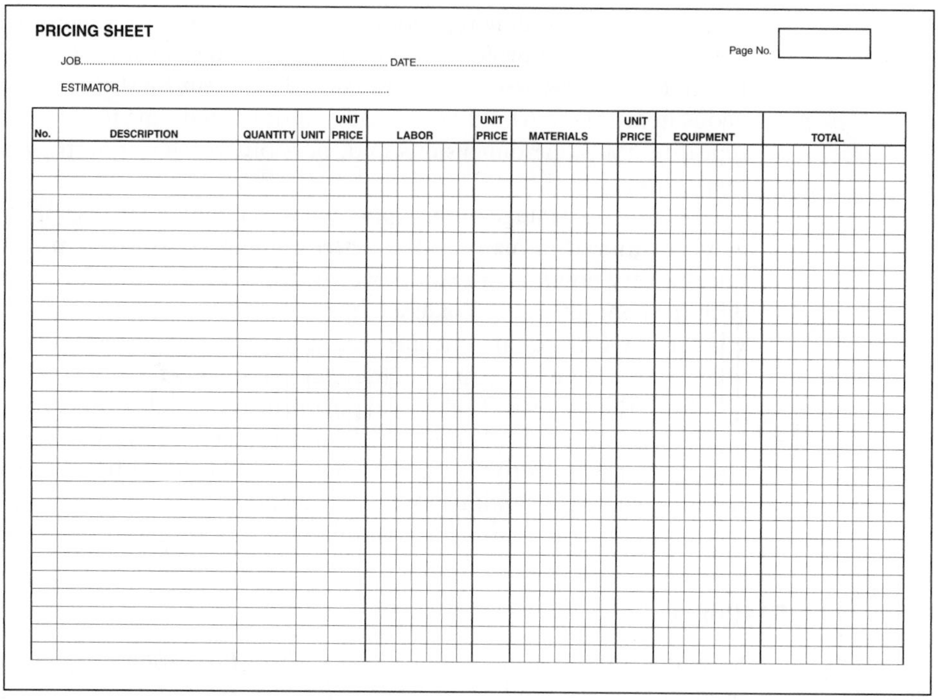

PRICING SHEET

JOB.................................... DATE....................

ESTIMATOR....................................

Page No.

| No. | DESCRIPTION | QUANTITY | UNIT | UNIT PRICE | LABOR | UNIT PRICE | MATERIALS | UNIT PRICE | EQUIPMENT | TOTAL |
|---|---|---|---|---|---|---|---|---|---|---|

**Figure 3.3** Sample Pricing Sheet

In this book the quantity takeoff is recapped onto separate sheets for pricing, so one form of stationery is used for the takeoff and a second for the **recap**. The process of recapping quantities onto a separate pricing sheet is adopted because it makes the task of pricing an estimate far more efficient since all related items are gathered in a concise manner. To price the takeoff items directly would entail a long and tedious process with a great deal of repetition, especially for estimates of larger projects. On small projects, however, there may be some advantage to directly pricing the takeoff. Figure 3.4 shows an example of a combined takeoff and pricing sheet that could be used for this purpose.

## Takeoff by Computer

The use of computer programs in the takeoff process can greatly increase the efficiency of the process. The estimator can simply select takeoff items from a database of standard work items and input the dimensions for each item; the system then performs all the required calculations and sorts the takeoff items according to the required presentation format. Most estimating systems allow the estimator to select a number of related items together (an assembly) then enter one set of dimensions that apply to all the items selected. Even greater productivity can be achieved if a group of related items is saved as a preset assembly. When such assemblies are in place, the estimator just recalls an assembly (without having to select the individual items again) and then inputs the dimensions that apply to that assembly.

Some estimators like to prepare a "paper takeoff" before entering dimensions into the computer. This version provides a comprehensive record of how the estima-

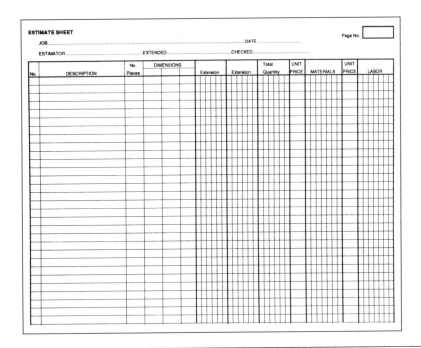

**Figure 3.4**    Stationery for Combined Takeoff and Pricing

tor proceeded and allows the takeoff to be checked in detail to ensure all the work has been measured properly. If this written takeoff is dispensed with, the takeoff production rate can certainly improve, but the ability to review the estimator's work will be reduced. However, as we have discussed, most computer estimating programs can generate reports that enable you to audit the takeoff and, as long as the estimator maintains explanatory notes with the takeoff, either on paper or in the computer, it will still be possible to conduct a detailed review of the takeoff.

## Formulas and Perimeter Centerline Calculations

The basic formulas used in estimating are given in Figures 3.5 and 3.6. The last formula shown on Figure 3.6 calculates the volume of a perimeter trench from the centerline length, width and depth of the trench. This is a formula that is used extensively in the takeoff process on many types of projects and, may be used many times on the same project takeoff.

Consider the example shown in Figure 3.7. The calculation of volumes of trench excavation shown in the figure's footing concrete and wall concrete all involve the same centerline (C.L.) perimeter length:

$$\text{Volume of trench} \quad = \quad \text{C.L. length} \times w_1 \times d_1$$

$$\text{Volume of footing} \quad = \quad \text{C.L. length} \times w_2 \times d_2$$

$$\text{Volume of wall} \quad = \quad \text{C.L. length} \times w_3 \times d_3$$

This identical centerline length will then be used in further calculations of footing and wall formwork areas. It is clear that a single centerline value can be used many times over in takeoff calculations, so adopting a systematic method of establishing centerline perimeters, such as that outlined in the following, can contribute to both the speed and accuracy of a takeoff.

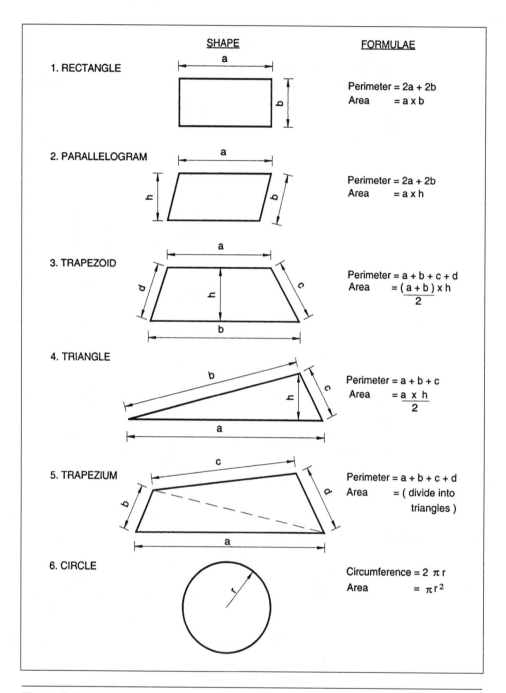

**Figure 3.5**  Perimeters and Areas of Plane Shapes

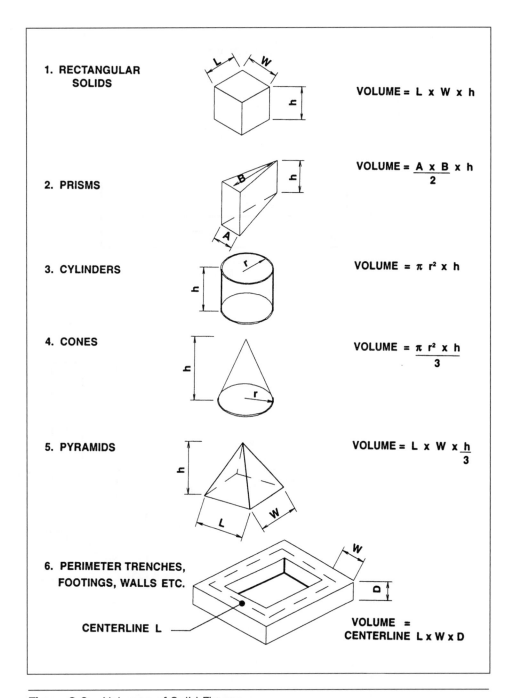

**1. RECTANGULAR SOLIDS**

VOLUME = L x W x h

**2. PRISMS**

VOLUME = $\dfrac{A \times B}{2} \times h$

**3. CYLINDERS**

VOLUME = $\pi\ r^2$ x h

**4. CONES**

VOLUME = $\dfrac{\pi\ r^2 \times h}{3}$

**5. PYRAMIDS**

VOLUME = L x W x $\dfrac{h}{3}$

**6. PERIMETER TRENCHES, FOOTINGS, WALLS ETC.**

CENTERLINE L

VOLUME = CENTERLINE L x W x D

**Figure 3.6**  Volumes of Solid Figures

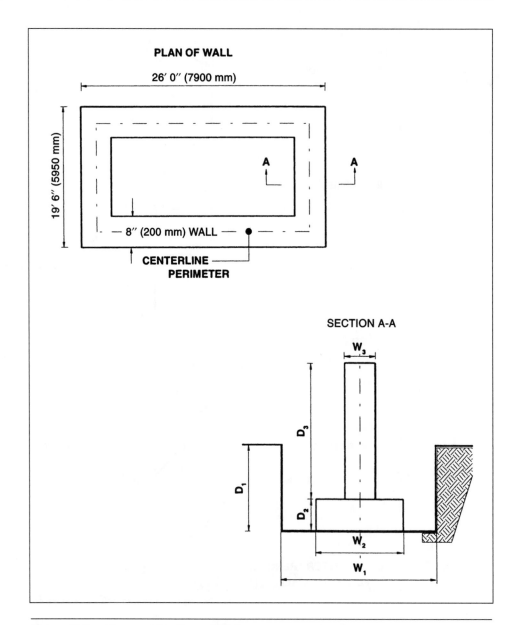

**PLAN OF WALL**

26' 0" (7900 mm)

19' 6" (5950 mm)

A

A

— 8" (200 mm) WALL —

CENTERLINE
PERIMETER

SECTION A-A

$W_3$

$D_3$

$D_1$

$D_2$

$W_2$

$W_1$

**Figure 3.7** Plan and Section of Wall

A perimeter length is calculated from the figured dimensions that are provided on the plan of the structure. In this example the dimensions of the exterior of the walls are given. The length of the outside perimeter can be determined thus:

$$2 \times 26' \ 0" = 52' \ 0" \quad (2 \times 7.90 \text{ m} = 15.8 \text{ m})$$
$$2 \times 19' \ 6" = \underline{39' \ 0"} \quad (2 \times 5.95 \text{ m} = \underline{11.9 \text{ m}})$$
$$\underline{91' \ 0"} \quad (\underline{27.7 \text{ m}})$$

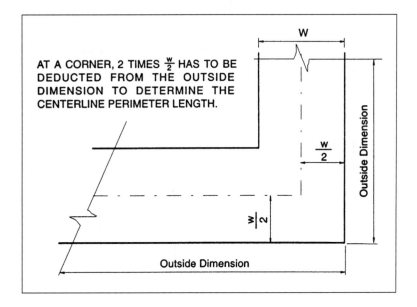

AT A CORNER, 2 TIMES $\frac{w}{2}$ HAS TO BE DEDUCTED FROM THE OUTSIDE DIMENSION TO DETERMINE THE CENTERLINE PERIMETER LENGTH.

**Figure 3.8**    Plan of Corner of Wall

To arrive at the centerline length it is necessary to deduct the thickness of the wall at each corner, that is, $4 \times 8" = 2'8"$ ($4 \times 200$ mm $= 800$ mm), which makes the total centerline 88'4" (26.9 m). The reason for making this adjustment at the corners can be seen from the enlarged detail of a corner shown in Figure 3.8.

When the first perimeter length was calculated, it was to the exterior corner, but to obtain the centerline perimeter amount, $\frac{w}{2}$ has to be deducted from each dimension. So, at this corner, 2 times $\frac{w}{2}$ has to be deducted and 2 times $\frac{w}{2}$ = the width of the wall, 8" in this case. The full calculation of the centerline perimeter would appear like this:

$$
\begin{array}{llll}
2 \times 26'\ 0" & = \underline{\phantom{x}52'\ 0"} & \text{or} & 2 \times 7.9\,\text{m} & = \underline{\phantom{x}15.80\,\text{m}} \\
2 \times 19'\ 6" & = \underline{39'\ 0"} & & 2 \times 5.95\,\text{m} & = \underline{11.90\,\text{m}} \\
& \phantom{=} 91'\ 0" & & & \phantom{=} 27.70\,\text{m} \\
\\
\text{minus } 4 \times 8" & = \underline{\langle\ 2'\ 8"\ \rangle} & & \text{minus } 4 \times .20\,\text{m} & \underline{\langle 0.80 \rangle\ \text{m}} \\
& \phantom{=} \underline{\underline{88'\ 4"}} & & & \phantom{=} \underline{\underline{26.90\,\text{m}}}
\end{array}
$$

In Figure 3.9 the wall is not a rectangle on plan but it cuts in at two corners. The external perimeter can still be obtained by adding 2 times the extreme length and 2 times the extreme width as shown. The centerline perimeter is determined, once again, by making an adjustment for four exterior corners as the effect of an interior corner is balanced by an exterior corner, which is verified on the sketch of a pair of corners shown on Figure 3.10.

Figures 3.11 and 3.12 show additional examples of perimeter calculations for a number of different building shapes.

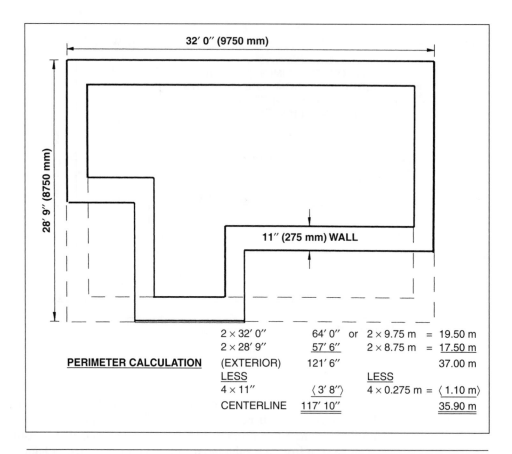

|  |  |  |  |  |
| --- | --- | --- | --- | --- |
| 2 × 32′ 0″ | 64′ 0″ | or | 2 × 9.75 m | = 19.50 m |
| 2 × 28′ 9″ | 57′ 6″ |  | 2 × 8.75 m | = 17.50 m |
| **PERIMETER CALCULATION** (EXTERIOR) | 121′ 6″ |  |  | 37.00 m |
| LESS |  |  | LESS |  |
| 4 × 11″ | ⟨ 3′ 8″⟩ |  | 4 × 0.275 m | = ⟨ 1.10 m⟩ |
| CENTERLINE | 117′ 10″ |  |  | 35.90 m |

**Figure 3.9**  Plan and Perimeter Calculation of Wall

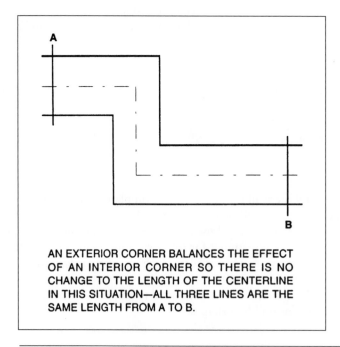

AN EXTERIOR CORNER BALANCES THE EFFECT OF AN INTERIOR CORNER SO THERE IS NO CHANGE TO THE LENGTH OF THE CENTERLINE IN THIS SITUATION—ALL THREE LINES ARE THE SAME LENGTH FROM A TO B.

**Figure 3.10**  Plan of Exterior/Interior Corner

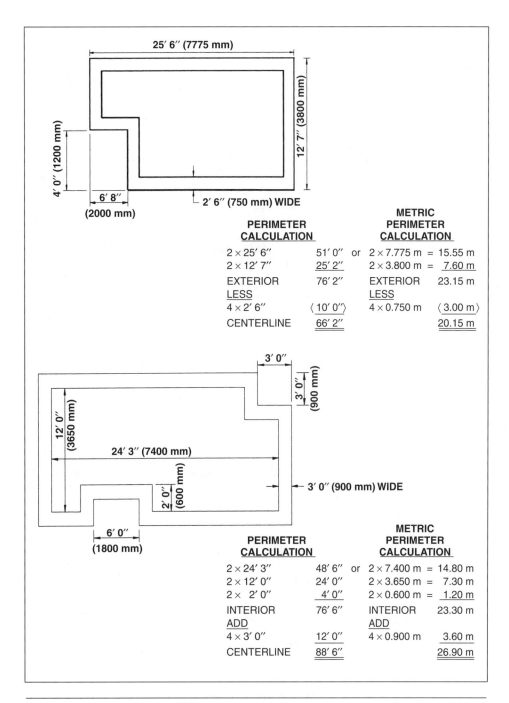

**Figure 3.11**   Wall Plans and Perimeter Calculations

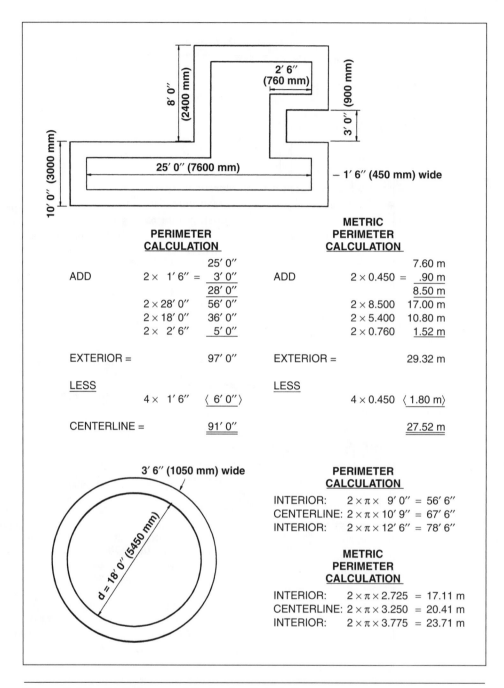

**Figure 3.12**  Wall Plans and Perimeter Calculations

## SUMMARY

- A quantity takeoff is defined as a process of measuring the work of a project in the form of a list of quantified work items measured in accordance with a uniform set of rules.
- The rules for measurement are contained in the "measuring notes" for each trade found in the following chapters. Collectively these rules amount to a "Method of Measurement" that can be shared by a group of estimators so that their takeoffs and the data they use are handled in a consistent manner.
- Note that quantities of work are measured "net in place" in the takeoff process and this is different from measuring quantities of materials.
- There are two systems of measurement used in the North American construction industry: the traditional English units and metric units.
- Estimators are advised to work entirely in English or metric units according to the units used on the project drawings since mixing the two systems of measurement in an estimate increase the likelihood of error.
- For reasons of consistency and objectivity, work is measured "net in place"; this means that quantities are calculated using the dimensions shown on the drawings with no adjustments made for waste factors and suchlike.
- Takeoff items comprise dimensions and a description; dimensions are always recorded in the order of length, width and height.
- Descriptions contain sufficient information for the estimator to price the work.
- The quantity takeoff should accurately reflect the amount of work in the project. While high accuracy is possible with well-detailed concrete items, a margin of error has to be allowed with excavation work as it is never so well-detailed on project drawings.
- The order of the takeoff will generally follow the sequence of construction, although some estimators prefer to measure concrete work before an assessment of earthwork requirements is undertaken.
- A strategy is needed for completing a takeoff of larger projects:
  a. Divide the project into manageable-size parts.
  b. Consider one part of the project at a time.
  c. Measure the work as a series of assemblies.
  d. Complete the takeoff for one assembly before passing on to the next.
- Specially printed stationery designed for each of the various estimating procedures is used to increase productivity and contribute to the accuracy of an estimate.
- Computers have become an important tool for the estimator. Their use can greatly increase the efficiency of the estimating process and also lead to improved accuracy.
- The estimator needs to be able to calculate the area of various shapes and calculate the volume of a variety of solid figures.
- It is also useful for the estimator to be able to quickly calculate the centerline perimeter of the plan of a trench/footing/wall around a building, given the dimensions of the building.

**REVIEW QUESTIONS**

1. What is the objective of a quantity takeoff?
2. Why is it recommended that estimators follow a standard method of measurement when preparing a takeoff?
3. What is the difference between an estimator's quantity takeoff and a foreman's material takeoff?
4. Estimators usually measure "net in place" quantities. What does this mean?
5. What are the two components that make up a takeoff item?
6. In what order are dimensions entered onto the takeoff?
7. Why should a figured dimension be used in preference to a dimension scaled off a drawing?
8. Why should an estimator include headings and side notes throughout a takeoff?
9. The cost of transportation of materials has to be allowed for in the estimate, but this item is not usually measured or even referenced in the takeoff. List six other similar items.
10. What level of accuracy should the estimator pursue in the measurement of work in a takeoff?
11. What is an assembly and how is the notion of assemblies used in the takeoff process?
12. Calculate the perimeter and the area of the parallelograms with the following dimensions (refer to Figure 3.5):

    a = 12'0"    a = 3600 mm
    b = 8'3"    b = 2000 mm
    h = 8'0"    h = 2400 mm

13. Calculate the perimeter and the area of the triangles with the following dimensions (refer to Figure 3.5):

    a = 10'0"    a = 2750 mm
    b = 9'6"    b = 2050 mm
    c = 3'0"    c = 300 mm
    h = 2'0"    h = 600 mm

14. Calculate the perimeter and the area of the circles with the following dimensions (refer to Figure 3.5):

    r = 14'0"    r = 4250 mm

15. Calculate the volume of the prisms with the following dimensions (refer to Figure 3.6):

    a = 4'0"    a = 1250 mm
    b = 19'6"    b = 5950 mm
    h = 12'0"    h = 3250 mm

16. Calculate the volume of the cylinders with the following dimensions (refer to Figure 3.6):

    r = 25'0"    r = 7500 mm
    h = 11'3"    h = 3625 mm

17. Calculate the volume of the cones with the following dimensions (refer to Figure 3.6):

    r = 4'9"    r = 1750 mm
    h = 12'0"    h = 3250 mm

18. Calculate the volume of the pyramids with the following dimensions (refer to Figure 3.6):

    L = 36'0"    L = 10,975 mm
    W = 32'6"    W = 9900 mm
    h = 10'0"    h = 3000 mm

# 4

# MEASURING SITEWORK, EXCAVATION AND PILING

## OBJECTIVES

*After reading this chapter and completing the review questions, you should be able to do the following:*

- Make an assessment of sitework requirements from project drawings and specifications.
- Identify sources in addition to plans and specifications that can be consulted when measuring the sitework on a project.
- Determine the likely soil conditions at a site by studying a soils report.
- Explain the use of swell factor and compaction factor when calculating volumes of excavation and backfill.
- Define bank measure and explain how it applies to sitework takeoffs.
- Explain why the sitework estimator needs a thorough knowledge of OSHA requirements for excavation work.
- Describe the use of an electronic digitizer and its function in sitework takeoffs.
- Explain how excavation work, backfill work and fill materials are measured in a takeoff.
- Describe and use the "grid method" of calculating cut and fill volumes.
- Describe and use the "section method" of calculating cut and fill volumes.
- Identify what a general contractor's estimator usually measures when the piling work of a project is subcontracted out.
- Describe how bearing piles are classified and measured in a takeoff.
- Describe how sheet piling is measured in a takeoff.
- Complete a manual takeoff of sitework.
- Complete a computer takeoff of sitework using the Precision Engineering software.

## Generally

Measuring sitework and excavation work is different from measuring most other work of a construction project because bid drawings usually provide very little detail about the specific requirements of sitework operations. The drawings will give details of the new construction required for the project, but information about what is currently to be found at the site of the proposed work may not be provided. Contractors are usually advised in the **"Instructions to Bidders"** to satisfy themselves as to the present condition of the site. Furthermore, the size and depth of foundations may be defined in detail on the drawings, but there is typically nothing disclosed about the dimensions and shape of the excavations required to accommodate these constructions. Before the estimator can measure the site work, he or she has to make an assessment of excavation requirements from data in the specifications and drawings and from information gathered on visits to the site. As the site conditions will also impact the pricing of many other items in the estimate, the estimator should make use of a checklist of items to consider on a site visit so that important information is not overlooked. This is discussed in Chapter 2 under "Site Visit."

## Soils Report

Another source of site information available on most large projects and also on some of the smaller jobs is the soils report. The report can sometimes be found bound with the specifications, or it may be accessible at the design consultant's office. Even when a soils report is included with the specifications, it is usually not a bid document. The estimator is cautioned to read the contract documents carefully to determine the true status of the soils report because, while some contracts contemplate extra compensation to a contractor who is misled by information contained in the report, other contracts may state that the contractor should not rely on information provided by the soils report but should make his or her own investigation of subsurface conditions. In this latter situation the prudent contractor has to assess the risk imposed by this contract condition and adjust the bid accordingly.

The soils report provides information about the subsurface conditions at the site obtained from bore holes or other investigations made by the soils engineer. The purpose of the report is to furnish data about the site that is necessary to undertake the design of the foundation system. Indeed, the report usually includes advice about the type of foundation systems suitable for the conditions encountered. However, the information contained in the report can also be very useful to the excavator as it not only discloses the type of soil that will probably be found at the site, but also provides an indication of the moisture content of the soil. Both of these factors can profoundly affect the cost of excavation work.

A soils report usually begins with an introduction identifying the name and location of the project, the name of the owner and the prime consultant, and states the objective of the study. Details of the investigation follow, including the date the site was examined, the method of investigation undertaken and general details of the bore hole procedures. Next the report addresses the nature of the subsurface conditions encountered and goes on to provide comments and recommendations about the foundation system together with detailed information about the test hole results.

See the section entitled "Subsurface Conditions" in Appendix A, "Extract from Typical Soils Report."

### Comments on the Sample Soils Report

The kind of information that the estimator would highlight and note from the soils report shown in Appendix A would include the following points:

1. **Topsoil** is 8" deep and is to be stripped from both the building area and the parking area.

2.  Most of the excavation on this project will be in the top silt-sand layer. This material is not usually difficult to excavate, but in wet conditions it can turn into a sticky mud that can bog down equipment.

3.  The moisture content of the silt-sand encountered was quite low and there was little ground water at the time of the investigation. However, the report indicates that "ground water levels are subject to wide fluctuations," so a contingency sum for dewatering may be necessary in the estimate.

4.  Soil compaction factors of a fairly high value are required. In order to obtain the factors specified, there will be a need for some rigorous compaction procedures and the water content of the soil will have to be carefully controlled.

5.  It is recommended that the sides of excavations (see section on Excavation Safety Considerations) be cut back to a slope of 1.5 Horizontal to 1.0 Vertical. This is a low slope ratio, which will result in wide cuts at the top of excavations. Shoring may be necessary if space is limited.

    The estimator should also keep in mind the OSHA (see section on Excavation Safety Conditions) requirements for cut-back of excavations of different depths.

6.  Good positive drainage of the finished grades is important on this project.

7.  Foundation concrete does not have to contain sulfate-resisting cement.

All of the soils report recommendations should be checked to determine if they are consistent with the contents of the project specifications. Note that it is the specifications that the contractor is required to follow, and they may vary from the recommendations of the soils engineer.

## Bank Measure, Swell and Compaction Factors

The soil that is extracted from an excavation is less dense than before it was excavated, so it will occupy more space than it did when it was in the ground. For example, if a hole of 10 cubic yards volume is excavated, the pile of soil removed from the hole might occupy 13 cubic yards; therefore, 13 cubic yards of material will have to be transported if it is required to be removed from the site. The difference between the volume of the hole and the volume of the material once it has been dug out is known as the **swell factor**.

A similar adjustment factor is required with regard to filling operations. If a hole of 10 cubic yards capacity is to be filled with gravel, 14 cubic yards of loose gravel may be required because the material will have to be compacted; that is, it will be more dense after it is deposited and there will also be some wastage of gravel to consider. Here the difference between the volume of the hole to be filled and the volume of fill material is referred to as the **compaction factor**.

In accordance with the general principle of measuring net quantities, excavation and backfill quantities are calculated using "**bank measure**." "Bank measure" amounts are obtained by using the dimensions of the holes excavated or filled with no adjustment made to the quantities obtained for swell or compaction of materials. Swell factors and compaction factors will be accounted for when the takeoff items are priced.

## Excavation Safety Considerations

The potential danger to workers in trenches and by the sides of excavations due to cave-ins of the earth embankments is a safety hazard that must be considered in every quantity takeoff of excavation work. OSHA's *Construction Safety and Health*

*Regulations* require that the sides of all earth embankments and trenches over 5'0" deep be adequately protected by a shoring system or by cutting back the sides of the excavations to a safe angle. As a consequence, the estimator must allow extra excavation for cutting back the face of excavations to a suitable angle wherever this is possible. Where restricted space or other circumstances prohibit cut-backs, the estimator may have to include for a system of shoring and bracing, but this is usually an inferior choice as it is a far more expensive alternative.

It is important that the estimator carefully studies the OSHA requirements for excavations since excavation rules are so strongly enforced by OSHA. The estimator will also need to consider particular state regulations regarding excavation safety requirements as specific details may vary from state to state. For instance, some states require that shoring systems, over a certain depth, be designed by and constructed under the supervision of a professional engineer. This requirement will have cost implications that cannot be ignored in the estimate.

## Use of Digitizers

Digitizers are electronic devices that enable the user to take measurements from drawings and input the data directly into a computer program. There are two main types of digitizers: sonic and tablet. Both of these types of digitizers employ a pointer or cursor to locate points and lines. With the sonic digitizer, the cursor emits a sonic code that is identified by two receivers that are used to calculate the precise location of the cursor. Using this system, any drawing can be scanned regardless of its size or the type of surface it is placed on. The sonic receivers have to be set up so that there is no obstruction between them and the cursor as it travels across the drawing, and the estimator has to ensure that measurements recorded are in accordance with the scale of the drawing. Clearly, drawings have to be drawn to scale for any type of digitizer to provide accurate data.

With tablet digitizers, drawings have to be laid out on an electronic tablet that functions to identify the specific location of the cursor as it moves over the surface of the tablet. Because of the size of construction project drawings, tablets as large as 42" × 60" (107 mm × 152 mm) and larger may have to be used if all the information on a drawing is to be accessible at one time.

The information gathered by the digitizer is then available for processing in computer programs so that lengths, areas, volumes, item counts and sophisticated calculations such as cut and fill volumes can be determined very swiftly. Digitizers can also be linked to estimating software systems and used as an alternative to the keyboard for inputting the data into the system. Some estimating systems can be operated directly from the digitizer thus eliminating any need to handle the keyboard. This setup provides a powerful tool for estimators because it allows large amounts of data to be quickly and accurately input into the computer without having to "key in" long lists of numbers.

One of the main disadvantages of the use of digitizers is that the accuracy of the system depends entirely on the accuracy of the drawings that are scanned. A fundamental takeoff principle is that the estimator should only rely on scaled dimensions as a last resort; sizes may be changed in the design process and are often carried out by modifying the figured dimensions without changing the actual size of objects shown on the drawings to comply with their new dimensions. This shortcutting results in drawings that wind up out of scale, and the digitizer system that scans these drawings generates erroneous output.

### Measuring Notes—Excavation and Backfill

1.  Excavations, backfill and fill material shall be measured in cubic yards or cubic meters "bank measure."
2.  If different types of materials will be encountered in the excavations, each type of material shall be described and measured separately.
3.  Excavations shall be classified and measured separately in the following categories:
    a. Site clearing
    b. Bulk excavation
    c. Basement excavations
    d. Trench excavations
    e. Pit excavations
4.  Hand excavation shall be measured separately.
5.  Quantities of different types of fill and backfill materials shall be kept separate.
6.  Fill and backfill materials shall be measured separately in the following categories:
    a. Bulk fill
    b. Backfill to basements
    c. Backfill to trenches
    d. Backfill to pits
    e. Gravel under slabs on grade
7.  An item of disposal of surplus soil shall be measured when excess excavated material is required to be removed from the site. Measurement of this item can be left until quantities are recapped ready for pricing, then the total volumes of excavation and common backfill can be used to determine the amount of surplus material.

    Alternatively, the price of disposal of surplus excavated materials can be included with the excavation price, in which case the description shall state the type of excavation "including removal of surplus soil."

## Calculation of Cut and Fill Using the "Grid Method"

Certainly the fastest and, probably, the most accurate way to calculate volumes of cut and fill over a site, when true scale drawings are available, is to use an electronic digitizer in conjunction with a software program specifically for this type of application (as discussed previously). Here we consider two alternative "manual" methods of obtaining the quantities of cut and fill beginning with the "grid method."

Calculation by the "grid method" requires a survey of the site showing the elevation of the existing grade at each intersection point on the grid. The elevation of the required new grade is also plotted at each intersection point, and from these two elevations the depth of cut or fill can then be obtained at each point. From here on, cut calculations are separate from fill calculations. To figure the volume of cut at an intersection point, the depth of cut at this point is multiplied by the area "covered" by that intersection point. Then adding together all the individual cut volumes computed in this way will give the total volume of cut on the site. Following the same process using the fill depths will establish the individual and total fill volumes for the site. The area "covered" by an intersection point means the area that point applies to as shown in Figure 4.1.

The accuracy of this method of calculating cut and fill volumes depends on the grid spacing; generally, the closer the grid spacing, the more accurate the results are.

The use of the "grid method" to calculate volumes of cut and fill requires the estimator to consider the depth of cut or fill at each point where the grid lines intersect (station) on the survey grid and then determine the "area covered" by that station.

At each station on the grid the elevation of the existing grade obtained from the site survey is noted in the top right quadrant. The elevation of the required new grade is noted in the top left quadrant. The difference between these grades gives the depth of fill or cut at this location. If it is a fill, the depth is noted in the bottom right quadrant; if a cut, the depth is noted in the bottom left quadrant.

| GRID INTERSECTION | |
|---|---|
| NEW GRADE ELEVATION | EXISTING GRADE ELEVATION |
| DEPTH OF CUT | DEPTH OF FILL |

Consider the accompanying survey grid. If the areas formed by the grid lines are "A" square feet, the "area covered" by point 1-A is one quarter of area "A," but point 1-B applies to two areas so the "area covered" by this point is two quarters of area "A," and so on as shown. The number of quarters of area "A" that the station point applies to is labeled the "frequency" when tabulating this data.

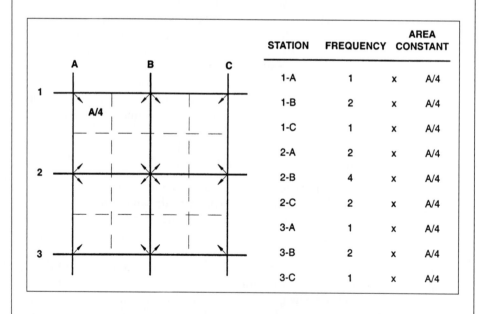

| STATION | FREQUENCY | | AREA CONSTANT |
|---|---|---|---|
| 1-A | 1 | x | A/4 |
| 1-B | 2 | x | A/4 |
| 1-C | 1 | x | A/4 |
| 2-A | 2 | x | A/4 |
| 2-B | 4 | x | A/4 |
| 2-C | 2 | x | A/4 |
| 3-A | 1 | x | A/4 |
| 3-B | 2 | x | A/4 |
| 3-C | 1 | x | A/4 |

**Figure 4.1**    Mass Excavation Calculations by Grid Method

However, a closer grid spacing leads to more calculations and a longer processing time. As this process is very repetitive, the processing time can be much reduced by using a computer program to perform the calculations.

See Figure 4.2 for a complete calculation of volumes of cut and fill over a site using this "grid method" with a computer spreadsheet program.

## Mass Excavation Calculations by the Grid Method (Grid 20' × 16')

**Grid Elevations**

|   | A |  | B |  | C |  | D |  | E |  |
|---|---|---|---|---|---|---|---|---|---|---|
| **1** | 4.2 | 6.5 | 4.4 | 5.0 | 4.6 | 3.0 | 4.8 | 1.9 | 5.0 | 2.2 |
|   | 2.3 |  |  | 0.6 |  | 1.6 |  | 2.9 |  | 2.8 |
| **2** | 4.4 | 5.1 | 4.6 | 3.2 | 4.8 | 2.8 | 5.0 | 4.5 | 5.2 | 5.2 |
|   | 0.7 |  |  | 1.4 |  | 2.0 |  | 0.5 | 0.0 | 0.0 |
| **3** | 4.6 | 3.6 | 4.8 | 2.0 | 5.0 | 5.3 | 5.2 | 7.1 | 5.4 | 7.9 |
|   |  | 1.0 |  | 2.8 |  | 0.3 |  | 1.9 |  | 2.5 |
| **4** | 4.8 | 1.9 | 5.0 | 4.0 | 5.2 | 8.2 | 5.4 | 10.0 | 5.6 | 10.3 |
|   |  | 2.9 |  | 1.0 |  | 3.0 |  | 4.6 |  | 4.7 |
| **5** | 5.0 | 3.0 | 5.2 | 3.8 | 5.4 | 6.4 | 5.6 | 7.0 | 5.8 | 7.5 |
|   |  | 2.0 |  | 1.4 |  | 1.0 |  | 1.4 |  | 1.7 |

**Tabulation of Results**

| Station | New Elevation | Existing Elevation | Depth Cut | Depth Fill | Frequency | Area Constant | Volume Cut | Volume Fill |
|---------|---------------|--------------------|-----------|-----------|-----------|---------------|------------|-------------|
| 1A | 4.2 | 6.5 | 2.3 | 0.0 | 1 | 80 | 184 | 0 |
| 1B | 4.4 | 5.0 | 0.6 | 0.0 | 2 | 80 | 96 | 0 |
| 1C | 4.6 | 3.0 | 0.0 | 1.6 | 2 | 80 | 0 | 256 |
| 1D | 4.8 | 1.9 | 0.0 | 2.9 | 2 | 80 | 0 | 464 |
| 1E | 5.0 | 2.2 | 0.0 | 2.8 | 1 | 80 | 0 | 224 |
| 2A | 4.4 | 5.1 | 0.7 | 0.0 | 2 | 80 | 112 | 0 |
| 2B | 4.6 | 3.2 | 0.0 | 1.4 | 4 | 80 | 0 | 448 |
| 2C | 4.8 | 2.8 | 0.0 | 2.0 | 4 | 80 | 0 | 640 |
| 2D | 5.0 | 4.5 | 0.0 | 0.5 | 4 | 80 | 0 | 160 |
| 2E | 5.2 | 5.2 | 0.0 | 0.0 | 2 | 80 | 0 | 0 |
| 3A | 4.6 | 3.6 | 0.0 | 1.0 | 2 | 80 | 0 | 160 |
| 3B | 4.8 | 2.0 | 0.0 | 2.8 | 4 | 80 | 0 | 896 |
| 3C | 5.0 | 5.3 | 0.3 | 0.0 | 4 | 80 | 96 | 0 |
| 3D | 5.2 | 7.1 | 1.9 | 0.0 | 4 | 80 | 608 | 0 |
| 3E | 5.4 | 7.9 | 2.5 | 0.0 | 2 | 80 | 400 | 0 |
| 4A | 4.8 | 1.9 | 0.0 | 2.9 | 2 | 80 | 0 | 464 |
| 4B | 5.0 | 4.0 | 0.0 | 1.0 | 4 | 80 | 0 | 320 |
| 4C | 5.2 | 8.2 | 3.0 | 0.0 | 4 | 80 | 960 | 0 |
| 4D | 5.4 | 10.0 | 4.6 | 0.0 | 4 | 80 | 1472 | 0 |
| 4E | 5.6 | 10.3 | 4.7 | 0.0 | 2 | 80 | 752 | 0 |
| 5A | 5.0 | 3.0 | 0.0 | 2.0 | 1 | 80 | 0 | 160 |
| 5B | 5.2 | 3.8 | 0.0 | 1.4 | 2 | 80 | 0 | 224 |
| 5C | 5.4 | 6.4 | 1.0 | 0.0 | 2 | 80 | 160 | 0 |
| 5D | 5.6 | 7.0 | 1.4 | 0.0 | 2 | 80 | 224 | 0 |
| 5E | 5.8 | 7.5 | 1.7 | 0.0 | 1 | 80 | 136 | 0 |
|   |   |   |   |   |   |   | 5200 cu. ft. | 4416 cu. ft. |
|   |   |   |   |   |   |   | 193 cu. yd. | 164 cu. yd. |

**Summary**

| Bulk Cut | — | 193 cubic yards |
|----------|---|-----------------|
| Bulk Fill | — | 164 cubic yards |
| Dispose Surplus | — | 29 cubic yards |

**Figure 4.2**  Sample Calculation Using Computer Spreadsheet Program

## Calculation of Cut and Fill Using the Section Method

This method of calculating volumes of cut and fill is mostly used with long, relatively narrow areas of cut and fill of the type encountered in road and railroad construction. Procedure:

1. At regular intervals (stations) along the centerline of the construction, survey the existing ground elevations on each side of the centerline.

    Alternatively, make use of a general survey of the site. On this survey, locate the centerline of the project and mark off stations at regular intervals along its length. Then determine the existing ground elevations to each side of the centerline at each station.
2. On graph paper, plot the cross-section at each station showing the existing grade and the required new grade.
3. From the plot of the cross-section, compute the area of cut and the area of fill at each station.
4. Using the station spacing and the area of cut at each station, calculate the total volume of cut. Similarly, from the station spacing and the area of fill at each station, the total fill volume can be obtained.

    The volume of cut or fill can be calculated by tabulating the data as shown on Figure 4.3 or by using the following formula:

$$\text{Volume} = \text{Station Spacing} \times \left( \text{Area 1} + \text{Area N} + \frac{\text{Sum of Other Areas}}{2} \right)$$

where N is the number of stations and Area N is the area at the last station.

Closer spacing of the stations can improve accuracy but the quality of the results obtained from this method mostly depends on the accuracy of the plotting of the section profiles. As with the "grid method," there are a number of computer programs available for use with the section method that can improve the accuracy and productivity of this procedure. Digitizers are a particularly useful tool for analyzing the profile of the sections to determine the areas of cut and fill.

See Figure 4.3 for a worked example of the section method of calculating volumes of cut and fill.

## Measuring Notes—Piling

Piling is often subcontracted out to companies specialized in this work, in which case the subtrade may provide a lump sum for the complete piling system. However, even when such a lump sum quote for the work is obtained, the general contractor will possibly have to perform some work in connection with this trade. Typical general contractor items in connection with piling include:

a. Layout of piles
b. Cutting off the tops of piles
c. Removing the excavated material produced in piling operations

In this example five stations have been laid out at 20-foot intervals across a proposed parking lot area. At each station the existing grade is surveyed and plotted on a cross-section profile, and the elevation of the new grade has also been plotted on the cross-section as shown:

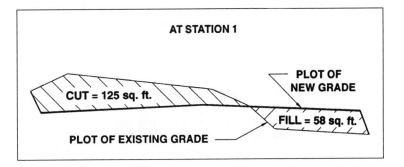

Cross-sections at stations 2 through 5 are also plotted, and the areas of cut and fill are calculated at each station. The areas of cut and fill calculated from all cross-sections are as follows:

| Station | Area of Cut (sq. ft.) | Area of Fill (sq. ft.) |
|---|---|---|
| 1 | 125 | 58 |
| 2 | 92 | 73 |
| 3 | 51 | 185 |
| 4 | 27 | 243 |
| 5 | 75 | 129 |

From this data the volumes of cut and fill can be calculated on a spreadsheet in the following way: the volume of cut = section spacing × average cut, and volume of fill = section spacing × average fill:

**Tabulation of Results**

| Station | Section Spacing | Actual Cut | Average Cut | Actual Fill | Average Fill | Volume Cut (cu. ft) | Volume Fill (cu. ft.) |
|---|---|---|---|---|---|---|---|
| 1 | | 125.0 | | 58.0 | | | |
| | 20.0 | | 108.5 | | 65.0 | 2,170.0 | 1,310.0 |
| 2 | | 92.0 | | 73.0 | | | |
| | 20.0 | | 71.5 | | 129.0 | 1,430.0 | 2,580.0 |
| 3 | | 51.0 | | 185.0 | | | |
| | 20.0 | | 39.0 | | 214.0 | 780.0 | 4,280.0 |
| 4 | | 27.0 | | 243.0 | | | |
| | 20.0 | | 51.0 | | 186.0 | 1,020.0 | 3,720.0 |
| 5 | | 75.0 | | 129.0 | | | |
| | | | | | | 5,400.0 | 11,890.0 |
| | | | | cubic yards | | 200.0 | 440.0 |

**Figure 4.3**    Mass Excavation Calculations by Section Method

If these, or any other work items, are required of the general contractor, they should be measured in the general contractor's takeoff.

Some piling subtrades are reluctant to offer a lump sum price for the work but will quote on a per pile or a per linear foot basis for the type of pile specified. In this case the contractor will have to measure the pile work specifying the type and size of pile required and measuring both the number and total length of each type of piling.

In the case in which the general contractor is to perform the piling work, a more detailed takeoff of bearing piles or sheet piling is required in accordance with the following rules of measurement.

### Bearing Piles, Generally

1. Bearing piles shall be described and measured in linear feet or linear meters, classified as to type, section and length. The unit of measurement covering all handling, pitching and driving shall be per linear foot (meter), measured from tip to cut-off level along the axis of the pile, except that the supply of the pile driving equipment shall be covered by a lump sum.
2. Batter piles, piles not fully embedded, piles driven in water or piles driven under other special circumstances shall be separately measured.
3. Where considered desirable, measurement of piling may be subdivided into (a) supply of piling materials and (b) handling and driving. Where this is done, the measurement of the supply item shall be on the basis of specified lengths as a minimum, that is, without deduction for cut-off.

*Classification of Bearing Piles*

4. Bearing piles shall be classified under one of the following headings, with separate descriptions and measurements for each:
   a. Round timber piles
   b. Sawn timber piles
   c. Steel piles
   d. Precast concrete piles
   e. Precast prestressed concrete piles
   f. Poured-in-place concrete piles (no permanent casing)
   g. Concrete-filled shell or pipe piles, closed end
   h. Open-end concrete-filled steel pipe piles
   i. Caisson piles
   j. Composite or combination type piles
   k. Special and patented type piles

*Timber Piles*

5. Timber piles shall be described and measured in linear feet or linear meters. Separate items are to be provided for piles up to 20 feet (6 m) long and in stages of 10 feet (3 m) thereafter.
6. Piles shall be described as to species of timber and size of butt and tip.
7. Where piles treated with preservative such as creosote are specified, the required preservative shall be given together with details of any special requirements such as pointing or shoeing of piles and ringing with steel bands, which shall be measured per pile.

8.   Cut-off of timber piling shall be measured separately per pile and it shall be clearly stated if cut-off is to be carried out below water level. Any treatment of pile butts is to be detailed. If piles are to be spliced, details shall be given and measurement made per splice.

*Steel Piles*

9.   Steel "H" piles or other shapes of all-steel bearing piles shall be described and measured according to section, size and weight. Separate items are to be provided for piles in 10-foot (3-m) stages of length. Splices, if any, shall be fully described. The circumstances under which splices will be permitted shall be stated, and splices shall be measured by number.

10.   Cutting-off of piles shall be measured by number, according to section. If piles are to be painted or otherwise treated or protected, this shall be fully described, and measurement made per linear foot (meter) of pile painted or treated.

*Precast Concrete Piles*

11.   Precast concrete piles shall be fully described and measured according to section and length in 10-foot (3-m) stages of length.

12.   The units of measurement for precast concrete piles are to be:
     a. Supply and delivery of piles, complete, based on the lengths ordered by the Engineer—per linear foot (meter).
     b. Handling and driving of piles, based on lengths below cut-off—per linear foot.
     c. Cutting-off piles—number.

*Poured-in-Place Concrete Piles*

13.   Poured-in-place concrete piles shall be measured in linear feet (meters), measured from the top of the concrete to the base of the excavated shaft or driven temporary casing tube. Where a special base or bulb is to be formed, an additional length allowance shall be given for the base, this length to be stated.

14.   Where tops of cast-in-place piles are to be trimmed, an item shall be listed, per pile.

15.   All details as to shell type and thickness, internal reinforcement and so on shall be given.

16.   Pile shoes, plates or points shall be fully described and measured separately; similarly any pile caps or dowels.

17.   Cutting-off of piles shall be measured by number according to pile size.

## Sheet Piling, Generally

1.   Sheet piling shall be described and measured in square feet or square meters, classified as to type, section and length. The unit of measurement covering all handling, threading and driving shall be in square feet or square meters (measured in plane elevation along the center line of the piling). The measured area shall be the net area as placed in position, from tip to cut-off level.

2. Piles driven under differing conditions, piles not fully embedded, piles in confined areas, piles driven to a batter, piles driven over water or under any other special circumstances shall be separately measured.

3. Where considered desirable, measurement of piling may be subdivided into (a) supply of piling materials and (b) handling, threading and driving. Where this is done, the measurement of the supply item shall be on the basis of specified length as a minimum, that is, without deduction for cut-off portions.

4. Measurement for withdrawing of piles is to be made as described under "Extraction of Piles."

5. Cutting-off of sheet piling shall be measured per linear foot or per linear meter in plane elevation along the centerline of the piling.

6. Splices, if any, shall be fully described and measured by number.

7. Timber sheet piling, concrete sheet piling and steel sheet piling shall all be measured according to the foregoing principles.

*Extraction of Piles*

8. If piles are to be extracted, separate items are to be provided for extraction of each type and section of pile, on the basis of linear foot (meter) measurement for bearing piles and square foot measurement for sheet piles.

## Drawings

Figure 4.4 is a set of drawings for a project comprising a single family house, and Figure 4.5 is a set of drawings for an office/warehouse project. The takeoff examples in this chapter and in the chapters that follow are based on these drawings.

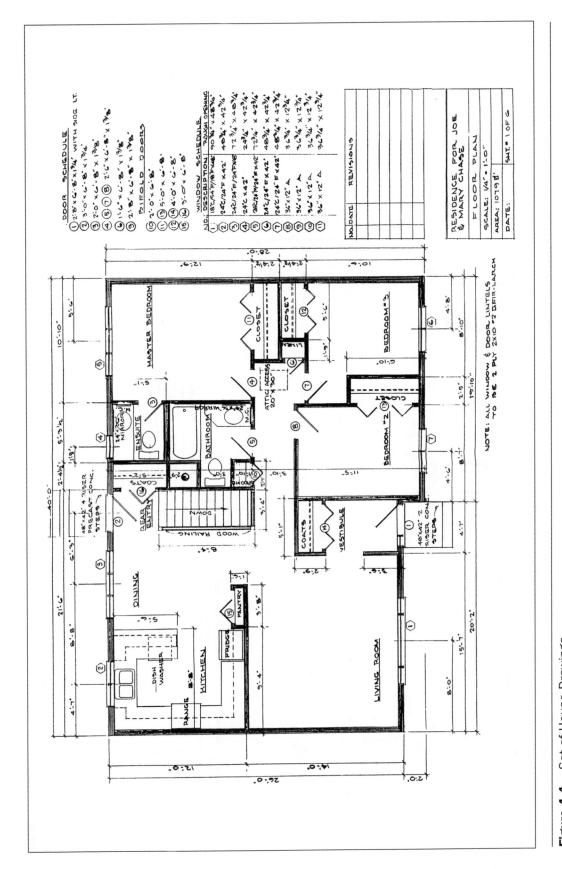

**Figure 4.4**    Set of House Drawings
(Drawings may not be to scale.)

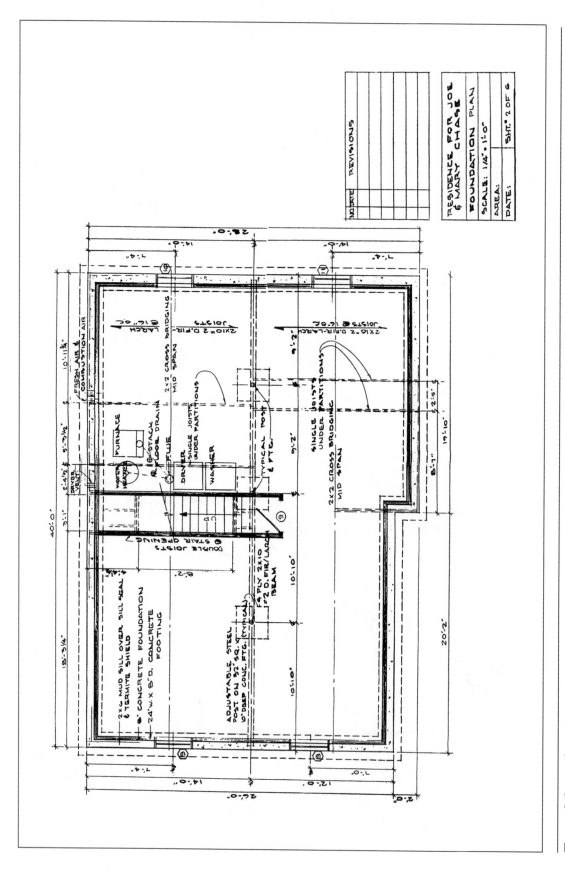

**Figure 4.4 continued**

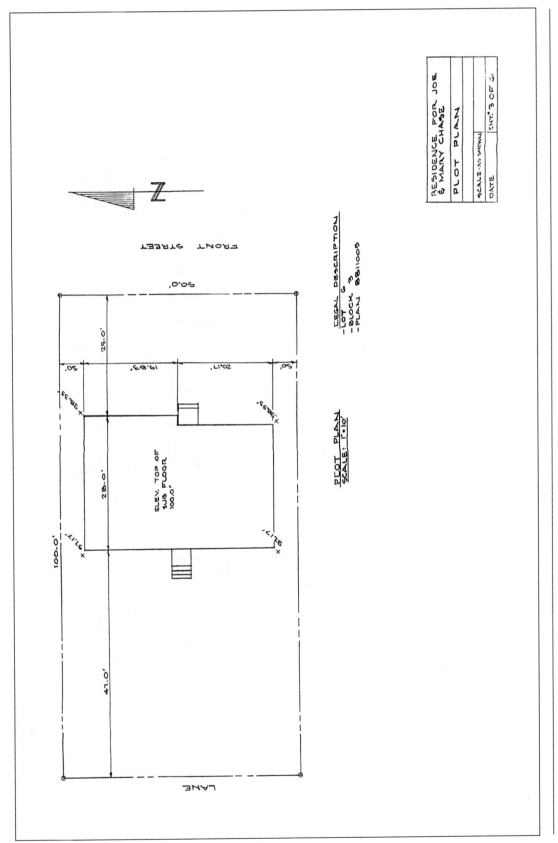

**Figure 4.4 continued**

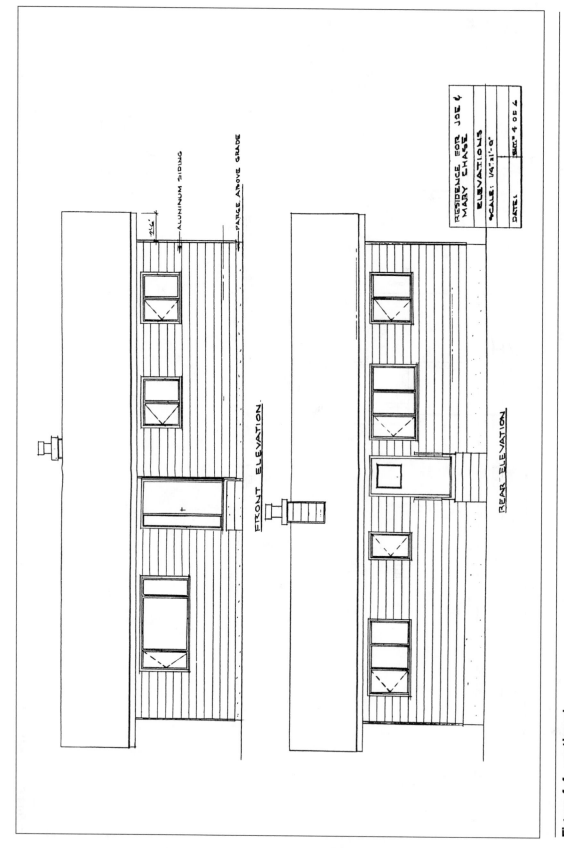

**Figure 4.4 continued**

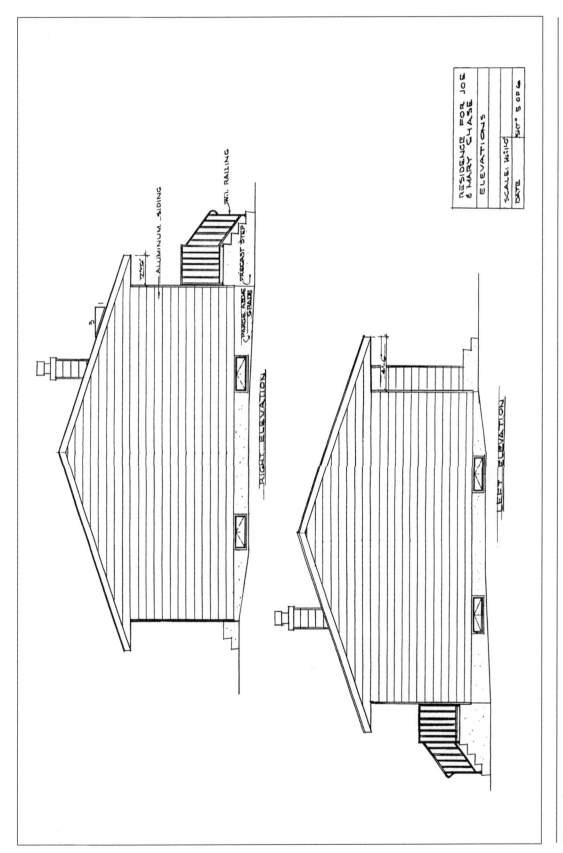

**Figure 4.4 continued**

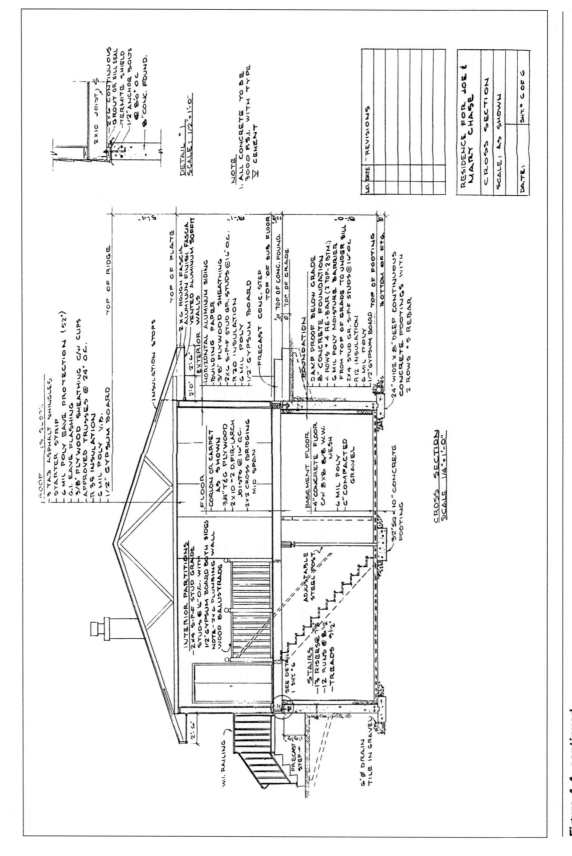

**Figure 4.4 continued**

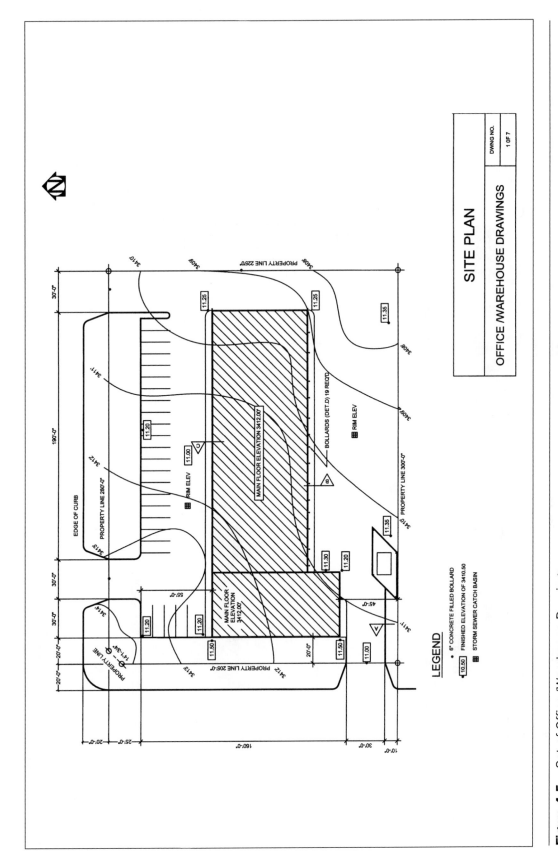

**Figure 4.5**    Set of Office/Warehouse Drawings
(Drawings may not be to scale.)

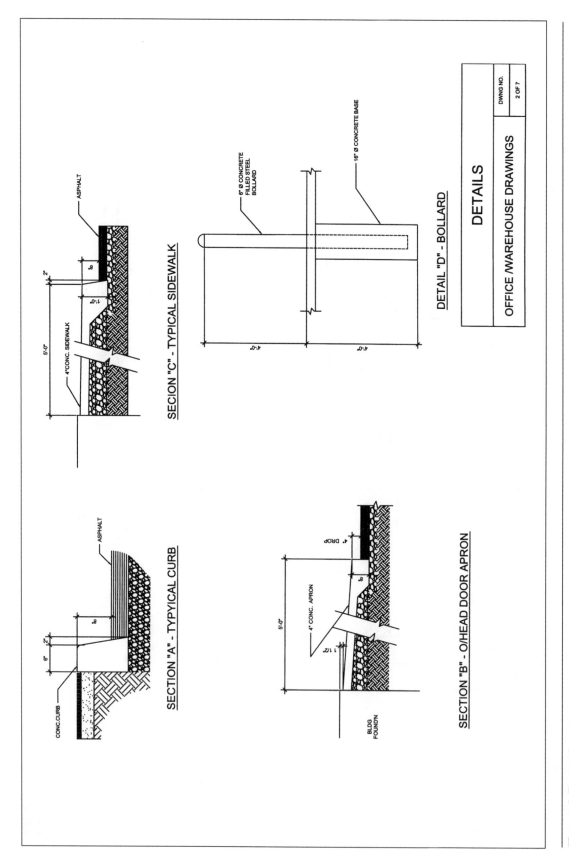

**Figure 4.5 continued**

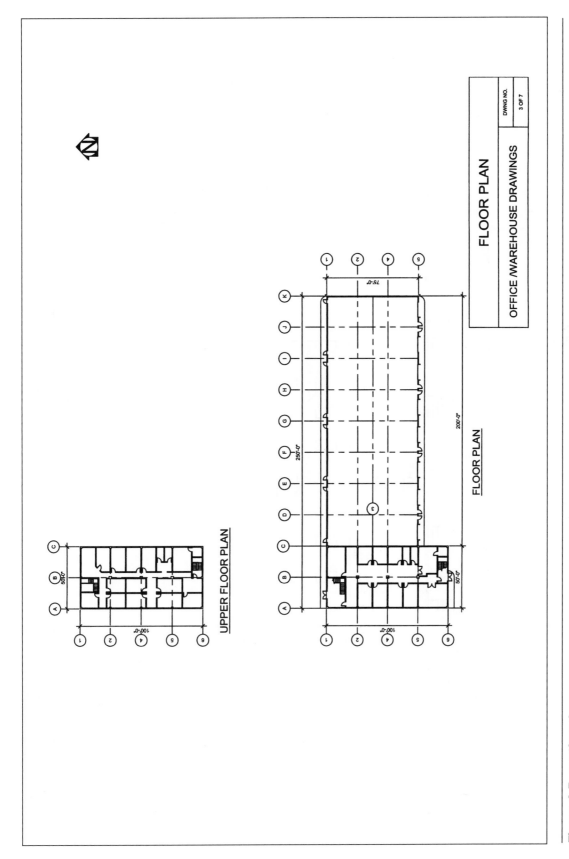

**Figure 4.5 continued**

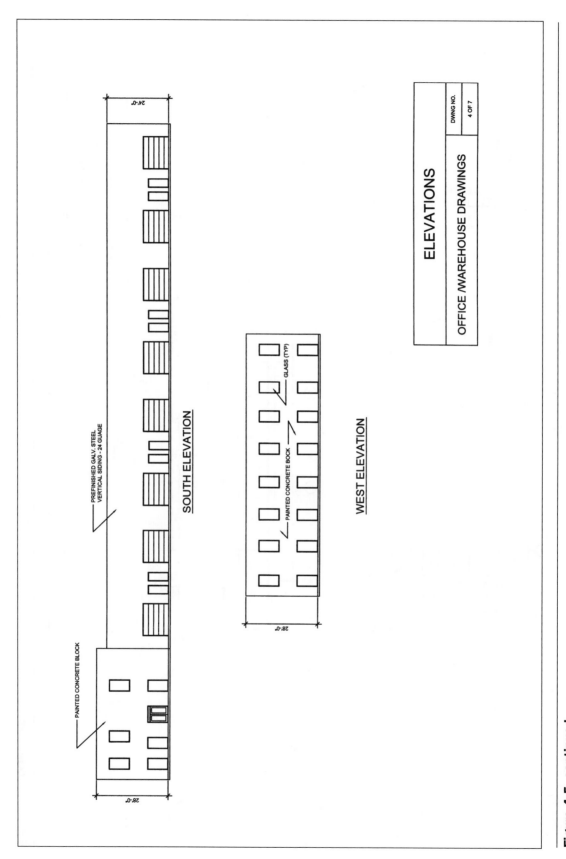

**Figure 4.5 continued**

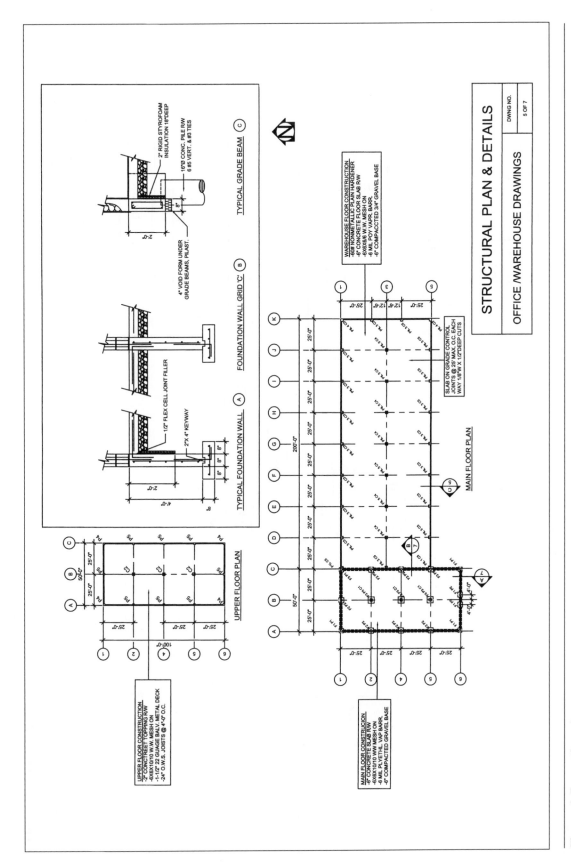

**Figure 4.5 continued**

### PAD FOOTINGS SCHEDULE

| | | | | | |
|---|---|---|---|---|---|
| F1 | 3'-6" X 3'-6" | 8" | 4 #5 E W | / | 4 #5 DWLS |
| F2 | 4'-0" X 4'-0" | 8" | 4 #5 E W | / | 4 #5 DWLS |
| F3 | 5'-0" X 5'-0" | 8" | 4 #5 E W | / | 4 #5 DWLS |

### PIERS SCHEDULE

| | | | | |
|---|---|---|---|---|
| P1 | 1'-4" X 1'-4" | 4'-0" | 4 #5 VERT | #3 T @ 12" |
| P2 | 1'-4" X 1'-4" | 4'-0" | 4 #5 VERT | #3 T @ 12" |
| P3 | 2'-0 " X 2'-0" | 2'-0" | 3 #5 VERT | #3 T @ 12" |
| P4 | 1'-4" X 1'-4" | N/A | 4 #5 VERT | / |
| P5 | 1'-4" X 1'-4" | N/A | 4 #5 VERT | / |

### PILASTERS SCHEDULE

| | | | | |
|---|---|---|---|---|
| PIL1 | 1'-4" X 1'-4" | 4'-0" | 4 #5 VERT | 5 #3 TIES |
| PIL2 | 1'-4" X 1'-4" | 4'-0" | 4 #5 VERT | 5 #3 TIES |
| PIL3 | 2'-0 " X 2'-0" | 2'-0" | 6 #5 VERT | 3 #3 TIES |
| PIL4 | 2'-0 " X 2'-0" | 2'-0" | 6 #5 VERT | 3 #3 TIES |
| PIL5 | 2'-0 " X 2'-0" | 2'-0" | 6 #5 VERT | 3 #3 TIES |

### COLUMNS SCHEDULE

| | | |
|---|---|---|
| C1 | 8"X8"X.250" | 4 ANCHOR BOLTS 3/4"ØX16" |
| C2 | 6"X6"X.250" | N/A |
| C3 | W12X40 | 4 ANCHOR BOLTS 1"ØX18" |
| C4 | W12X40 | 4 ANCHOR BOLTS 1"ØX18" |

### STRUCTURAL DETAILS

OFFICE /WAREHOUSE DRAWINGS

DWNG NO. 6 OF 7

PIER/PILASTER/ANCHOR BOLT PLANS

**Figure 4.5  continued**

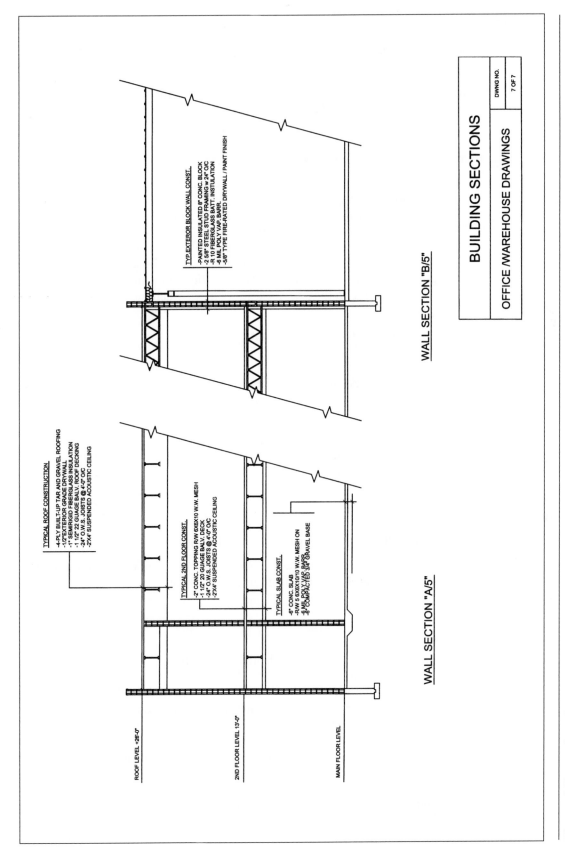

**Figure 4.5  continued**

## Sitework Takeoff

### Example 1—The House Project

Refer to Figure 4.6a for excavation and backfill takeoff. The site of this house has 6 inches of topsoil, and operations will begin by stripping the topsoil from the lot and stockpiling it on site. After the construction of the house the topsoil will be spread over the lawn area as a base for future sodding.

### Comments on the Takeoff Shown as Figure 4.6a

1. No calculations are required for the stripping topsoil item so this can be measured directly.
2. The average ground level over the area of the house is calculated from the elevations at the four corners of the house.
3. The depth of the basement excavation will extend from the ground level over the area of the house down to the level of the underside (u/s) of the gravel below the basement slab. The elevation of the underside of the gravel is not provided on the drawings, so we have to determine it from the information given:

| | |
|---|---:|
| The elevation of top of the main floor: | 100.0 |

(this is indicated on the site plan)

| | |
|---|---:|
| Minus: the thickness of the floor sheathing: | – ¾" |
| the depth of the joists: | – 9¼" |
| the sill plate thickness: | – 1½" |
| the height from footing to joists: | – 8'0" |
| the depth of the gravel: | – 6" |
| | – 9'5½" |

So elevation 100.0 – 9'5½" gives an elevation of 90'6½".

Then the difference between the average ground level and the elevation of the bottom of the gravel gives the depth of the basement excavation:

| | |
|---|---:|
| | 97'3" |
| | –90'6½" |
| depth of excavation: | 6'8½" |

4. The length and width of the basement excavation is based on the size of the house, so we start with the dimensions of the house to the outside of the walls shown on the basement plan.
   There is a 2'0" jog in the front wall of the house, which (theoretically) reduces the size of the basement excavation but can be ignored as the excavator will probably not make any adjustment for it.
5. A 1'0" wide work space is allowed outside of the foundation walls, and an allowance is also made for cutting back the sides of the excavation. (See Figure 4.6b.)

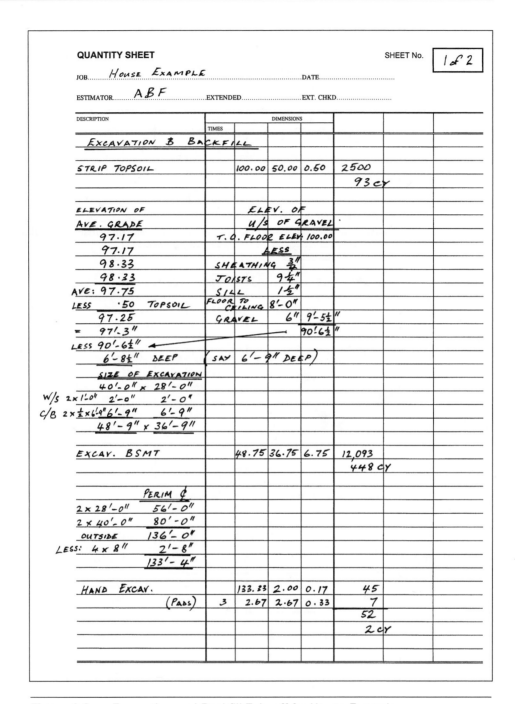

**Figure 4.6a**  Excavation and Backfill Takeoff for House Example

6.  The top of the perimeter footings, the top of the pad footings and top of the gravel under the basement slab are all level, but the perimeter footings extend 2" deeper than the gravel and the pad footings are 4" deeper. To obtain this extra depth an item of hand excavation is taken.

**QUANTITY SHEET**

SHEET No. 2 of 2

JOB    HOUSE EXAMPLE    DATE

ESTIMATOR    ABF    EXTENDED    EXT. CHKD

| DESCRIPTION | TIMES | DIMENSIONS | | | | | |
|---|---|---|---|---|---|---|---|
| EXTR. PERIM: 136'-0" | | | | | | | |
| ADD | | | | | | | |
| 4 x 2 x 1'-0"  8'-0" | | | | | | | |
| ℄ OF PIPE:  144'-0" | | | | | | | |
| | | | | | | | |
| 6" DIA. DRAIN TILE | | 144.00 | | | 144 LF | | |
| | | | | | | | |
| DRAIN GRAVEL | | 144.00 | 2.00 | 2.00 | 576 | | |
| | | | | | 21 CY | | |
| | | | | | | | |
| BACKFILL BSMT. | | AS EXCAV. | | | 12,093 | | |
| (BLAG:)   DDT | | 20.17 | 26.00 | 6.75 | ‹3540› | | |
| " | | 19.83 | 28.00 | 6.75 | ‹3748› | | |
| " | | DRAIN GRAVEL | | | ‹576› | | |
| | | | | | 4229 | | |
| | | | | | 157 CY | | |

**Figure 4.6a continued**

7.  The volume of the hand excavation for the perimeter footing = the centerline × the width of the footing × the 2" depth.

   The centerline calculation shows the perimeter on the outside of the wall to be 136'0" less the adjustment for the corners of 2'8" gives 133'4". This centerline length will be used a number of times in the takeoff.

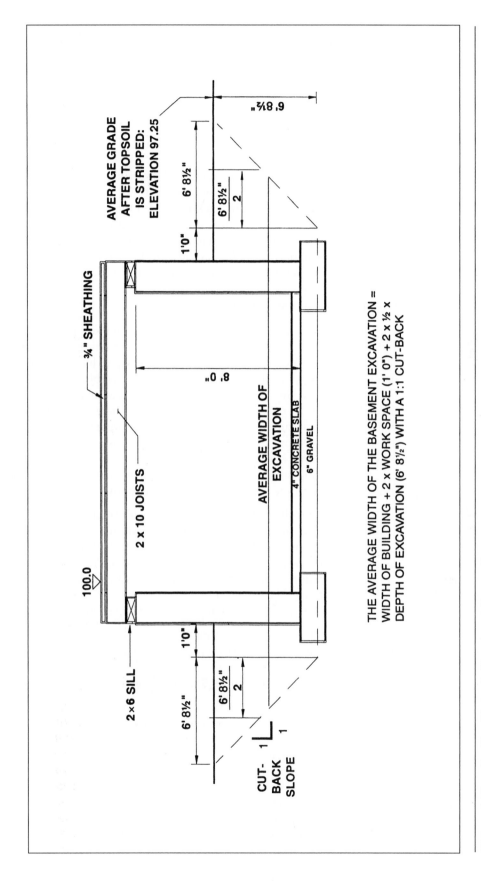

**Figure 4.6b**    Section Through Basement Excavation

8. A footing drain is located 1'0" outside of the foundation wall so the length of this drain is calculated from the outside perimeter of the house:

136'0"

Plus the adjustment for corners:

4 corners × 2/ 1.0 at each corner gives:

8'0"

144'0"

9. The footing drain is required to be surrounded by drain gravel so an area 2'0" × 2'0" is allowed around the pipe.
10. Finally, the amount of backfill around the foundations is calculated by deducting from the volume of basement excavation the volume displaced by the building and the volume displaced by the drain gravel.

    Note that the abbreviation "DDT" (Figure 4.6, 2 of 2) indicates a deduction and the amount of the deduction is placed in brackets.

### Example 2—The Office/Warehouse Project

For this second example of a sitework takeoff, we will be making use of the Precision Estimating Basic Edition software that accompanies this text. So that the student can see all the steps in a computerized estimate process, we have provided the estimator's hand notes that show how takeoff data are assembled before being input into the computer (see Figure 4.7a). More experienced estimators may not need to keep takeoff notes in as much detail as those shown here, but for the beginning estimator takeoff notes are very useful as a way of organizing the estimating process. These notes also provide a record of the dimensions entered into the computer and how these amounts were determined.

The following information was gathered from a visit to the site of the proposed office/warehouse project and from the specifications:

1. The site is clear of trees and any other obstructions.
2. The grade slopes as indicated by the contours on the site plan.
3. The ground has a sparse grass covering with about 6 inches of topsoil.
4. The soil at the site is a cohesive clay-sand mixture.
5. Excavations will not extend below the water table but a rainfall will entail dewatering of excavations.
6. Fill below slabs and paving shall be approved granular material.

The takeoff begins with the measurement of topsoil to be stripped from the building area and the paved parking areas around the building.

### Comments on the Takeoff Notes Shown as Figure 4.7a

1. Because of the slope of the site, part of the building area will have to be cut and part filled. By interpolating between the contours shown on the site plan, the elevation of the existing grade at each corner of the building can be found (see Figure 4.7b).

   As the top of the main floor elevation is 3412.0 and there is 6" of concrete and 6" of gravel below this, the elevation of the underside of the gravel is 3411.0. The contour where the cut meets the fill (the cut/fill line) will be 3411.5 because 6" of topsoil is removed before cut and fill operations begin. The approximate location of this cut/fill line is also shown on Figure 4.7b.

| NOTES | TIMES | DIMENSIONS | | |
|---|---|---|---|---|

**TAKEOFF NOTES**    SHEET No. $\boxed{1 \text{ of } 3}$

PROJECT:  OFFICE/WAREHOUSE    DATE:

ESTIMATOR:  A B F

| NOTES | | | TIMES | | DIMENSIONS | |
|---|---|---|---|---|---|---|
| BUILDING EXCAV. & BACKFILL | | | | | | |
| STRIP TOPSOIL 6" DEEP | | | | | | |
| (BUILDG & PARKING AREAS) | | | | 280.00 | 200.00 | 6" |
| (X - OVER) | | | | 50.00 | 30.00 | 6" |
| (X - OVER) | | | | 40.00 | 30.00 | 6" |
| SITE AREA | | | | | | |
| 50'-0" x 45'-0" | | | | | | |
| 200'-0" 100'-0" | | | | | | |
| 30'-0" 55'-0" | | | | | | |
| 280'-0" x 200'-0" | | | | | | |
| CUT & FILL ZONES | | | | | | |

| ZONE ① | ZONE ② | ZONE ③ | ZONE ④ | ZONE ⑤ |
|---|---|---|---|---|
| 11.50 | 11.50 | 13.00 | 11.50 | 11.50 |
| 12.60 | 11.50 | 11.50 | 10.30 | 9.50 |
| 13.00 | 10.90 | 11.50 | 11.30 | 9.00 |
| 11.50 | 33.90 | 36.00 | 11.50 | 10.30 |
| 48.60 | | | 44.60 | 40.30 |
| AVE: 12.15 | 11.30 | 12.00 | 11.15 | 10.08 |
| LESS: 11.50 | 11.50 | 11.50 | 11.50 | 11.50 |
| 0.65 CUT | 0.20 FILL | 0.50 CUT | 0.35 FILL | 1.43 FILL |

| NOTES | | | | TIMES | | DIMENSIONS | |
|---|---|---|---|---|---|---|---|
| BULK EXCAV. | | ZONE ① | | | 82.00 | 50.00 | 0.65 |
| | | ZONE ③ | | ½ | 86.00 | 64.00 | 0.50 |
| 100.00 | | | | | | | |
| 64.00 | | | | | | | |
| AVE: 82.00 | | | | | | | |
| GRAVEL FILL | | ZONE ② | | ½ | 36.00 | 50.00 | 0.20 |
| | | ZONE ④ | | | 86.00 | 43.00 | 0.35 |
| 75.00 | | ZONE ⑤ | | | 114.00 | 75.00 | 1.43 |
| -64.00 | | | | | | | |
| 11.00 | 200.00 | | | | | | |
| +75.00 | -86.00 | | | | | | |
| 86.00 | 114.00 | | | | | | |
| AVE: 43.00 | | | | | | | |

**Figure 4.7a**    Excavation and Backfill Takeoff Notes for Office/Warehouse Example

All elevations greater than 3411.5, which are to the left of this line on the sketch, will have to be cut and all elevations less than 3411.5, to the right, need to be filled.

To facilitate the calculation of volumes of cut and fill the area has been divided into five areas (zones). The average grade of each zone is then calculated, and from this the depth of cut or fill for the zone is obtained.

TAKEOFF NOTES SHEET No. 2 of 3

PROJECT: OFFICE/WAREHOUSE  DATE:

ESTIMATOR: A B F

| NOTES | | | | | TIMES | DIMENSIONS | | |
|---|---|---|---|---|---|---|---|---|
| EXCAV. FOR OFFICE | | | | | | | | |
| EXCAV. DEPTH | PADS F1 | PADS F2 | PADS F3 | STRIP FTGS | | | | |
| 4'-8" | 3'-6" | 4'-0" | 5'-0" | 2'-0" | | | | |
| LESS <1'-0"> | ADD w/s 2'-0" | 2'-0" | 2'-0" | 2'-0" | | | | |
| 3'-8" | 5'-6" | 6'-0" | 7'-0" | 4'-0" | | | | |
| EXCAV. TRE. | (F1) | | | | 5 | 5.50 | 5.50 | 3.67 |
| | (F2) | | | | 6 | 6.00 | 6.00 | 3.67 |
| OFFICE PERIM. | (F3) | | | | 2 | 7.00 | 7.00 | 3.67 |
| 2 x 100' 200'-0" | (STRIP FTG) | | | | | 219.83 | 4.00 | 3.67 |
| 2 x 50' 100'-0" | | | | | | | | |
| 300'-0" | | | | | | | | |
| LESS 4 x 8" < 2'-8"> | | | | | | | | |
| LESS 297'-4" | | | | | | | | |
| 5 x 5'6" 27'-6" | | | | | | | | |
| 6 x 6'0" 36'-0" | | | | | | | | |
| 2 x 7'0" 14'-0" <77'-6"> | | | | | | | | |
| 219'-10" | | | | | | | | |
| EXCAV. PITS | (F3) | | | | 3 | 7.00 | 7.00 | 1.67 |
| GRADE & TRIM | | | | | 3 | 7.00 | 7.00 | |
| B/FILL TRE. | | | | | | AS TRE. EXCAV. | | |
| | DDT (F1) | | | | <5 | 3.50 | 3.50 | 0.67> |
| STRIP FTG | " (F2) | | | | <6 | 4.00 | 4.00 | 0.67> |
| LESS 297'-4" | " (F3) | | | | <2 | 5.00 | 5.00 | 0.67> |
| 5 x 3'-6" 17'-6" | " (STRIP FTG) | | | | < | 245.83 | 2.00 | 0.67> |
| 6 x 4'-0" 24'-0" | " (WALL) | | | | < | 297.33 | 0.67 | 3.00> |
| 2 x 5'-0" 10'-0" <51'-6"> | " (PILS. 1) | | | | <4 | 0.67 | 0.67 | 3.00> |
| 245'-10" | " (PILS.2) | | | | <9 | 1.33 | 0.67 | 3.00> |

**Figure 4.7a  continued**

2. On this project the specifications do not allow excavated material to be used as fill under slabs, so "granular" fill is measured. This will be an imported gravel material that meets specification requirements.
3. Because the office has a different foundation system from the warehouse, the excavation for the office is kept separate from that of the warehouse.

**TAKEOFF NOTES**

SHEET No. 3 of 3

PROJECT: *Office / Warehouse*      DATE:

ESTIMATOR: *A B F*

| NOTES | | TIMES | DIMENSIONS | | |
|---|---|---|---|---|---|
| B/FILL PITS | | AS | PIT EXCAV. | | |
| DDT (F3) | ⟨ 3 | 5·00 | 5·00 | 0·67 ⟩ |
| " (P3) | ⟨ 3 | 2·00 | 2·00 | 1·00 ⟩ |

EXCAV. FOR WAREHOUSE

| EXCAV. DEPTH | | TRENCH LENGTH | |
|---|---|---|---|
| 2'-0" | 2 x 200' | 400'-0" | |
| ADD 4" | + | 75'-0" | |
| 2'-4" | | 475'-0" | |
| LESS ⟨1'-0"⟩ | LESS 2 x 8" ⟨1'-4"⟩ | | |
| 1'-4" | | 473'-8" (LENGTH OF GRADE BEAM) | |
| | LESS 2 L 1" | | |
| 2'-4" ⟨4'-5"⟩ | | (ADJUSTMENT FOR OVERLAP | |
| 469'-3" | | WITH OFFICE EXCAVATION) | |

| | | | | | |
|---|---|---|---|---|---|
| EXCAV. TRE | | | 469.25 | 2·67 | 1·33 |
| (PILS) | 16 | 4.00 | 1.33 | 1·33 |

PILASTERS

| | | | |
|---|---|---|---|
| 2'-0" x 1'-4" | | | |
| w/ 2 x 1' 2'-0" | | | |
| 4'-0" x 1'-4" | | | |

| | | | | |
|---|---|---|---|---|
| GRADE 3 TRIM | | 469.25 | 2·67 | |
| | 16 | 4.00 | 1·33 | |

| | | | | | |
|---|---|---|---|---|---|
| EXCAV. PITS (PIL 5) | 7 | 4.00 | 4.00 | 1·00 | |

| | | | | |
|---|---|---|---|---|
| GRADE 3 TRIM | 7 | 4.00 | 4·00 | |

| | | | | | |
|---|---|---|---|---|---|
| B/FILL TRE. | | AS | TRENCH EXCAV | | |
| DDT (G.B.) | ⟨ | 473.67 | 0·67 | 1·33 ⟩ |
| " (PILS 1) | ⟨ 2 | 1·33 | 0·67 | 1·00 ⟩ |
| " (PILS 3) | ⟨ 16 | 2·00 | 1·33 | 1·00 ⟩ |
| " (PILS 4) | ⟨ 2 | 1·33 | 1·33 | 1·00 ⟩ |

| | | | | | |
|---|---|---|---|---|---|
| B/FILL PITS | | AS | PIT EXCAV. | | |
| DDT (PILE CAPS) | ⟨ 7 | 2·00 | 2·00 | 1·00 ⟩ |

**Figure 4.7a continued**

4. The foundations around the office extend 4'8" below the floor level but the ground level is 1'0" below the top of floor (6" concrete + 6" of gravel), so the depth of the excavations is 3'8".

5. As this depth of excavation is shallow (less than 5'0") and the soil is cohesive, a minimum workspace of 1'0" is allowed from the face of footings.

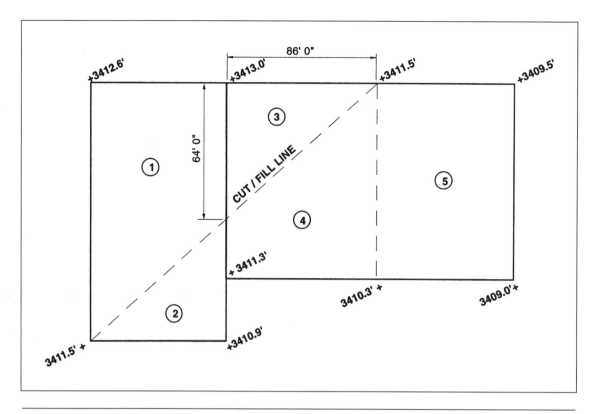

**Figure 4.7b**   Site Elevation for Office/Warehouse

6. Although the excavation around the perimeter of the office will be a single trench, the trench will be wider where the pad footings are located, so excavation for the pads is measured first, then excavation between the pads is calculated by taking the full perimeter and deducting excavations for the pads.

7. Excavation for isolated pads (pits) is measured separately as this will probably be more expensive than a continuous trench. Note that the depth of the inside pads (F3) is only 2'8" below the top of floor less 1'0" (concrete slab and gravel) gives 1'8" depth of excavation.

8. Backfill quantities are obtained by deducting from the excavation volumes the space taken up by the footings, the walls and the pilasters up to the ground level.

9. A 2'0" deep grade beam with a 4" deep void form below it is required to the north, south and east sides of the warehouse, this calls for a 1'4" deep trench. Note the centerline calculation only accounts for two corners as this is not a full perimeter calculation.

10. The perimeter trench is widened at the pilasters as shown in Figure 4.7c.

**Precision Estimating Takeoff**

Using Precision Estimating, we will input the information we have assembled in the preceding notes to complete the takeoff of the Office/Warehouse project. First you need to install the software. On the CD attached to the inside of the back cover of this book you will find a file called "Getting Started with Estimating," that explains

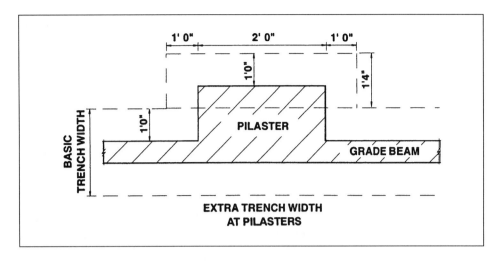

**Figure 4.7c**    Dimensions at Grade Beam Pilaster

how to install the Precision Estimating software and provides basic information on how to use this program.

Once you have installed Precision Estimating, you can follow the step-by-step procedure that follows to perform the computer-takeoff of the project work.

1.  Start up Precision Estimating and click on the New Estimate button found on the left end of the tool bar.
    ■ The New Estimate window will open to allow you to enter the file name. See Figure 4.8.

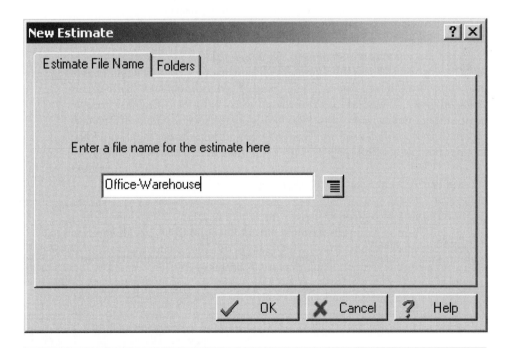

**Figure 4.8**    New Estimate Window

2. Before you close this window, click on the Folders tab. Here you can verify the name of the folder in which the estimate is to be stored and the name of the folder containing the database to be used for this estimate. See Figure 4.9.
   - We are using the "Sample Commercial GC" database, which is one of the sample databases provided with this software.
3. Click OK when you are satisfied with the Folders information.
   - The Spreadsheet window will open and the screen will be ready for you to begin the takeoff. At this stage you can input some information about this estimate using the Estimate Information window.
4. Click on the Takeoff item on the menu bar and select Estimate Information to open the Estimate Information window. See Figure 4.10.
   - You will find that the name of the project field in this window has been prefilled by default with the estimate name. All the other fields in this window are optional. Entering the Job Size here is useful as it enables the program to keep track of the cost per square foot of the estimate.
5. Enter "24,250" as the Job size and "sf" as the units.
   - This sample database has only a limited number of items, so we need to set up some new phases and items for this takeoff.
6. On the menu bar, click on Database and select Phases from the list. The Database Phase window will now open.
7. On this window, click Add and enter the following data:

   Phase:              2201.00
   Description:        Earthwk: Remove Topsoil
   Unit description:   CY

8. Click OK to create this new phase. See Figure 4.11.
9. Repeat these steps to set up the phases listed on Figure 4.12. Now we can set up the new items in these phases.

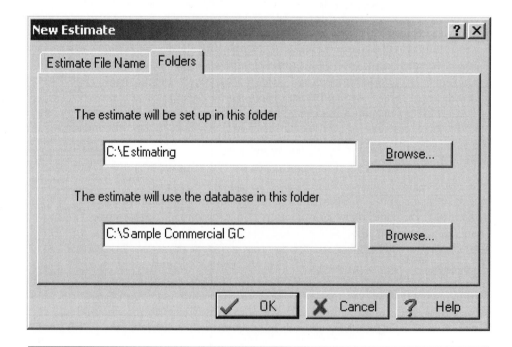

**Figure 4.9**   New Estimate Window—Folders

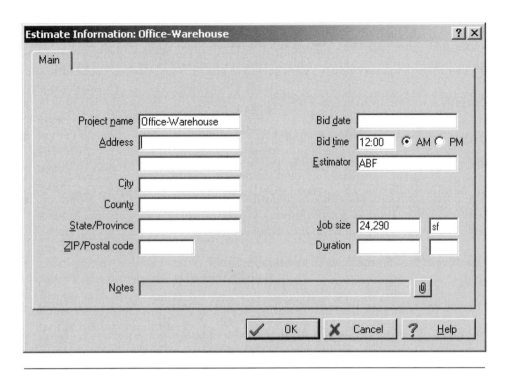

**Figure 4.10**    Estimate Information Window

10. On the menu bar, click on database and select Items from the list. The Database Item window will now open.

11. Click on the Add button and enter the following data:

Phase:        2201.00       Item: 10
Description:     Earthwk: Remove Topsoil
Takeoff Unit:     CY
Formula:     CY L × W × D"/27 (You can select this formula from the list that is accessible from this field.)

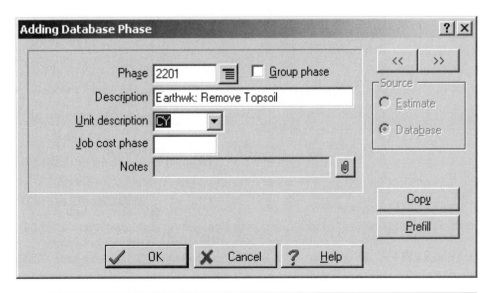

**Figure 4.11**    Database Phase Window

| Phase Code | Description | Unit |
|---|---|---|
| 2205.00 | Earthwk: Excavate Trench | CY |
| 2210.00 | Earthwk: Excavate Pits | CY |
| 2220.00 | Earthwk: Bulk Excavation | CY |
| 2220.40 | Earthwk: Grade & Trim | sf |
| 2221.50 | Earthwk: Gravel Fill | CY |
| 2222.00 | Earthwk: Dispose Surplus | CY |
| 2225.00 | Backfill Trenches | CY |
| 2230.00 | Backfill Pits | CY |

**Figure 4.12**    New Phases

- We will not need to enter prices at this stage, but we do have to select at least one Cost Category. For this item we will select both labor and equipment.
12. Click on New Category and select Labor.
13. Click the second New Category and this time select Equipment.
14. Click OK to create the new item. See Figure 4.13.
15. Repeat these steps to set up the items listed on Figure 4.14.

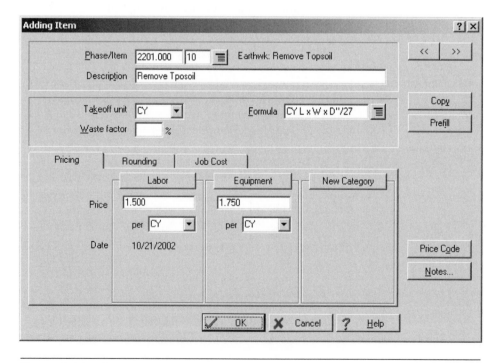

**Figure 4.13**    Database Item Window

| Phase Code | Item Code | Description | Unit | Formula | Cost Categories |
|---|---|---|---|---|---|
| 2205.00 | 10 | Excavate Trench | CY | CY L × W × D'/27 | L, E |
| 2210.00 | 10 | Excavate Pits | CY | CY L × W × D'/27 | L, E |
| 2220.00 | 10 | Bulk Excavation | CY | CY L × W × D'/27 | L, E |
| 2220.40 | 10 | Grade & Trim | sf | SF L × W | L, E |
| 2221.50 | 10 | Gravel Fill | CY | CY L × W × D'/27 | L, M, E |
| 2222.00 | 10 | Dispose Surplus | CY | none | L, E |
| 2225.00 | 10 | Backfill Trenches | CY | CY L × W × D'/27 | L, E |
| 2230.00 | 10 | Backfill Pits | CY | CY L × W × D'/27 | L, E |

**Figure 4.14**   New Items

16. Include a waste factor of 20% in item 2221.50 Gravel Fill and check the Apply Waste % box in the material cost category that can be found on the Rounding tab in the Database Item window (see Figure 4.13).
    - Now we are ready to begin the takeoff. There are various ways to proceed. You could pick an item, enter the dimensions for that item, and then move on to the next item. Alternatively, some people like to pick a number of items and enter the dimensions later.
17. Click on the Quick Takeoff button on the menu bar and select the group phase SITEWORK. See Figure 4.15.
18. Click on the Remove Topsoil phase. The items under this phase will now open.

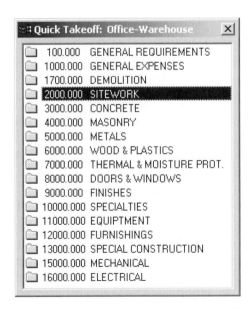

**Figure 4.15**   Quick takeoff—Database Group Phases

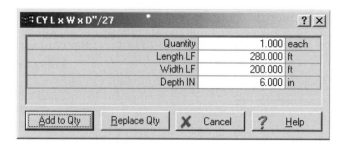

**Figure 4.16**  Enter Dimensions for Formula

19. Double click on the Remove Topsoil item found under the phase; this will place the item into the spreadsheet, and then you can enter the dimensions for that item.
    - Dimensions can be entered by means of a formula accessed with a mouse right click (see Figure 4.16) or by using the calculator also accessed with a mouse right click.
    - Continue until all of the items on the takeoff notes have been entered. See Figure 4.17.

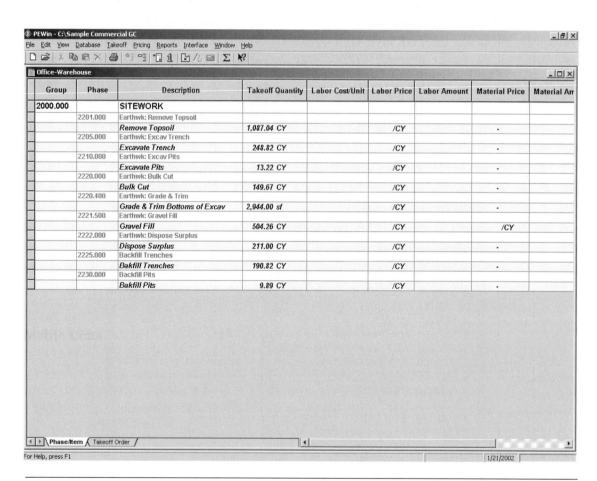

**Figure 4.17**  Spreadsheet

■ One further item, Dispose Surplus, has been measured. This quantity is obtained by adding together all the excavation amounts including the Bulk Cut (411.71CY) and deducting the total amount of material used for backfill (200.71CY).

## SUMMARY

■ The estimator has to make an assessment of sitework requirements based on what is shown on the plans and in the specifications, but there is usually very little detail of excavation requirements provided on the bid drawings.

■ The estimator's site visit will be another useful source of information about sitework requirements for the project. A checklist should be used to ensure that all of the important items of information are obtained on the visit.

■ A soils report will usually be available with large projects. It should be carefully studied to determine the soil conditions likely to be encountered in the excavation work.

■ Because soil expands when it is excavated, a swell factor is used to calculate the increased volume of material after it has been excavated.

■ Similarly, a compaction factor is used to calculate the total amount of material required to backfill excavations with compacted soils or gravels.

■ Excavation and backfill quantities are measured using bank measure quantities with no allowance for swell factors or compaction factors at the time of the takeoff.

■ It is important that the estimator carefully studies the OSHA requirements for excavations to ensure that the sitework takeoff reflects these requirements.

■ The efficiency of the sitework takeoff, especially cut and fill operations, can be much increased by using a digitizer, an electronic device that enables you to take measurements from drawings and input this data directly into a computer program.

■ In general, items of excavation, backfill and fill materials are measured in cubic yards or cubic meters bank measure.

■ The "grid method" may be used to calculate volumes of cut and fill over a site. This requires a survey showing the elevation of the existing grade at each intersection point on the grid. The elevation of the required new grade is also plotted at each intersection point and from these two elevations the depth of cut or fill can then be obtained at each point. From here on, cut calculations are separate from fill calculations. To figure the volume of cut at an intersection point, the depth of cut at this point is multiplied by the area "covered" by that intersection point. Then adding together all of the individual cut volumes computed in this way will give the total volume of cut on the site.

■ Cut and fill volumes can also be calculated using the "section method" as follows:

   a. At regular intervals (stations) along the centerline of the construction, survey the existing ground elevations on each side of the centerline.

   b. On graph paper, plot the cross-section at each station showing the existing grade and the required new grade.

   c. From the plot of the cross-section, compute the area of cut and the area of fill at each station.

   d. Using the station spacing and the area of cut at each station, calculate the total volume of cut. Similarly, from the station spacing and the area of fill at each station, the total fill volume can be obtained.

   e. The volume of cut or fill can then be calculated.

■ Piling work is usually subcontracted, in which case the general contractor's work in connection with this trade would be restricted to:

    a. Layout of piles
    b. Cutting off tops of piles
    c. Removing the excavated material produced in piling operations.

■ Bearing piles are measured in linear feet or in linear meters and are classified as follows:
    a. Round timber piles
    b. Sawn timber piles
    c. Steel piles
    d. Precast concrete piles
    e. Precast prestressed concrete piles
    f. Poured-in-place concrete piles (no permanent casing)
    g. Concrete-filled shell or pipe piles, closed end
    h. Open-end concrete-filled steel pipe piles
    i. Caisson piles
    j. Composite or combination type piles
    k. Special and patented type piles

■ Sheet piling is measured in square feet or square meters.

■ Sitework takeoffs can be done manually or via computer.

## REVIEW QUESTIONS

1. Why is the process of measuring excavation different from measuring the work of other trades?
2. Why is it important for the estimator to carefully examine the site of the work before the estimate is prepared?
3. What is the purpose of a soils report and how can the information it contains assist the estimator?
4. Explain the meaning of the following terms as they relate to the estimating process:
    a. Bank measure
    b. Swell factor
    c. Compaction factor
5. Why is it important for an estimator to be knowledgeable of the OSHA requirements regarding excavations?
6. Describe the use of digitizers and explain the difference between the two main types of digitizers.
7. Why are excavations measured separately in different categories?
8. Why is hand excavation always measured separately?
9. What is required before calculations of cut and fill volumes can be made using the "grid method?"
10. Discuss the advantages and disadvantages of calculating volumes of cut and fill by using the "grid method" rather than by using the "section method."
11. What piling work is often performed by the general contractor even when there is a piling subcontractor employed on the project?
12. Why is it critical to accurately calculate the centerline of a perimeter foundation to a building?
13. Figure 4.18 shows the plan of a site that is 60 m × 40 m overall, with a grid laid out at 10 m on center. The elevation (in meters) of the existing grade is shown at each intersection point on this grid. Calculate the volume of cut and the volume of fill in cubic meters required to level this site to a new grade of 100.0 m over the entire area.

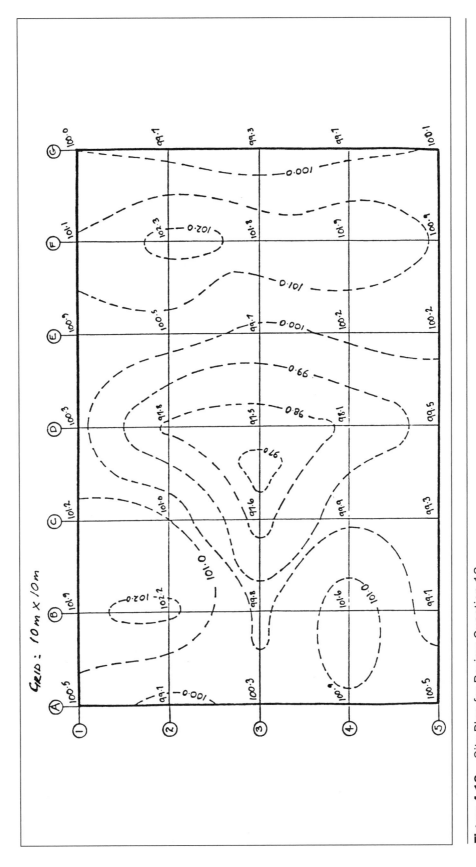

**Figure 4.18**   Site Plan for Review Question 13

# 5

# MEASURING CONCRETE WORK

## OBJECTIVES

*After reading this chapter and completing the review questions, you should be able to do the following:*

■ Explain how concrete work and associated items are measured in a takeoff.
■ Measure concrete items from drawings and specifications.
■ Make an assessment of formwork requirements from the details of concrete given.
■ Explain how concrete and formwork are classified in the takeoff process.
■ Describe how formwork to pilasters, bulkheads, edges and suchlike is measured.
■ Describe how formwork for grooves, keyways, notches and chamfers is measured using appropriate units.
■ Describe how forms to circular columns are measured.
■ Describe how concrete finishes are measured.
■ Describe how welded wire mesh is measured.
■ Describe how reinforcing steel is measured.
■ Complete a manual takeoff of concrete work and associated items.
■ Complete a computer takeoff of concrete work using the Precision Estimating software.

### Concrete Work Generally

Preparing a quantity takeoff of concrete work requires the estimator to measure a combination of items—some of which are shown on the drawings while others are to be inferred from the drawings. When drawings are well prepared, items of concrete such as footings, walls and columns are clearly shown, but nowhere on the drawings will the estimator find details of the formwork required for this work. The assessment of formwork requirements will be based on the estimator's knowledge of what is required for each of the different concrete components. Similarly, cement finishes and items such as screeds and curing are not generally explicit on the drawings, so the estimator must apply the specification requirements to what is shown on the drawings to determine the extent of these implicit items.

Because the concrete is detailed on the drawings, it makes sense to begin the takeoff by measuring the volume of concrete in an item. Then, after the concrete dimensions are defined, consider the formwork requirements, followed by the finishes that are needed and so on. Also, in accordance with the basic principles previously discussed regarding taking off by assembly, a good practice is to measure all work associated with one concrete assembly before passing on to consider the next assembly.

For example, if there are concrete footings, walls and columns to consider on a project, we would begin with the footings. First we would ascertain the dimensions of footing concrete from the drawings, then reuse these dimensions to calculate the area of formwork and the length of keyway required and conclude this assembly by measuring anything else associated with the footings. After the work on the footings is measured, we would turn our attention to the walls and measure all the items associated with them. Finally, we would deal with the columns in the same way.

This approach allows the estimator to focus on one piece of work at a time and fully understand all of its requirements before moving on to the next item. Alternative approaches such as measuring all the concrete volumes and then all the formwork areas require the estimator to be constantly jumping about from one type of item to another. He or she may have to come back to footings three or four times with this method, which is not an efficient way to proceed.

## Measuring Notes—Concrete

1.  Concrete shall be measured in cubic yards or cubic meters net in place. The volume of concrete is calculated from the dimensions given on the drawings with no adjustment for "add-on" factors. Additional material required because of spillage, expanding forms and wastage will be accounted for later in the pricing process by means of a waste factor added to concrete items.

    One exception to this general rule applies to the situation in which concrete is to be placed on rock or shale. In this case, the "overbreak" in the excavation is generally required to be filled with concrete and, as the volume of "overbreak" is not indicated, the estimator will have to add an assessed amount to the quantity of concrete to allow for this "overbreak."

2.  No adjustment to the quantity of concrete is made for reinforcing steel and insets that displace concrete. Also, no deductions are made for openings in the concrete that do not exceed one cubic foot or 0.05 cubic meter of volume.

3.  Concrete shall be classified and measured separately in the following categories:

    | | |
    |---|---|
    | a. Underpinning | j. Suspended slabs |
    | b. Pile caps | k. Floor toppings |
    | c. Isolated footings | l. Stairs and landings |
    | d. Continuous footing | m. Curbs |
    | e. Retaining walls | n. Manholes |
    | f. Grade beams | o. Equipment bases |
    | g. Columns and pedestals | p. Roads |
    | h. Beams | q. Sidewalks |
    | i. Slabs on grade | r. Other structures not listed |

4.  Different mixes of concrete shall be measured separately in each of the categories listed in item 3. For instance, where a high rise is being estimated, if

the columns up to the second floor are specified as 6000 psi concrete, from the second up to the tenth floor they are 5000 psi concrete, and above the tenth floor they are 4000 psi concrete, a separate quantity of each strength of concrete is required.

Often on a project all the concrete for a certain use, for example footings, is specified as the same mix, in which case there is no need to note the mix on the takeoff as it will not have to be considered until the recap and pricing stage is reached.

Mixes are further complicated by the different types of cement that could be specified. For instance, concrete in contact with soil may have to be made with type V, sulfate-resisting cement. Furthermore, air entrainment may or may not be required, to say nothing of super-plasticizers. The many combinations of variables result in a multitude of possible concrete mixes. To simplify the pricing of concrete, the items listed on the recap are priced for the cost of placing (labor and equipment) only. A separate list of the amounts of the various mixes is prepared and against each item on this list the price of the particular mix of concrete is entered. See Chapter 11 for more details of concrete work recaps.

5.  Where columns and walls extend between the floors of a building, they shall be measured from the top surface of the slab below up to the undersurface of the slab or beam above as illustrated in Figure 5.1.

6.  Beams may be measured separately from slabs, but if they are to be poured monolithically with the slabs, the quantity of concrete in the beams should be added to the slab concrete for pricing.

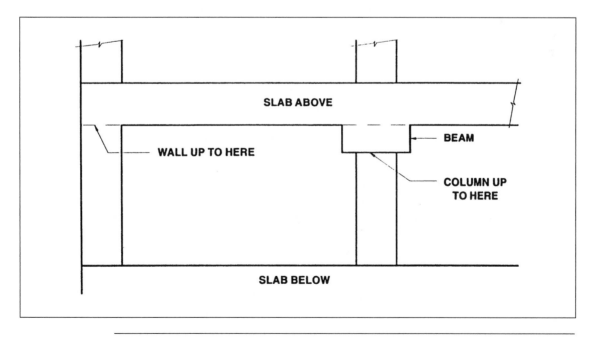

**Figure 5.1**  Measuring Concrete Walls and Columns

## Measuring Formwork Generally

The formwork operations involve a number of activities including fabricating and erecting the forms, stripping, moving, cleaning and oiling the forms for reuse. All of these activities and the materials involved are allowed for in the pricing of the forms. The estimator measures the surface area of the concrete that comes into contact with the forms; this is known as the *contact area*.

Because only the area of formwork is measured, the estimator does not have to be concerned about the design of the forms at the time of the takeoff. All that needs to be established is which surfaces of the concrete require forms. In the past, estimators have agonized over such things as whether the bottom of an opening or the sloped top surface of a wall needs to be formed. If discussion with your colleagues does not provide an answer, the prudent estimator will always exercise caution and allow the forms. .

## Measuring Notes—Formwork

1. Formwork shall be measured in square feet or square meters of contact area, that is, the actual surface of formwork that is in contact with the concrete.
2. Formwork is classified into the same categories as listed for concrete. As an illustration, consider a project with concrete footings, walls and columns, forms to footings; forms to walls and forms to columns would each be described and measured separately.

   There are, however, a number of factors that may have no effect on the price of the concreting operations but do affect the price of formwork and, therefore, should be noted. For example, the volume of concrete in all walls, whether they are straight or curved, will have the same price, but the price of forms to curved walls will differ from the price of straight walls so the forms to curved surfaces must be kept separate.
3. Bulkheads and edge forms shall be measured separately within these categories, so, if there are construction joints required for long lengths of walls, the area of bulkheads to form these construction joints would be measured separately from the wall forms. Similarly, if there are **pilasters** projecting from the walls, the area of the pilasters would be calculated and noted separately from the wall forms. See Figure 5.2 for different categories of formwork in a wall system.
4. Forms to slab edges are measured separately from forms to beams and forms to walls, even where the edge forms may be extensions of beam or wall forms (Figure 5.3).
5. Where there is an opening in a form system, no deduction is made from the total area of the forms if the size of the opening is less than 100 square feet (10 m$^2$). Examples of such openings would include openings for windows in walls, stairway openings or elevator shaft openings in suspended slabs. The estimator must distinguish between what are openings and what are cut outs. Openings less than 100 square feet (10 m$^2$) are not deducted, but all cut outs would be deducted. (See Figure 5.4.)
6. Items of formwork that are lineal in nature, such as grooves, notches, keyways, chamfers and the like are described stating their size, and measured in linear feet or meters. Narrow strips of formwork, less than 1-foot wide (300 mm), can also be measured in this fashion.

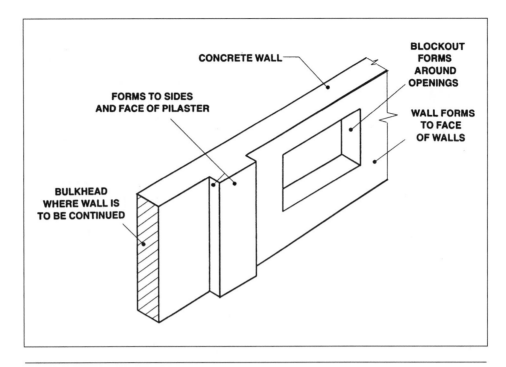

**Figure 5.2**    Formwork Categories

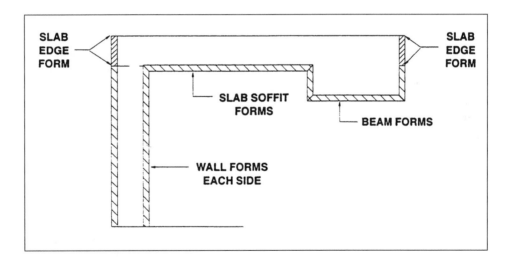

**Figure 5.3**    Types of Forms

7. Forms to circular columns are described giving the diameter of the column and measured in linear feet or meters to the height of the column. Capitals, which are the widening at the top of circular columns, are described and enumerated. Some estimators find it useful to draw on their takeoff small sketches of complex items such as capitals to clarify exactly what is being measured.

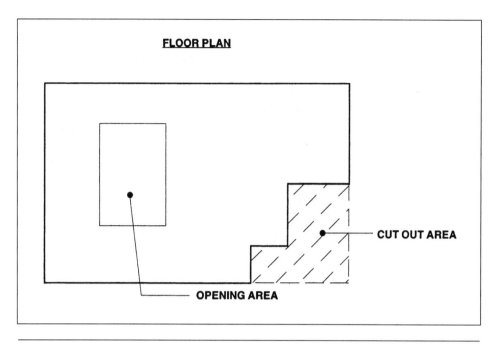

**Figure 5.4**   Openings and Cut Outs in Formwork

## Measuring Notes—Finishes and Miscellaneous Work

### Slab Finishes

1.  Slab finishes are measured in square feet or square meters of plan area, and vertical finishes are measured in square feet or square meters of exposed surface of concrete.
2.  Finishes are first classified in terms of what is being finished: slabs, walls, columns, stairs, sidewalks and so on. They are then separated by type of finish: wood float, sack rubbed, bush hammered, steel trowel, broom finish and so on. In addition, if floor hardeners are required, they are measured separately describing the type of hardener specified.

### Screeds and Curing Slabs

3.  Screeds can be measured in square feet or square meters as the plan area of the slab, or some estimators prefer to measure the length of the individual screed strips. This length can easily be obtained by dividing the area of the slabs by the spacing of the screed strips.
4.  Curing slabs is also measured in plan area, as, too, are vapor barriers, but because these membranes are lapped at the edges, an additional 10% is added to the square feet or square meter area.

### Welded Wire Mesh

5.  Welded wire mesh is likewise measured in plan area with an additional 10% included for laps. If different types of mesh are required in different parts of the project, each type should be fully described as it is taken off. However, if there is only one specification for mesh, the takeoff can merely describe W.W.M. and the type of mesh considered later at the pricing stage.

### Inserts and Waterstops

6. Inserts into the concrete such as anchor bolts are described and enumerated in the takeoff.

7. Linear inserts such as waterstops are measured in linear feet or meters.

8. Construction joints in walls are often required to incorporate waterstops, grooves and sealants, each of which is measured in linear feet or meters. The takeoff of this type of construction joint will comprise bulkhead forms in square feet or square meters; then waterstop, forming grooves and sealant, each measured separately, in linear feet or meters.

    Here again, a small sketch incorporated into the takeoff is useful in clarifying this kind of complex detail.

### Expansion Joints

9. Expansion joints are measured in linear feet or meters stating the size of joint and the type of material used. If a sealant is required in connection with the joint, it shall be measured separately in linear feet or meters stating the size and type of sealant.

### Nonshrink Grout

10. Grout to anchor bolts and base plates is measured in cubic feet or cubic meters and different types of materials are measured separately.

### Saw Cutting Slabs

11. Saw cuts are measured in linear feet or meters stating the size of the cut. Here it is important to note whether old or new concrete is to be cut and what is to be cut; slabs, walls, columns and so on should be stated.

### Reinforcing Steel

12. Fabricating and placing reinforcing steel is often subcontracted by general contractors. The subcontractor would takeoff and price this trade.

13. Bar reinforcing steel is taken off in linear feet or meters of bar and converted to a weight by multiplying the lengths measured by the unit weight per foot or meter for that size of bar.

## Example 1—House

### Comments on the Takeoff Shown as Figure 5.5

1. The specifications call for the same strength concrete (3000 psi) to be used for the footings, walls and slab-on-grade so the strength is not mentioned in the takeoff but is left until the pricing stage to consider.

2. The centerline length of the perimeter footing has previously been calculated with the excavation work, and the width and height of the footing are shown on the drawings; these figures, therefore, can be entered directly into the dimension columns.

3. The reinforcing bars in the footings and walls are continuous from corner to corner. At the corners the bars are spliced using L-shaped bars with 2'0" legs as shown in the small sketch on the takeoff.

**QUANTITY SHEET**

JOB _HOUSE EXAMPLE_  DATE ____

ESTIMATOR _ABF_  EXTENDED ____  EXT. CHKD ____

SHEET No. | 1 of 2 |

CONCRETE WORK

| DESCRIPTION | TIMES | DIMENSIONS | | | CONC. L×W×D | FORM STRIP FTGS. 2×L×D | FORM PAD FTGS. 2(L+W)×D |
|---|---|---|---|---|---|---|---|
| CONC. FTGS. | | 133.33 | 2.00 | 0.67 | 179 | 179 | — |
| | 3 | 2.67 | 2.67 | 0.83 | 18 | — | 27 |
| | | | | | 197 | 179 SF | 27 SF |
| | | | | | 7 CY | | |
| FORM 2×4 KEYWAY | | 133.33 | | | 133 LF | | |
| #5 REBAR | 2 | 133.33 | | | 267 | | |
| (CORNER L's) | 6×2 | 4.00 | | | 48 | | |
| | | | | | 315 × 1.043 | | |
| | | | | | 329 lbs | | |

| | | | | | CONC. L×W×D | FORM WALL 2×L×D | BULKHEADS 2(L+D)×W |
|---|---|---|---|---|---|---|---|
| CONC. WALLS | | 133.33 | 0.67 | 8.00 | 715 | 2133 | — |
| (WINDOWS) DDT | 4 | 3.08 | 0.67 | 1.08 | ⟨9⟩ | — | 22 |
| | | | | | 706 | 2133 SF | 22 SF |
| | | | | | 26 CY | | |
| #4 REBAR | 4 | 133.33 | | | 533 | | |
| (CORNER L's) | 6×4 | 4.00 | | | 96 | | |
| | | | | | 629 × 0.668 | | |
| | | | | | 420 lbs | | |

40'-0" × 28'-0"

LESS

2 × 8"   1'-4"   1'-4"

38'-8" × 26'-8"

| | | | | | CONC. L×W×D | FINISH L×W | |
|---|---|---|---|---|---|---|---|
| CONC. S.O.G. | | 38.67 | 26.67 | 0.33 | 340 | 1031 | |
| DDT | | 20.17 | 2.00 | 0.33 | ⟨13⟩ | ⟨40⟩ | |
| | | | | | 327 | 991 SF | |
| | | | | | 12 CY | | |

**Figure 5.5**    House Example—Concrete Work

4. Bulkhead forms are measured to the perimeter of the window openings and they are the same width as the wall.
5. The basement slab-on-grade extends to the inside of the exterior walls so the thickness of the walls is deducted from the outside dimensions to obtain the dimensions of the slab.

**QUANTITY SHEET**                          SHEET No.  | 2 of 2 |

JOB......... *HOUSE EXAMPLE* ......................... DATE.............................

ESTIMATOR......... *ABF* ..........EXTENDED.......................EXT. CHKD...................

| DESCRIPTION | TIMES | DIMENSIONS | | | | | |
|---|---|---|---|---|---|---|---|
| *SCREED & CURING* | | *As* | *FINISH* | | 991 SF | | |
| | | | | | | | |
| *6 MIL POLY. V.B.* | | *As* | *FINISH + 10%* | | 1090 SF | | |
| | | | | | | | |
| *8 × 8 × 8/8 WWM* | | | *DITTO* | | 1090 SF | | |
| | | | | | | | |
| *GRAVEL UNDER* | | 38.67 | 26.67 | 0.50 | 516 | | |
| *S.O.G.*      *DDT.* | | 20.17 | 2.00 | 0.50 | ⟨ 20⟩ | | |
| | | | | | 496 | | |
| | | | | | 18 CY | | |
| | | | | | | | |
| *2 × 6 SILL PLATE* | | 136.00 | | | 136 BF | | |
| | | | | | | | |
| | | 136'-0" | | | | | |
| | | 8'-0" | | | | | |
| | | = 17 | | | | | |
| *CORNERS* | | + 6 | | | | | |
| | | 23 | | | | | |
| *½" × 9" A.B.s* | | 23 | | | 23 No. | | |

**Figure 5.5  continued**

6.  The plan area of the slab is measured for items of finishing, screeds and curing. Items of screeds and curing are often not noted on the takeoff but are added later when the finishes are listed on the recap. The quantity of screeds and curing is easy to determine as it is equal to the sum of all the areas of floor finishing.

7.  The vapor barrier under the slab is also measured as the plan area, but the quantity is increased by 10% to allow for laps in the material.

8.  A convenient place to measure the gravel under the slab-on-grade is directly after the concrete as it is the same area as the concrete—only the thickness is different.

9.  It is also convenient to measure the sill plate at this stage, but this item will be recapped and priced in the rough carpentry section.

## Example 2—Office/Warehouse Building

### Comments on Concrete Work Takeoff Notes Shown as Figure 5.6a

1.  On this project the concrete strength specified for all concrete is 3000 psi and concrete generally is to be normal weight except lightweight concrete is to be used for the topping on the metal deck.

2.  The sizes of the pad foundations are taken directly from the drawings and the length of the continuous strip footing between the pads has been previously calculated with the excavation work, so there are no side calculations needed at this stage.

3.  The footings to the perimeter of the office will be poured together so the concrete in pad footings and continuous footings can be added together. However, the forms to the sides of the continuous strip footings are kept separate from the forms of the pad footings because there will probably be more work per square foot, and thus a higher cost, for the pad footings.

    Note that the area of the pad footing forms where they meet the ends of the continuous strip footings is deducted as there is no need for pad footing forms in these locations.

4.  Pilaster concrete is added to the wall concrete as this material is all placed together, but forms to pilasters are measured separately from wall forms in accordance with the rules of measurement.

    Separating the different types of forms allows the estimator to price pilaster forms at a premium as incorporating pilasters reduces the forming crews productivity, that is, a plane wall without pilasters is easier to build, thus less expensive, than a wall with pilasters. So the area of the face of the pilasters is deducted from the area of the wall forms, and then the face and sides of the pilasters are taken off separately. See Figure 5.6b for an illustration of this detail.

5.  Bulkhead forms for construction joints to the walls have not been measured in this example as they are not considered to be necessary here.

    Sometimes, however, the specifications limit the length of wall that can be placed in a single concrete pour, or the project supervisor may require bulkheads to divide the walls into separate pours. So the estimator may wish to discuss this item with the proposed supervisor if this person is available. Estimators should always be encouraged to take a team approach to their work, but tight deadlines and the fact that other staff are frequently not

| TAKEOFF NOTES | | | | SHEET No. 1 of 6 |

PROJECT: OFFICE/WAREHOUSE          DATE:

ESTIMATOR: A B F

| NOTES | TIMES | DIMENSIONS | | |
|---|---|---|---|---|
| CONCRETE WORK | | | | |
| OFFICE | | | | |
| 3000 psi CONT. FTG.          (F₁) | 5 | 3.50 | 3.50 | 0.67 |
| (F₂) | 6 | 4.00 | 4.00 | 0.67 |
| (F₃) | 2 | 5.00 | 5.00 | 0.67 |
| (STRIP FTG.) | | 245.83 | 2.00 | 0.67 |
| DDT (ENDS OF STRIP FTG.) | 2 x 13 | 2.00 | — | 0.67 > |
| FORM PAD FTGS | 5 | 14.00 | — | 0.67 |
| | 6 | 16.00 | — | 0.67 |
| | 2 | 20.00 | — | 0.67 |
| FORM CONT. FTG. | 2 | 245.83 | — | 0.67 |
| FORM 2 x 4 KEYWAY | | 297.33 | | |
| 3000 psi WALLS | | 297.33 | 0.67 | 4.00 |
| PILASTERS (P₁) | 4 | 0.67 | 0.67 | 4.00 |
| (P₂) | 9 | 1.33 | 0.67 | 4.00 |
| FORM WALLS | 2 | 297.33 | — | 4.00 |
| DDT (P₁) | 4 | 1.33 | — | 4.00 > |
| " (P₂) | 9 | 1.33 | — | 4.00 > |
| FORM PILASTERS | 4 | 1.33 | — | 4.00 |
| | 9 | 2.67 | — | 4.00 |
| 3000 psi PAD FTG. (F₅) | 3 | 5.00 | 5.00 | 0.67 |
| FORM PAD FTGS. | 3 | 20.00 | — | 0.67 |

2 x 8"  1'-4"  1'-4"  2'-8"

**Figure 5.6a**  Office/Warehouse Example—Concrete Work

**TAKEOFF NOTES**

PROJECT: OFFICE / WAREHOUSE

DATE:

ESTIMATOR: ABF

| NOTES | | TIMES | DIMENSIONS | | |
|---|---|---|---|---|---|
| 3000 psi PEDESTALS | (P3) | 3 | 2.00 | 2.00 | 2.00 |
| FORM PEDESTALS | | 3 | 8.00 | — | 2.00 |
| ¾" DIA. A.B. 16" LONG | (C1) | 3 | 4 | | |
| GROUT BASE PLATES | | 3 x 4 | 1.00 | 1.00 | 0.08 |
| 2" RIGID INSULATION | | | 221.00 | — | 2.00 |
| | | 2 x 10 | 0.67 | — | 2.00 |
| OFFICE PERIM. | | | | | |
| 2 x 50'  100'-0" | | | | | |
| +  100'-0" | | | | | |
| +  25'-0" | | | | | |
| LESS  225'-0" | | | | | |
| 3 x 2 x 8" ⟨ 4'-0" ⟩ | | | | | |
| 221'-0" INSIDE PERIM. | | | | | |
| WAREHOUSE | | | | | |
| 3000 psi PILE CAPS | (PIL 2) | | 1.33 | 1.33 | 4.00 |
| | (PIL 5) | 7 | 2.00 | 2.00 | 2.00 |
| FORM PILE CAPS | | | 5.33 | — | 4.00 |
| | | 7 | 8.00 | — | 2.00 |
| 4 x 1'-4" = 5'-4" | | | | | |
| 3000 psi GRADE BEAM | | | 473.67 | 0.67 | 2.00 |
| | (PIL 1) | 2 | 1.33 | 0.67 | 2.00 |
| | (PIL 3) | 16 | 2.00 | 1.33 | 2.00 |
| | (PIL 4) | 2 | 1.33 | 1.33 | 2.00 |

**Figure 5.6a  continued**

**TAKEOFF NOTES**

SHEET No. 3 of 6

PROJECT: OFFICE / WAREHOUSE

DATE:

ESTIMATOR: ABF

| NOTES | TIMES | DIMENSIONS | | |
|---|---|---|---|---|
| FORM GRADE BM. | 2 | 473.67 | — | 2.00 |
| DDT (PIL 1) | < 2 | 1.33 | — | 2.00 > |
| " (PIL 3) | < 16 | 2.00 | — | 2.00 > |
| " (PIL 4) | < 2 | 2.67 | — | 2.00 > |
| | | | | |
| 4" × 8" VOID FORM | | 473.67 | | |
| | | | | |
| FORM PILASTERS | 2 | 2.00 | — | 2.00 |
| 2'-0" | 16 | 4.67 | — | 2.00 |
| 2×1'4" 2'-8" | 2 | 2.67 | — | 2.00 |
| 4'-8" | | | | |
| INSIDE PERIM. | | | | |
| OUTSIDE = 475'-0" | | | | |
| LESS | | | | |
| 2 × 2 × 8"    2'-8" | | | | |
| 472'-4" | | | | |
| | | | | |
| 2" RIGID INSULN. | | 472.33 | — | 1.50 |
| (PIL 1) | 2 | 0.67 | — | 1.50 |
| (PIL 3) | 2×16 | 1.33 | — | 1.50 |
| | | | | |
| 1" A.B. 18" LONG    (C3 + C4) | 26 | 4 | | |
| | | | | |
| GROUT BASE PLATES | 26 | 1.17 | 0.67 | 0.08 |
| | | | | |
| SLAB ON GRADE | | | | |
| OFFICE            WAREHOUSE | | | | |
| LESS  50'-0" × 100'-0"    200'-0" × 75'-0" | | | | |
| 2 × 8" <1'-4"> <1'-4">  LESS  < 8"> <1'-4"> | | | | |
| 48'-8" × 98'-8"    199'-4" × 73'-8" | | | | |

**Figure 5.6a  continued**

TAKEOFF NOTES

SHEET No. 4 of 6

PROJECT: OFFICE/WAREHOUSE

DATE:

ESTIMATOR: ABF

| NOTES | | TIMES | DIMENSIONS | | |
|---|---|---|---|---|---|
| 3000 psi S.O.G. | (OFF.) | | 48.67 | 98.67 | 0.50 |
| | (W/HOUSE) | | 199.33 | 73.67 | 0.50 |
| | | | | | |
| TROWEL FINISH SLAB | | | 48.67 | 98.67 | |
| | | | 199.33 | 73.67 | |
| | | | | | |
| NON METALIC HARDNER | (W/HOUSE) | | 199.33 | 73.67 | |
| SCREEDS | | | AS TROWEL FIN. | | |
| CURING | | | AS TROWEL FIN | | |
| 6 MIL POLY V.B. | | | AS TROWEL FIN + 10% | | |
| 6 x 6 x 10/10 WWM | | | AS OFF. FIN + 10% | | |
| 6 x 6 x 6/6 WWM | | | AS W/HOUSE FIN + 10% | | |
| GRAVEL UNDER S.O.G. | | | AS CONC. S.O.G. | | |
| ½" x 6" EXPN. JNT. FILLER | (OFF.) | 2 | 48.67 | | |
| | | 2 | 98.67 | | |
| | (PILS) | 2 x 9 | 0.67 | | |
| | (W/HOUSE) | 2 | 199.33 | | |
| | | 2 | 73.67 | | |
| | (PILS) | 2 x 16 | 1.33 | | |
| | (PILE CAPS) | 10 | 8.00 | | |
| | (PIL 3) | 3 | 1.33 | | |
| | | | | | |
| FORM CONSTN. JNT. | | | 48.67 | — | 0.50 |
| | | 7 | 73.67 | — | 0.50 |

**Figure 5.6a  continued**

TAKEOFF NOTES                                                    SHEET No. 5 of 6

PROJECT:  OFFICE / WAREHOUSE                    DATE:

ESTIMATOR:  ABF

| NOTES | TIMES | DIMENSIONS | | |
|---|---|---|---|---|
| ¼" × 1½" SAWCUTS (ON GRID LINES) | 3 | 48.67 | | |
| | | 98.67 | | |
| | | 199.33 | | |
| | 7 | 73.67 | | |
| | | | | |
| SECOND FLOOR | | | | |
|     3000 psi SLAB ON METAL DECK | | 48.67 | 98.67 | 0.25 |
|         (LIGHT-WEIGHT CONC.) | | | | |
| TOPPING   2" | | | | |
| DECK  1½" = ¾" | | | | |
| 2      2¾"   SAY 3" DEEP | | | | |
| | | | | |
|     TROWEL FINISH SLAB | | 48.67 | 98.67 | |
| | | | | |
|     SCREEDS | | DITTO | | |
| | | | | |
|     CURING | | DITTO | | |
| | | | | |
|     6" × 6" × 10/10  WWM | | DITTO + 10% | | |
| | | | | |
| | | | | |
|     FORM EDGES OF | | | | |
|     SLAB ON METAL DECK  (OPENINGS) | 4 | 16.00 | — | 0.33 |
| | | | | |
| STAIRS | | | | |
|     3000 psi IN STAIR TREADS (1½" DEEP) | | 8.00 | 4.00 | 0.13 |
| | 2×6 | 3.50 | 0.92 | 0.13 |
| | | | | |
|     FINISH  STAIRS | | 8.00 | 4.00 | |
| | 2×6 | 3.50 | 0.92 | |

**Figure 5.6a**    Office/Warehouse Example—Concrete Work

**TAKEOFF NOTES**

SHEET No. 6 of 6

PROJECT: *Office / Warehouse*                    DATE:

ESTIMATOR: *A B F*

| NOTES | TIMES | DIMENSIONS | | |
|---|---|---|---|---|
| *EXTERIOR CONCRETE* | | | | |
| 3000 psi CURBS | | 125.00 | 0.58 | 1.00 |
| | | 320.00 | | |
| | | 45.00 | | |
| | | 86.00 | | |
| | 2 | 26.00 | | |
| | 2 | 32.00 | 0.58 | 1.00 |
| | | | | |
| FORM CURBS | 2 | 692.00 | — | 1.00 |
| | | | | |
| FINISH TOP OF CURBS | | 692.00 | | |
| | | | | |
| 3000 psi SIDEWALKS | | 250.00 | 5.00 | 0.33 |
| (THICKEN) | | 250.00 | 0.92 | 0.67 |
| (APRON) | | 200.00 | 5.00 | 0.33 |
| (THICKEN) | | 200.00 | 0.83 | 0.33 |
| FORM EDGE OF CURBS | | 250.00 | — | 1.00 |
| | | 200.00 | — | 0.67 |
| FINISH SIDEWALKS | | 250.00 | 5.00 | |
| | | 200.00 | 5.00 | |
| CURE + SCREEDS | | DITTO | | |
| | | | | |
| GRAVEL UNDER SLABS | | 250.00 | 5.00 | 0.50 |
| | | 200.00 | 5.00 | 0.50 |
| | | | | |
| 6" DIA. BOLLARDS | | 19 | | |

**Figure 5.6a  continued**

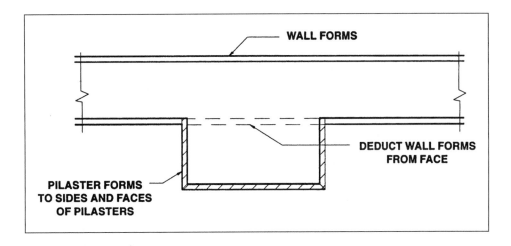

**Figure 5.6b**  Part Plan of Grade Beam and Pilasters

available, do make it difficult to integrate these discussions into the estimating schedule.

6. Rigid insulation is applied to the inside face of the outside walls of the office but not to the wall between the office and the warehouse, so only 25'0" of the east wall of the office is included in the perimeter calculation. For the same reason, only three corners are adjusted for to obtain the inside perimeter.

7. As noted with the excavation takeoff, the warehouse is measured separately from the office because it is a different form of construction.

8. Void forms made of cardboard or expanded plastic strips are placed at the bottom of grade beam forms to prevent contact between the concrete and the subgrade. This precaution is taken in an attempt to avoid frost heave on the underside of the beam.

9. The slab-on-grade is measured to the inside of walls, so the 8-inch wall thicknesses are deducted from the outside dimensions to obtain the inside dimensions.

10. The slab finish together with an item for screeds and curing are measured as the plan area of the slab. If screeds are required to be priced by the linear foot, the length of screeds can be obtained by dividing the floor area by the spacing of the screeds.

11. Welded wire mesh and vapor barrier are also measured as the plan area of slab, but quantity is increased by 10% to allow for laps.

12. You will observe that there are few, if any, specifications noted in this takeoff. As we have previously discussed, the estimator will have examined the specifications and made notes of requirements before commencing the takeoff. If the specifications called for certain grade beams to have 3000 psi

concrete and others to have 4000 psi concrete, the takeoff descriptions would reflect this and the quantity of each type of grade beam would be measured separately.

13. Gravel is measured after the concrete slab-on-grade as before, using the same plan area as the slab multiplied by the thickness of gravel.

14. A joint filler is placed between the slab and the grade beams, so the perimeter of each slab is calculated to arrive at the quantity of this material.

15. Slab construction joints are frequently not shown on drawings, but specifications, and sometimes limitations on the number of finishers available, can restrict the amount of slab that can be placed in one pour. Consequently, the contractor will need to form bulkheads at construction joints to separate slab pours. This is another instance in which discussions with field staff can be very useful. In this example bulkheads have been measured along the grid lines of the project, which divides the slab into areas of 625 square feet in the office and 937.5 square feet in the warehouse. These areas are typically required to be poured checkerboard fashion by the specifications, so adjacent areas cannot be poured together.

16. Contraction control joints (sawcuts) are specified to be located on the grid lines, so they are measured in linear feet.

### Precision Estimating Takeoff

Because the database we are using has only a small number of items, we need to begin by setting up some new phases and items for the concrete work.

1. Set up the new phases shown on Figure 5.7.
2. Set up the new items shown on Figure 5.8.
   - Where there is a waste factor % specified for an item, remember to check the "Apply Waste %" to the material cost factor on the Rounding tab.
   - Here we could get the program to round the quantities of materials if this is required. In this example we will leave the rounding as "none."
   - There is also a Taxable box on the rounding tab. Leave this box unchecked since we will be adding taxes directly to material prices in which they are applicable.
3. Some of the phases and items that came with the sample database do not fit into the numbering system we are using for the database items we have added, so we can tidy up the database by deleting those phases and items we are not going to use. First we will delete the items, then we will delete the phases that contain these items. On the menu bar click on Database and select Items from the list (see Figure 5.9).
   - The Database Item window will now open at the first item in the database.
4. Click on the list button located to the right of the phase code.
   - The list will open at the first item, but we need to collapse this list to Group Phases so that we can find a specific item.
5. Click on the Collapse All button, then double click on the CONCRETE group phase.

| Phase Code | Description | Unit |
|------------|-------------|------|
| 2144.00 | Gravel @ Slab | CY |
| 3112.00 | Forms: Pile Caps | SF |
| 3114.00 | Forms: Grade Beams | SF |
| 3114.50 | Forms: Voids | SF |
| 3115.00 | Forms: Walls | SF |
| 3119.00 | Forms: Pilasters | SF |
| 3123.00 | Forms: Pedestals | SF |
| 3127.00 | Forms: Edgeform Slabs | SF |
| 3135.00 | Forms: Edgeform Susp Slab | SF |
| 3139.00 | Forms: Curbs | SF |
| 3159.00 | Forms: Strip & Oil | SF |
| 3160.00 | Screeds | SF |
| 3214.00 | Rebar: Slab-on-Grade | SF |
| 3305.00 | Conc: Pile Caps | CY |
| 3306.00 | Conc: Footings | CY |
| 3307.00 | Conc: Walls | CY |
| 3309.00 | Conc: Pedestals | CY |
| 3310.00 | Conc: Slab-on-Grade | CY |
| 3310.50 | Conc: Sidewalks | CY |
| 3314.00 | Conc: Stairs | CY |
| 3318.00 | Conc: Grade Beams | CY |
| 3319.00 | Conc: Curbs | CY |
| 3320.00 | Conc: Suspended Slabs | CY |
| 3350.00 | Conc: Curing | SF |
| 3380.00 | Finish: General | SF |
| 3901.00 | Misc: Grout | CF |
| 3950.00 | Misc: Anchor Bolts | No. |
| 3955.00 | Misc: Insulation | SF |
| 3960.00 | Misc: Expansion Joint | LF |
| 3961.00 | Misc: Sawcut Concrete | LF |
| 3965.00 | Misc: Bollards | No. |
| 7100.00 | Vapor Barrier | SF |

**Figure 5.7**    New Phases

| Phase Code | Item Code | Item Description | Unit | Formula | Waste Factor | Cost Categories |
|---|---|---|---|---|---|---|
| 2144.00 | 10 | Gravel Under SOG | CY | CY L × W × D"/27 | 20% | L, M, E |
| 3110.10 | 20 | Form Pad Footings | SF | SF Footing Forms | 5% | L, M |
| 3110.10 | 30 | Form Continuous Footings | SF | — | 5% | L, M |
| 3112.00 | 10 | Form Pile caps | SF | SF Footing Forms | 5% | L, M |
| 3114.00 | 10 | Form Grade Beams | SF | — | 5% | L, M |
| 3114.50 | 10 | Form 4" × 8" Voids | LF | — | 5% | L, M |
| 3115.00 | 10 | Form Walls | SF | — | 5% | L, M |
| 3119.00 | 10 | Form Pilasters | SF | — | 5% | L, M |
| 3123.00 | 10 | Form Pedestals | SF | — | 5% | L, M |
| 3127.00 | 10 | Form Edge of SOG | SF | SF L × H | 5% | L, M |
| 3135.00 | 10 | Form Edge of Susp Slab | SF | SF L × H | 5% | L, M |
| 3139.00 | 10 | Form Curbs | SF | — | 5% | L, M |
| 3159.00 | 10 | Strip & Oil: Misc | SF | — | 10% | L, M |
| 3160.00 | 10 | Slab Screed | SF | SF L × W | 5% | L, M |
| 3214.00 | 10 | 6 × 6 × 10/10 WWM | SF | SF L × W | 5% | L, M |
| 3214.00 | 20 | 6 × 6 × 6/6 WWM | SF | SF L × W | 5% | L, M |
| 3305.00 | c30 | 3000 psi Pile Caps | CY | CY L × W × D'/27 | 3% | L, M, E |
| 3306.00 | c30 | 3000 psi Continuous Footings | CY | CY L × W × D'/27 | 3% | L, M, E |
| 3306.00 | c30 | 3000 psi Pad Footings | CY | CY L × W × D'/27 | 3% | L, M, E |
| 3307.00 | c30 | 3000 psi Walls | CY | CY L × W × D'/27 | 3% | L, M, E |
| 3309.00 | c30 | 3000 psi Pedestals | CY | CY L × W × D'/27 | 3% | L, M, E |
| 3310.00 | c30 | 3000 psi Slab-on-Grade | CY | CY L × W × D"/27 | 3% | L, M, E |
| 3310.50 | c30 | 3000 psi Sidewalks | CY | CY L × W × D'/27 | 3% | L, M, E |
| 3314.00 | c30 | 3000 psi Stairs | CY | CY L × W × D"/27 | 3% | L, M, E |
| 3318.00 | c30 | 3000 psi Grade Beams | CY | CY L × W × D'/27 | 3% | L, M, E |
| 3319.00 | c30 | 3000 psi Curbs | CY | CY L × W × D'/27 | 3% | L, M, E |
| 3320.00 | c30 | 3000 psi Lt Wt Susp Slabs | CY | CY L × W × D"/27 | 3% | L, M, E |
| 3350.00 | 10 | Curing Slabs | SF | SF L × W | 10% | L, M |
| 3380.00 | 10 | Trowel Finish | SF | SF L × W | — | L, E |
| 3380.00 | 15 | Stair Finish | SF | SF L × W | — | L |
| 3380.00 | 20 | Sidewalk Finish | SF | SF L × W | — | L, E |
| 3380.00 | 30 | Nonmetallic Floor Hardener | SF | SF L × W | 10% | L, M |
| 3380.00 | 50 | Finish Curbs | LF | — | — | L |
| 3901.00 | 10 | Grout Base Plates | CF | — | 10% | L, M |
| 3950.00 | 10 | Install ³⁄₄" Anchor Bolts | No. | — | — | L |
| 3950.00 | 20 | Install 1" Anchor Bolts | No. | — | — | L |
| 3955.00 | 10 | 2" Rigid Insulation | SF | SF L × H | 10% | L, M |
| 3960.00 | 10 | ¹⁄₂" × 6" Expn Joint Filler | LF | — | 10% | L, M |
| 3961.00 | 10 | ¹⁄₄" × 1¹⁄₂" Sawcuts | LF | — | — | L, E |
| 3965.00 | 10 | 6" Dia. Bollards | No. | — | 5% | L, M |
| 7100.00 | 10 | 6 Mil Poly Vapor Barrier | SF | SF L × W | 5% | L, M |

**Figure 5.8**   New Items

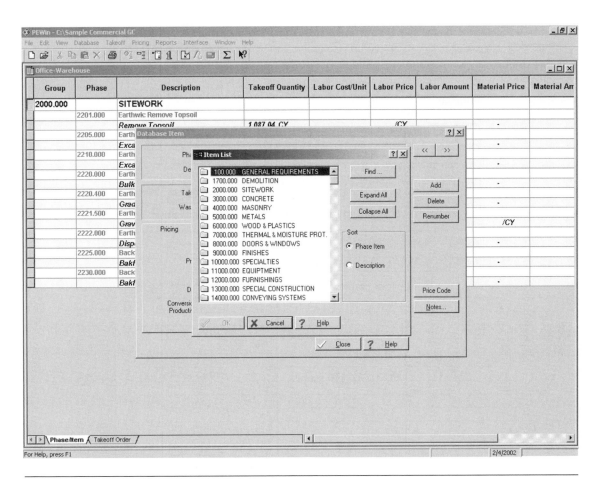

**Figure 5.9**    Database Item Window

6. Double click on Phase 3111.50, Forms: Strip/Oil, then double click on item 10 in this phase (see Figure 5.10).
   - The Database Item window will now open at this item.
7. Click on Delete, then confirm this by clicking on Yes.
   - You can delete item 3310.14 c30, Footing Conc 3000 psi, by following the same procedure.
8. Now we can delete the phases that contained these items. On the menu bar, click on Database and select Phases from the list.
   - The Database Phase window will now open at the first phase in the database.
9. Click on the list button located to the right of the phase code.
10. Double click on the CONCRETE group phase.
11. Double click on Phase 3111.50, Forms: Strip & Oil.
    - The Database Phase window for this phase will now open.
12. Click on Delete, then confirm this by clicking on Yes.
    - You can delete Phase 3310.14, Conc: Footings, by following the same procedure.
    - The database is now ready for the concrete work takeoff.
13. Use the Quick Takeoff to takeoff all the items listed on the takeoff notes beginning with 3000psi Concrete Footings.

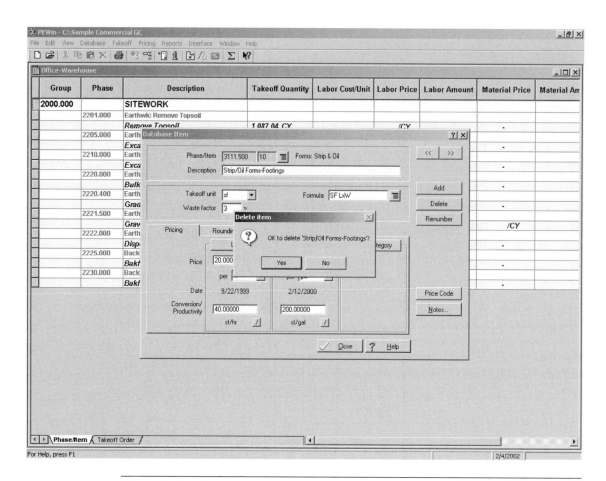

**Figure 5.10** Deleting an Item from the Database Item Window

- Figure 5.11 shows the spreadsheet after the takeoff is completed.
- Do not be concerned about prices at this stage—we will consider the pricing later.

## SUMMARY

- Preparing a quantity takeoff of concrete works requires the estimator to measure a combination of items such as concrete, formwork, concrete finishes and so on.
- Whereas concrete is clearly shown on the project drawings, few details of formwork are provided. The estimator, therefore, needs to make an assessment of formwork requirements from the details of the concrete given.
- In a takeoff of concrete work, the estimator should work through the project dealing with one assembly at a time measuring all the work associated with one assembly before passing on to the next.
- Concrete is measured in cubic yards or cubic meters net in place.
- Concrete is classified in terms of the use to which the concrete is put.
- Different mixes of concrete are measured separately.
- Formwork generally is measured in square feet or square meters of contact area net in place.

PEWin - C:\Sample Commercial GC

File  Edit  View  Database  Takeoff  Pricing  Reports  Interface  Window  Help

Office-Warehouse

| Group | Phase | Description | Takeoff Quantity | Labor Cost/Unit | Labor Price | Labor Amount | Material Price | Material |
|-------|-------|-------------|------------------|-----------------|-------------|--------------|----------------|----------|
| 2000.000 | | SITEWORK | | | | | | |
| | 2144.000 | Gravel @ Slab | | | | | | |
| | | Gravel under SOG | 402.90 CY | 20.00 /CY | 20.00 /CY | 8,058 | /CY | |
| | 2201.000 | Earthwk: Remove Topsoil | | | | | | |
| | | Remove Topsoil | 1,087.04 CY | 20.00 /CY | 20.00 /CY | 21,741 | - | |
| | 2205.000 | Earthwk: Excav Trench | | | | | | |
| | | Excavate Trench | 248.82 CY | 20.00 /CY | 20.00 /CY | 4,976 | - | |
| | 2210.000 | Earthwk: Excav Pits | | | | | | |
| | | Excavate Pits | 13.22 CY | 20.00 /CY | 20.00 /CY | 264 | - | |
| | 2220.000 | Earthwk: Bulk Cut | | | | | | |
| | | Bulk Cut | 149.67 CY | 20.00 /CY | 20.00 /CY | 2,993 | - | |
| | 2220.400 | Earthwk: Grade & Trim | | | | | | |
| | | Grade & Trim Bottoms of Excav | 2,944.00 sf | 20.00 /sf | 20.00 /CY | 58,880 | - | |
| | 2221.500 | Earthwk: Gravel Fill | | | | | | |
| | | Gravel Fill | 504.26 CY | 20.00 /CY | 20.00 /CY | 10,085 | /CY | |
| | 2222.000 | Earthwk: Dispose Surplus | | | | | | |
| | | Dispose Surplus | 211.00 CY | 20.00 /CY | 20.00 /CY | 4,220 | - | |
| | 2225.000 | Backfill Trenches | | | | | | |
| | | Bakfill Trenches | 190.82 CY | 20.00 /CY | 20.00 /CY | 3,816 | - | |
| | 2230.000 | Backfill Pits | | | | | | |
| | | Bakfill Pits | 9.89 CY | 20.00 /CY | 20.00 /CY | 198 | - | |
| 3000.000 | | CONCRETE | | | | | | |
| | 3110.100 | Forms: Footings | | | | | | |
| | | Pad Footing Forms | 143.20 sf | 20.00 /sf | 20.00 /sf | 2,864 | 0.78 /sf | |
| | | Continuous Footing Forms | 329.41 sf | 20.00 /sf | 20.00 /sf | 6,588 | 0.78 /sf | |
| | | Keyway in Footing | 297.33 lf | 0.40 /lf | 20.00 /hr | 119 | 0.90 /lf | |
| | 3112.000 | Forms: Pile Caps | | | | | | |
| | | Form Pile Caps | 133.30 sf | 20.00 /sf | 20.00 /sf | 2,666 | 0.78 /sf | |
| | 3114.000 | Forms: Grade Beams | | | | | | |
| | | Form Grade Beams | 1,815.00 sf | 20.00 /sf | 20.00 /sf | 36,300 | 0.78 /sf | |
| | 3114.500 | Forms: Voids | | | | | | |
| | | 4" x 8" Void Forms | 473.67 lf | 20.00 /lf | 20.00 /lf | 9,473 | 0.78 /lf | |
| | 3115.000 | Forms: Walls | | | | | | |
| | | Form Walls | 2,310.00 sf | 20.00 /sf | 20.00 /sf | 46,200 | 0.78 /sf | |

Phase/Item  Takeoff Order

For Help, press F1                                                    2/16/2002

**Figure 5.11a**    Spreadsheet

- Formwork is classified in terms of use just as concrete is. Forms to curved surfaces are measured separately.
- Because only the area of forms in contact with the concrete is measured, the estimator does not have to be concerned about details of the design of the forms at the time of the takeoff.
- Formwork to pilasters, bulkheads, edges and suchlike are measured separately.
- Items of formwork that are linear in nature, such as grooves, keyways, notches and chamfers, are described stating their size and measured in linear feet or meters.
- Forms to circular columns are described giving the diameter of the column and measured in linear feet or meters to the height of the column.
- Concrete finishes are measured in square feet or square meters.
- Slab finishes, screeds and curing slabs are all measured in square feet or square meters of plan area.
- Welded wire mesh reinforcing is also measured in square feet or square meters of plan area with further percentage added to allow for the laps in the mesh.
- Fabricating and placing reinforcing steel is often subcontracted by general contractors, in which case the subcontractors would take off and price this trade.
- Concrete work takeoffs can be done manually or via computer.

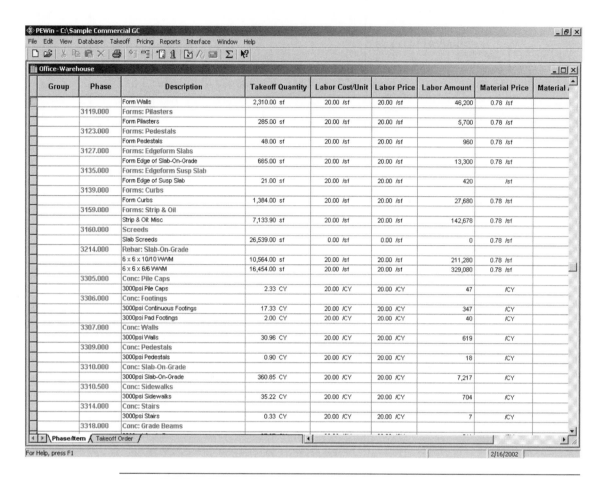

**Figure 5.11b** Spreadsheet continued

## REVIEW QUESTIONS

1. Why is it recommended that the estimator begin a takeoff of concrete work by measuring the quantity of concrete in an item rather than the amount of formwork?
2. Why is it also preferable to measure the concrete and all other work associated with an item before going on to measure the concrete in another item?
3. Why is nothing added to the concrete quantities for wastage at the time of the takeoff?
4. What size does an opening in concrete have to be before a deduction is made from the volume of concrete to account for the opening?
5. Why is concrete classified and measured separately in different categories?
6. Make a copy of Figure 5.1 and identify on the figure the following formwork:
   a. Wall forms
   b. Slab edge forms
   c. Column forms
   d. Beam forms
   e. Slab soffit forms
7. What, precisely, is measured in a formwork takeoff?
8. Why are pilaster forms measured separately from wall forms when the concrete in walls and pilasters is measured together?

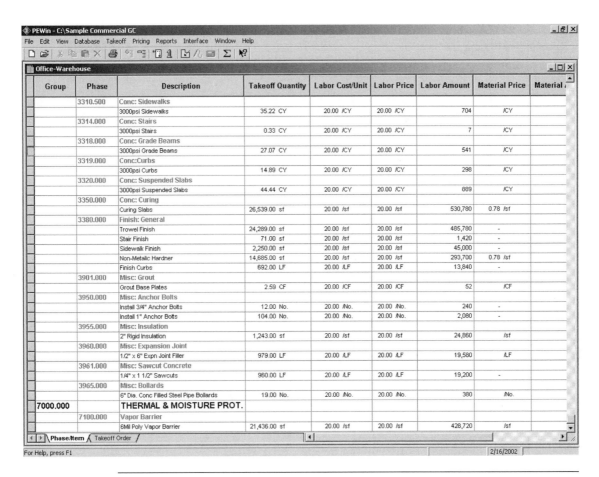

PEWin - C:\Sample Commercial GC

File  Edit  View  Database  Takeoff  Pricing  Reports  Interface  Window  Help

Office-Warehouse

| Group | Phase | Description | Takeoff Quantity | Labor Cost/Unit | Labor Price | Labor Amount | Material Price | Material |
|---|---|---|---|---|---|---|---|---|
| | 3310.500 | Conc: Sidewalks | | | | | | |
| | | 3000psi Sidewalks | 35.22 CY | 20.00 /CY | 20.00 /CY | 704 | /CY | |
| | 3314.000 | Conc: Stairs | | | | | | |
| | | 3000psi Stairs | 0.33 CY | 20.00 /CY | 20.00 /CY | 7 | /CY | |
| | 3318.000 | Conc: Grade Beams | | | | | | |
| | | 3000psi Grade Beams | 27.07 CY | 20.00 /CY | 20.00 /CY | 541 | /CY | |
| | 3319.000 | Conc:Curbs | | | | | | |
| | | 3000psi Curbs | 14.89 CY | 20.00 /CY | 20.00 /CY | 298 | /CY | |
| | 3320.000 | Conc: Suspended Slabs | | | | | | |
| | | 3000psi Suspended Slabs | 44.44 CY | 20.00 /CY | 20.00 /CY | 889 | /CY | |
| | 3350.000 | Conc: Curing | | | | | | |
| | | Curing Slabs | 26,539.00 sf | 20.00 /sf | 20.00 /sf | 530,780 | 0.78 /sf | |
| | 3380.000 | Finish: General | | | | | | |
| | | Trowel Finish | 24,289.00 sf | 20.00 /sf | 20.00 /sf | 485,780 | - | |
| | | Stair Finish | 71.00 sf | 20.00 /sf | 20.00 /sf | 1,420 | - | |
| | | Sidewalk Finish | 2,250.00 sf | 20.00 /sf | 20.00 /sf | 45,000 | - | |
| | | Non-Metalic Hardner | 14,685.00 sf | 20.00 /sf | 20.00 /sf | 293,700 | 0.78 /sf | |
| | | Finish Curbs | 692.00 LF | 20.00 /LF | 20.00 /LF | 13,840 | - | |
| | 3901.000 | Misc: Grout | | | | | | |
| | | Grout Base Plates | 2.59 CF | 20.00 /CF | 20.00 /CF | 52 | /CF | |
| | 3950.000 | Misc: Anchor Bolts | | | | | | |
| | | Install 3/4" Anchor Bolts | 12.00 No. | 20.00 /No. | 20.00 /No. | 240 | - | |
| | | Install 1" Anchor Bolts | 104.00 No. | 20.00 /No. | 20.00 /No. | 2,080 | - | |
| | 3955.000 | Misc: Insulation | | | | | | |
| | | 2" Rigid Insulation | 1,243.00 sf | 20.00 /sf | 20.00 /sf | 24,860 | /sf | |
| | 3960.000 | Misc: Expansion Joint | | | | | | |
| | | 1/2" x 6" Expn Joint Filler | 979.00 LF | 20.00 /LF | 20.00 /LF | 19,580 | /LF | |
| | 3961.000 | Misc: Sawcut Concrete | | | | | | |
| | | 1/4" x 1 1/2" Sawcuts | 960.00 LF | 20.00 /LF | 20.00 /LF | 19,200 | - | |
| | 3965.000 | Misc: Bollards | | | | | | |
| | | 6" Dia. Conc Filled Steel Pipe Bollards | 19.00 No. | 20.00 /No. | 20.00 /No. | 380 | /No. | |
| **7000.000** | | **THERMAL & MOISTURE PROT.** | | | | | | |
| | 7100.000 | Vapor Barrier | | | | | | |
| | | 6Mil Poly Vapor Barrier | 21,436.00 sf | 20.00 /sf | 20.00 /sf | 428,720 | /sf | |

Phase/Item  Takeoff Order

For Help, press F1    2/16/2002

**Figure 5.11c**    Spreadsheet continued

9. What size opening in an area of formwork would be deducted from the gross area of the forms?

10. List three examples of formwork items that would be measured in linear feet.

11. How is the circular columns formwork measured?

12. Takeoff the earthwork and the concrete work required for the construction of the carport shown on Figure 5.12, noting the following:

   a. Remove topsoil to a depth of 300 mm from the site of the carport extending 1 meter outside of the slab-on-grade.

   b. Replace topsoil around the outside of the structure at completion up to the original grade.

   c. Excavate a trench for the grade beam walls and allow a 600-mm wide workspace each side of the wall. This space is backfilled at completion.

   d. All concrete shall be 30 MPa. with air-entrainment.

   e. The slab-on-grade shall receive a steel trowel finish.

   f. Reinforcing steel is supplied and installed by subcontractor (Jones) for the price of $5,000.

   g. Structural steel columns are supplied and installed by subcontractor (Cortez) for the price of $2,500. Note that this subtrade supplies anchor bolts but they are installed by the general contractor.

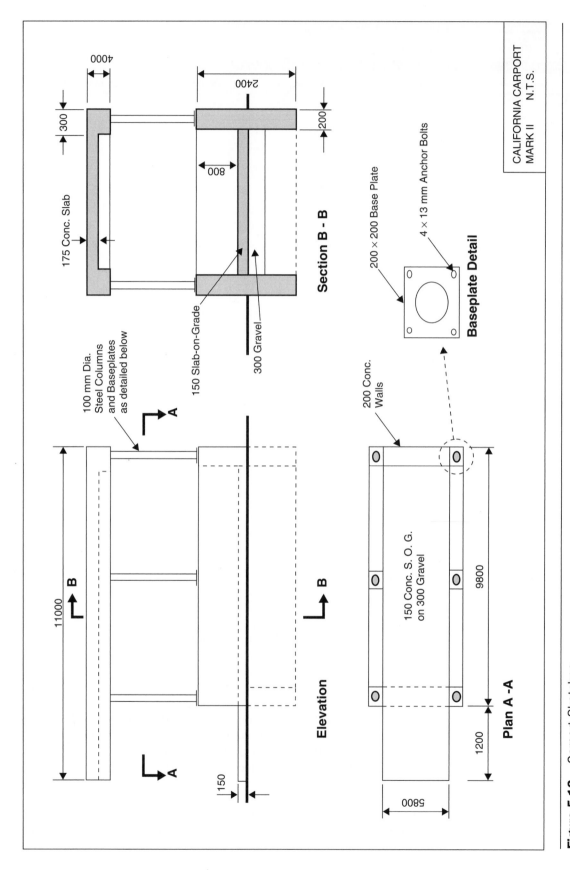

**Section B - B**

4000

2400

300

200

175 Conc. Slab

800

150 Slab-on-Grade

300 Gravel

100 mm Dia. Steel Columns and Baseplates as detailed below

CALIFORNIA CARPORT
MARK II     N.T.S.

200 × 200 Base Plate

4 × 13 mm Anchor Bolts

**Baseplate Detail**

200 Conc. Walls

**Elevation**

11000

150

**Plan A -A**

150 Conc. S. O. G.
on 300 Gravel

9800

1200

5800

**Figure 5.12**  Carport Sketches

# 6

# MEASURING MASONRY WORK

## OBJECTIVES

*After reading this chapter and completing the review questions, you should be able to do the following:*

- Explain how masonry work and associated items are measured in a takeoff.
- Measure masonry items from drawings and specifications.
- Measure items associated with masonry from drawings and spcifications.
- Explain how masonry items are classified in the takeoff process.
- Describe the main factors affecting the measurement of brick masonry.
- Explain what is meant by measuring "extra over."
- Use conversion factors to calculate quantities of bricks, blocks and masonry mortar.
- Given the size of masonry units and the thickness of the mortar joints between them, calculate the value of conversion factors to determine the number of bricks/blocks and the volume of mortar.
- Complete a manual takeoff of masonry work and associated items.
- Complete a computer takeoff of masonry work using the Precision Estimating software.

## Masonry Work Generally

Masonry work includes construction with clay bricks, concrete bricks and blocks, clay tiles and natural and artificial stone. This work is now typically performed by subcontractors who supply the labor, materials and equipment required to complete the work. In this chapter we will deal with the measurement of only standard clay brick and concrete block masonry, but the principles expressed can be applied to many types and sizes of masonry products. Estimating the costs of masonry work involving products other than standard bricks and concrete blocks can proceed in the same way as described hereunder, but the work is mostly the task of specialists rather than general contractors and, therefore, is beyond the scope of this text.

As we have previously stated with regard to other trades, a thorough knowledge of masonry construction is a definite prerequisite to the performance of good estimating. Particular attention has to be paid to items such as masonry bond and mortar joint treatment, which have considerable effect on material usage, waste factors and mortar consumption. A minimum knowledge of terminology is required before masonry specifications can be properly interpreted to determine such things as the type of units required and their requisite features, together with aspects of the numerous accessories associated with masonry work. We have not attempted to explore any of the many features of masonry construction practices in this text. The student is advised to consult and study one or more of the many texts that deal with this specific subject.

## Measuring Masonry Work

In accordance with the general principles followed in the measurement of the work of other trades, masonry work is measured "net in place" and the necessary allowances for waste and breakage are considered later in the estimating process when this work is priced. The units of measurement for masonry are generally the number of masonry pieces such as concrete blocks or, in the case of bricks, the number of thousands of clay bricks. Calculating the number of masonry units involves a two-stage process:

1. The area of masonry is measured.
2. A standard factor is applied to determine the number of masonry units required for area measured.

A number of other items associated with unit masonry that are detailed on the drawings or described in the specifications also have to be measured, including mortar, wire **"ladder" reinforcement** to joints, metal ties used between walls and extra work and materials involved in control joists, sills, lintels and arches. All of these items are considered in the masonry takeoff as detailed in the following text. Loose insulation and rigid insulation to masonry is also included in this work, and in some places the masonry trade includes membrane barriers, insulation and similar materials behind masonry when the masonry is the last work installed.

Estimators need to be familiar with the definition of the scope of work for the masonry trade in their geographical location. This trade scope is often defined for a geographical area by the local masonry association and may also be found in the bid depository rules of construction associations in many cities. This definition lists the items of work that are included in the masonry trade and those items of work that are excluded from the work of the masonry trade. Knowledge of this scope of work is essential to the person estimating the masonry work for the specialized masonry trade contractor. Knowledge of the masonry scope, together with the scopes of all other subtrades, is also valuable to the general contractor's estimator who is receiving subtrade prices.

A typical masonry scope of work definition is illustrated in Chapter 13. Estimators should be aware that this is merely an example of a scope as defined by a certain jurisdiction at a certain time; definitions vary according to location and time.

The general contractor estimator's role in handling problems associated with a subtrade's scope of work is also discussed later in Chapter 13. The contractor's estimator will, however, need to address items of work that are connected with the masonry trade but that are specifically excluded from the subtrade quote. If these items of work are not to be included as part of the subtrade's work, their cost will have to be estimated with the general contractor's work.

A typical example of the kind of work item that is related to masonry work but may not be included in the quote from a subtrade is the item of temporary bracing masonry work during construction. This item is excluded from the scope of the masonry trade in some localities, but the cost of the work involved in temporary bracing has to be estimated and included in the bid as it can be a substantial amount, especially where the project includes large expanses of masonry walls exposed to possible high winds during the course of construction.

## Brick Masonry

Bricks are made of different materials and manufactured by different methods, and they can be used in many different ways in the construction process. All of these factors will influence the price of the masonry, but the main factors affecting the measurement of brick masonry are the size of the brick units, the size of the joints between bricks, the wall thickness and the pattern of brick bond utilized. Bricks are available in a great number of sizes but the Common Brick Manufacturers Association has adopted a standard size with the nominal dimensions of $2\frac{1}{4}$ by $3\frac{3}{4}$ by 8 inches (57 mm by 95 mm by 203 mm). Bricks of this size are referred to as "standard bricks," and this is the brick size used in the takeoff examples that follow.

All of the variables previously noted will also impact on the measurement of the mortar needed for brick masonry. From the size of the bricks and the thickness of the mortar required, the amount of mortar per 100 square feet of wall can be readily calculated; examples follow. To avoid having to perform these calculations for all of the combinations of sizes involved, many estimators refer to texts that offer tables of factors to use in calculating numbers of masonry units and volumes of mortar per 100 square feet of wall area. For example, from Walker's *The Building Estimators Reference Book*, the amount of mortar required for the bricks described above is specified as 5.7 cubic feet (0.16 cubic meters) per thousand bricks.* Information about quantities of bricks per square foot and mortar requirements over a wide range of brick sizes are also available from brick manufacturers and from organizations such as the Brick Institute of America (BIA).

Sometimes designated courses of bricks are required to be laid in a different way than the rest of the courses in a wall. For example, a **soldier course** may be required above openings or a **brick-on-edge course** may be required at the sill of openings. These special courses can be measured in linear feet as **"extra over"** the brickwork in which they appear. This means that the extra cost of labor and materials over and above the cost of standard brickwork is calculated and added for the length of these special courses measured. Measurement as "extra over" avoids having to calculate and deduct the number of bricks laid in the standard fashion and adding the number of bricks laid in a soldier course, which is clearly a more cumbersome takeoff method.

## Concrete Blocks

There are fewer variables to consider with concrete blocks than with clay bricks, but blocks do come in different sizes and the thickness of joints between blocks can vary. See Figure 6.1 for standard block and bond beam sizes for an 8-inch (203 mm) thick wall. Concrete blocks are also available for different wall thicknesses (see Figure 6.2).

---

*The Building Estimator's Reference Book*, published by Frank R. Walker Company, 27th Edition, Chicago, 2003.

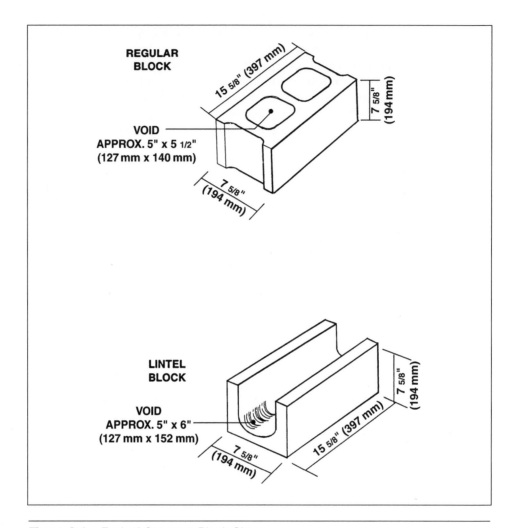

**Figure 6.1**    Typical Concrete Block Sizes

| NOMINAL WALL THICKNESS | BLOCK THICKNESS |
|:---:|:---:|
| 4" (102 mm) | $3\frac{5}{8}$" (92 mm) |
| 6" (152 mm) | $5\frac{5}{8}$" (143 mm) |
| 8" (203 mm) | $7\frac{5}{8}$" (194 mm) |
| 10" (254 mm) | $9\frac{5}{8}$" (244 mm) |
| 12" (305 mm) | $11\frac{5}{8}$" (295 mm) |

**Figure 6.2**    Nominal and Actual Block Thicknesses

Bond beams are measured as "extra over" the blockwork in which they are required. This means that, first, the number of blocks in a wall is calculated as if they are all standard blocks. Then, where bond beam blocks are required, instead of adding a number of bond beam blocks and deducting a number of standard blocks, labor and material prices per linear foot of bond beam are calculated to account for the extra cost of the bond beam blockwork over and above the cost of standard blockwork. This "extra over" price could be expanded to include the cost of concrete and reinforcing steel in the bond beam, but in the examples that follow, concrete and rebar to bond beams is measured separately for individual pricing.

## Conversion Factors

As mentioned, tables are published in reference works that list quantities of bricks or blocks per square foot area of masonry. The number of "standard bricks" per square foot of wall for different wall thicknesses, joint thicknesses and various bonds is given in *The Building Estimator's Reference Book*, which states, for instance, that a 4-inch (102 mm) wall with ¼-inch (6 mm) brick joints and laid in **running bond** will have 7 bricks per square foot of wall area.

Tables also list the number of standard-size concrete blocks and the amount of mortar required per 100 square feet of wall. From the reference source just cited, for a wall of 8-inch nominal thickness built with standard blocks of size 7⅝ by 7⅝ by 15⅝ inches and a joint thickness of ⅜-inches, the tables indicate that 112.5 blocks together with 6 cubic feet of mortar are required for 100 square feet of wall area. (Expressed in metric units we have: For a 102-mm wide standard brick wall with 6-mm joings in running bond, there will be 75 bricks per square meter. For a 203-mm wide standard block wall with 9.5-mm mortar joints, there will be 1210 blocks per 100 square meters, and 1.8 cubic meters of mortar is required per 100 square meters of wall.)

The estimator may have to calculate the conversion factors when nonstandard-size brick or block units are to be used. So, if bricks of size 3⅝ by 3⅝ by 15⅝-inches with ⅜-inch wide joints are to be used, the conversion factor may be calculated in the following way:

$$\frac{1 \text{ square foot (144 sq. inches)}}{(3\tfrac{5}{8} + \tfrac{3}{8}) \times (15\tfrac{5}{8} + \tfrac{3}{8})} = 2.25 \text{ brick/square foot}$$

(Note: Only the dimensions of the face of the brick are required; the brick thickness does not affect the calculation.)

For the volume of mortar per square foot of wall area, the volume of brick plus joints is calculated less the volume of brick per square foot of wall area. (Note: Here the brick thickness—3⅝—does have to be taken into account.)

$$\text{vol. of brick + joints } = 3\tfrac{5}{8} \times (3\tfrac{5}{8} + \tfrac{3}{8}) \times (15\tfrac{5}{8} + \tfrac{3}{8})$$

$$= 232 \text{ cu. inches/brick}$$

and since 1 cubic foot = 1728 cubic inches

$$\text{vol. of brick + joints } = \frac{232}{1728}$$

$$= 0.134 \text{ cu. feet/brick}$$

$$\text{vol. of brick alone} = \frac{3\frac{5}{8} \times 3\frac{5}{8} \times 15\frac{5}{8}}{1728}$$

$$= 0.119 \text{ cu. feet/brick}$$

therefore, the volume of mortar = 0.134 – 0.119 = 0.015 cu. feet per brick. Hence, the volume of mortar per square foot of wall area = 0.015 × 2.25 = 0.035 cu. feet per square foot, or, 0.129 cubic yards of mortar per 100 square feet of wall area.

However, this mortar quantity will be too low because, in addition to mortar wasted and spilled, in the process of laying bricks some mortar is squeezed into the voids in the bricks and, sometimes, mortar is left to bulge from the joints. Therefore the mortar factor could be increased by as much as 40% to allow for the additional quantity required.

## Measuring Notes—Masonry

### Generally

1. All quantities shall be measured "net in place" and no deductions shall be made for openings less than 10 square feet (1 square meter).
2. Masonry work circular on plan shall be measured separately.
3. Masonry work shall include scaffolding and hoisting and shall be kept separate in the following categories:
   a. Facings
   b. Backing to facings
   c. Walls and partitions
   d. Furring to walls
   e. Fire protection
4. Cleaning exposed masonry surfaces shall be measured in square feet or square meters and fully described.
5. Silicone treatment of masonry surfaces shall be measured in square feet or square meters, stating the number of coats.
6. Expansion joints or control joints in masonry shall be measured in linear feet or meters and fully described. These joints are filled with a material that should be specified and caulking is often required, all of which should be included in the description and priced as with the other components of the joint system.
7. Mortar shall be measured in cubic yards or cubic meters, with details of any admixtures required.
8. Colored mortar shall be measured separately.
9. Wire reinforcement in masonry joints shall be measured in linear feet or meters and fully described. "Ladder" or **"truss" reinforcement** can be used; the specification should define which is required.
10. Building in anchor bolts, sleeves, brackets and similar items shall be enumerated and fully described. These items together with the lintels and similar items mentioned in the following notes may be supplied and installed by the masonry trade, but trade rules in many localities state that miscellaneous metals components are supplied by the miscellaneous metals subtrade and only installed by the masonry contractor.
11. Building in lintels, sills, copings, flashings and similar items shall be measured in linear feet and fully described. Again trade rules may impact on this item, for example, flashings are often required to be supplied by the roofing

and sheet metal trade and merely installed by the masonry trade. The type of material involved should be noted as the installation cost can vary with different materials.

12. Where **weep holes** are required to be formed using rubber tubing and such like they shall be enumerated and fully described.

13. Rigid insulation to masonry work shall be measured in square feet describing the type and thickness of material.

### Brick Masonry

14. Bricks shall be measured in units of a thousand bricks describing the type and dimensions of bricks.

15. Facing brick shall be measured separately.

16. Bricks required to be laid in any other pattern than running bond, for example, soldier courses or brick-on-edge courses, shall be measured in linear feet as "extra over" the brickwork in which they are located.

17. **Brick ties** shall be enumerated and fully described.

### Concrete Block Masonry

18. Concrete block masonry units shall be enumerated stating the type and size of blocks.

19. Special units required at corners, jambs, heads, sills and other similar locations shall be described and enumerated separately.

20. Loose fill or foam insulation to blockwork cores shall be measured in cubic yards or cubic meters and fully described.

21. Bond beams (lintel blocks) shall be measured in linear feet or meters as "extra over" the blockwork in which they are located to allow for the extra cost of materials and labor for bond beam blocks.

22. Concrete to core fills and bond beams shall be measured in cubic yards or cubic meters stating the strength and type of concrete.

23. Reinforcing steel to core fills and bond beams shall be measured in linear feet or meters stating the size and type of rebar.

Trade rules in some locations require the subcontractor supplying reinforcing steel to the project to supply the rebar for the masonry work. This rebar is then installed by the masonry subtrade or the general contractor, whoever is completing the masonry work.

### Masonry Work—House (Brick Facings Alternative)

#### Comments on Takeoff Shown as Figure 6.3a

1. See Figures 6.3b and 6.3c for details of an alternative exterior finish of brick facings to the front of the house.

2. Facing bricks are to be applied to the front left side of the house as shown on the sketches. The front width plus the two returns gives an overall brickwork length of 23'10".

3. Note the net area of brickwork is required so all window openings are deducted.

4. The standard factors of 7 bricks per square foot and 5.7 cubic feet of mortar per 1000 bricks are used to determine quantities of bricks and mortar.

| QUANTITY SHEET | | | | | | | SHEET No. | $1 of 1$ |
|---|---|---|---|---|---|---|---|---|
| JOB _House Example_ | | | | | DATE | | | |
| ESTIMATOR _ABF_ | | | EXTENDED | | EXT. CHKD | | | |

| DESCRIPTION | TIMES | DIMENSIONS | | | | | | |
|---|---|---|---|---|---|---|---|---|
| MASONRY | | | | | | | | |
| FACING BRICKS ALTERNATIVE | | | | | | | | |
| 19'10" | | | | | | | | |
| 2 x 2'-0"  4'-0" | | | | | | | | |
| 23'-10" | | | | | | | | |
| | | | | | | | | |
| FACING BRICKS | | 23.83 | — | 8.83 | 210 | | | |
| (WINDOWS)  DDT | 2 | 4.08 | — | 3.58 | < 29 > | | | |
| | | | | | 181 sf x 7 = | | | 1267 № |
| | | | | | | | | |
| MORTAR | | —— | | —— | 1.27 M. BKS x 5.7 = | | | 7 CF. |
| | | | | | | | | |
| BRICK TIES | | —— | | —— | $\frac{181}{2}$ | | = | 91 № |
| ( SPACING 2' x 1' ) | | | | | | | | |
| | | | | | | | | |
| 3" x 3" x ¼" L | | 23.83 | | | 24 | | | |
| (OVER WINDOWS) | 2 | 6.08 | | | 12 | | | |
| 4'-1" | | | | | 36 LF | | | |
| EACH SIDE 2 x 1'-0" 2-0" | | | | | | | | |
| 6'-1" | | | | | | | | |
| | | | | | | | | |
| ½" DRILLED ANCHORS | 12 | | | | 12 № | | | |
| 23'-10" | | | | | | | | |
| 2'-0" | | | | | | | | |
| = 12 | | | | | | | | |
| | | | | | | | | |
| 12" WIDE FLASHING | | | AS | ANGLE | 36 LF | | | |
| | | | | | | | | |
| WEEP HOLES | 7 | | | | 7 № | | | |
| 23'-10" | | | | | | | | |
| 4'-0" | | | | | | | | |
| = 6+1 | | | | | | | | |

**Figure 6.3a**  House Example—Masonry Work

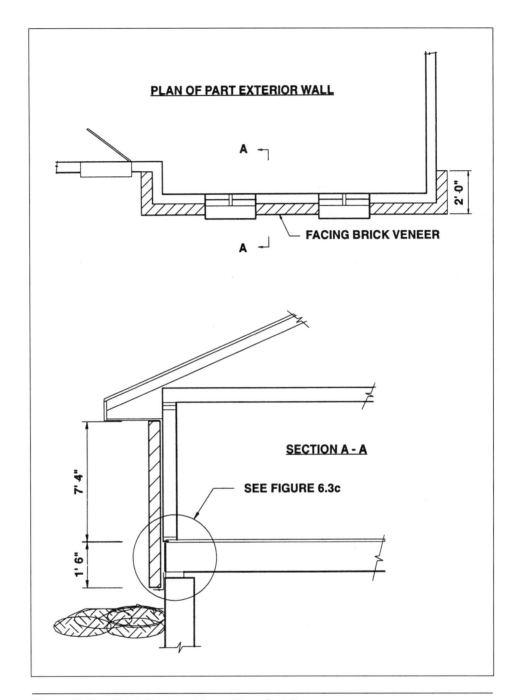

**PLAN OF PART EXTERIOR WALL**

A

2' 0"

FACING BRICK VENEER

A

**SECTION A - A**

SEE FIGURE 6.3c

7' 4"

1' 6"

**Figure 6.3b**    Brick Facings Alternative—Details

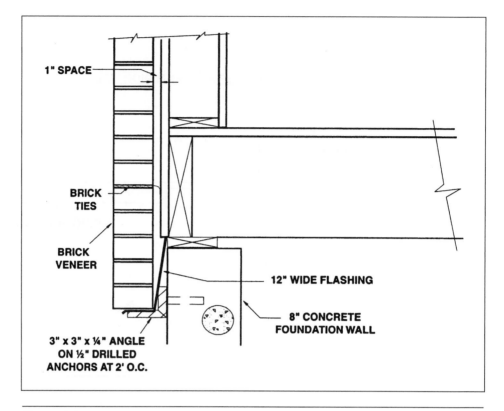

**Figure 6.3c**    Brick Facing—Detail at Foot of Facings

5.  Brick ties are spaced 2 feet horizontally and 1 foot vertically; dividing the wall area by the product of the spacing (2 × 1) gives the number of ties required.
6.  The steel angle lintel over the windows extends 1 foot each side of the opening, so 2 feet are added to the opening width to give the length of lintel required.
7.  The shelf angle supporting the full length of the brick wall is attached to the footing by means of anchors drilled into the concrete foundation wall. The number of these inserts is found using their spacing.
8.  Weep holes at 4'0" spacing are built into the lowest brick course, which sits on top of the shelf angle. These allow water that gets behind the facings to drain to the outside of the brickwork.

## Masonry Work—Office/Warehouse Building

### Comments on Takeoff Notes Shown as Figure 6.4

1.  See Figure 6.1 for sizes of concrete blocks and bond beams measured in this takeoff.
2.  Blockwork is confined to the office section of this project. The warehouse is finished with insulated cladding, which will be covered in a subcontractor's quote.
3.  The exterior perimeter (300 feet) is used to calculate the quantity of blocks. Areas are deducted for openings over 10 square feet, and areas are added to allow for the extra blocks in pilasters.

| TAKEOFF NOTES | | | SHEET No. 1 of 2 |
|---|---|---|---|

PROJECT: *OFFICE / WAREHOUSE*    DATE:

ESTIMATOR: *A B F*

| NOTES | TIMES | DIMENSIONS | | |
|---|---|---|---|---|
| **MASONRY** | | | | |
| EXTERIOR WALLS: 2 x 50'-0"  100'-0" | | | | |
| 2 x 100'-0"  200'-0" | | | | |
| 300'-0" | | | | |
| 8" LT. WT. CONC. BLOCKS | | 300.00 | 28.00 | |
| (112.5 Blks/100 SF)  (WINDOWS)  DDT | ‹31 | 4.00 | 8.00 | › |
| (DOORS)  DDT | ‹2 | 6.00 | 8.00 | › |
| (PILASTERS)  ADD | 8 | 1.33 | 28.00 | |
| "  ADD | 4 | 0.67 | 28.00 | |
| TYPE N MORTAR  (6 CF/100 SF) | | DITTO | | |
| LOOSE VERMICULITE INSULN  (27.6 CF/100 SF) | | DITTO | | |
| 8" LADDER REINF.  (75 LF/100 SF) | | DITTO | | |
| (EVERY 2ND. COURSE) | | | | |
| VERTICAL CONTROL JOINTS | | | | |
| IN 8" BLOCK WALL  (20' SPACING) | 2x4 | 28.00 | | |
| | 2x2 | 28.00 | | |
| LINTEL BLOCKS  (1-COURSE HIGH) | 4 | 300.00 | | |
| | 3 | 16.67 | | |
| DOOR/WINDOW UNIT  DOOR:  6'-0" | 29 | 6.67 | | |
| 2 x 4'-0"  8'-0"  + 2'-8" | | 8.67 | | |
| 6'-0"  8'-8" | | | | |
| 2 x 1'-4"  2'-8" | | | | |
| 16'-8"  CONC. IN LINTEL BLOCKS | | 1452.11 | 0.42 | 0.50 |
| WINDOWS:  4'-0" | | | | |
| 2 x 1'-4"  2'-8" | | | | |
| 6'-8"  CONC. IN CORE FILLS | 75 | 0.42 | 0.46 | 28.00 |

**Figure 6.4**  Office/Warehouse Example—Masonry Work

4. There is a situation on the drawings regarding block pilasters that needs clarification. In the center of the south wall on the office, a pilaster is indicated on the plan of the upper floor (see Figure 4.5, drawing A-2 of 4). On the lower floor plan there is a doorway in the center of this wall. The designer should be contacted to verify what is required here.

   Let us say that the answer we receive indicates that a full-height pilaster is required on this wall; the takeoff must reflect this clarification.

5. The conversion factors used in the takeoff are obtained from *The Building Estimator's Reference Book*, which indicates that the cavities of 8" block walls can be filled at the following rate: 14.5 square feet of wall per 4 cubic feet bag of insulation.

| TAKEOFF NOTES | | | | SHEET No. 2 of 2 | |
|---|---|---|---|---|---|
| PROJECT: | | | | DATE: | |
| ESTIMATOR: | | | | | |

| NOTES | | | | DIMENSIONS | |
|---|---|---|---|---|---|
| | | TIMES | | | |
| MASONRY PARTITIONS | | | | | |
| 25'-0"  2 × 14'-10"  29'-8" | | | | | |
| 15'-6"  10'-8" | | | | | |
| 40'-6"  40'-4" | | | | | |
| 14'-2' | | | | | |
| 10'-2"  8" LT. WT. CONC BLOCKS | | | 40.50 | 13.00 | |
| 5'-0" | | | 45.33 | 26.00 | |
| 2×8'  16'-0" | | | 40.33 | 13.00 | |
| 45'-4"  (DOORS)  DDT | | < 6 | 3.00 | 7.00 | > |
| TYPE N MORTAR (.6 CF/100 SF) | | | DITTO | | |
| 8" LADDER REINF. (75 LF/100 SF) | | | DITTO | | |
| LINTEL BLOCKS | | 6 | 5.67 | | |
| 3'-0" | | | | | |
| 2×1'-4'  2-8' | | | | | |
| 5'-8"  CONC. IN LINTEL BLOCKS | | 6 | 5.67 | 0.42 | 0.50 |

**Figure 6.4 continued**

At this rate the number of bags required to fill the cavities

$$\text{of 100 square feet of wall} = \frac{100}{14.5} = 6.9 \text{ bags per 100 SF}$$

6.9 bags × 4 cubic feet per bag = 27.6 cubic feet per 100 SF

6. Wire "ladder" reinforcing is required in every second course of blockwork. The length of reinforcing can be determined by dividing the area of block-

work by the spacing of reinforcing, which is 1'4" (two 8" courses) in this case. The number of linear feet of "ladder" reinforcing required

$$\text{per 100 square feet of wall} = 100/1.33$$
$$= 75 \text{ LF/100 SF}$$

7. It is assumed that the masonry specifications call for vertical control joints in the block walls at 20' intervals. This requires four joints on the long sides of the building and two joints on the short sides.
8. Two double-height bond beams are shown to the perimeter of the building: one at the second floor level and the other at the roof level. Bond beams consist of lintel blocks filled with concrete.
9. Bond beams are also measured over all openings. The bond beam is wider than the opening by one block (16") on each side.
10. On the structural drawing, S-1 dowels are shown to extend up from the grade beam into core fills in the masonry wall at 4'0" on center. Core fills refer to the practice of filling the vertical core of the blocks with mortar. The size of the hollow core of a concrete block is indicated on Figure 6.1.
11. Reinforcing steel would be required to bond beams and to vertical core fills. In this case we will assume that this rebar is to be supplied and installed by the rebar subtrade so the work does not have to be measured with the masonry section.

**Precision Estimating Takeoff**

A number of the items we need for the masonry takeoff can be found in the sample database, but we need to set up some additional phases and items for the masonry work.

1. Set up the following new phase: 4101.10   Control Joints   LF
2. Set up the new items shown on Figure 6.5.
   - There are waste factors to include, but there is no need to use the Round Quantities option with these items.
3. We will be using item 4080.10 LD8, Horiz Wall Reinf 8" Ladder, for measuring the ladder reinforcing, but we want to change the takeoff unit from MLF (1000 linear feet) to LF. On the menu bar click on Database and select Items from the list. The Database Item window will now open.
4. Click on the list button to the right of the Phase/Item code.
5. Click the Collapse All button to get a list of Group Phases.
6. Double click on MASONRY.

| Phase Code | Item Code | Item Description | Unit | Formula | Waste Factor | Cost Categories |
|---|---|---|---|---|---|---|
| 7210.05 | v10 | Vermiculite Insulation | CY | CY L × W × D'/27 | 10% | L, M |
| 4101.10 | 08 | PVC Control Joint in 8" Wall | LF | — | 5% | L, M |
| 4070.10 | 10 | 3000 psi Fill Lintel Blocks | CY | CY L × W × D'/27 | 5% | L, M |
| 4070.10 | 20 | 3000 psi Core Fill Blocks | | CY L × W × D'/27 | 5% | L, M |

**Figure 6.5**   New Items

7. Double click on phase 4080.10, then double click on item LD8 to bring up the 8" Ladder item.
8. Change the takeoff unit to LF.
   - To simplify pricing later, it is also advisable to change the order units to LF at this time
9. Change the labor order unit to LF. (The labor order unit is found below the labor price.)
10. Change the material order unit to LF. (The material order unit is found below the material price.)
    - Now we are ready to complete the masonry takeoff.
11. Use the Quick Takeoff to takeoff all the items listed on the takeoff notes for the masonry work beginning with the 8" light weight concrete blocks.
    - Figure 6.6 shows the spreadsheet after the takeoff is completed.
    - Again, do not be concerned about prices at this stage—we will consider the pricing of the masonry work later.

| Group | Phase | Description | Takeoff Quantity | Labor Cost/Unit | Labor Price | Labor Amount | Material Price | Material |
|---|---|---|---|---|---|---|---|---|
| | | Finish Curbs | 692.00 LF | 20.00 /LF | 20.00 /LF | 13,840 | - | |
| | 3901.000 | Misc: Grout | | | | | | |
| | | Grout Base Plates | 2.59 CF | 20.00 /CF | 20.00 /CF | 52 | /CF | |
| | 3950.000 | Misc: Anchor Bolts | | | | | | |
| | | Install 3/4" Anchor Bolts | 12.00 No. | 20.00 /No. | 20.00 /No. | 240 | - | |
| | | Install 1" Anchor Bolts | 104.00 No. | 20.00 /No. | 20.00 /No. | 2,080 | - | |
| | 3955.000 | Misc: Insulation | | | | | | |
| | | 2" Rigid Insulation | 1,243.00 sf | 20.00 /sf | 20.00 /sf | 24,860 | /sf | |
| | 3960.000 | Misc: Expansion Joint | | | | | | |
| | | 1/2" x 6" Expn Joint Filler | 979.00 LF | 20.00 /LF | 20.00 /LF | 19,580 | /LF | |
| | 3961.000 | Misc: Sawcut Concrete | | | | | | |
| | | 1/4" x 1 1/2" Sawcuts | 960.00 LF | 20.00 /LF | 20.00 /LF | 19,200 | - | |
| | 3965.000 | Misc: Bollards | | | | | | |
| | | 6" Dia. Conc Filled Steel Pipe Bollards | 19.00 No. | 20.00 /No. | 20.00 /No. | 380 | /No. | |
| **4000.000** | | **MASONRY** | | | | | | |
| | 4060.100 | Mortar: All Types | | | | | | |
| | | Mortar Type "N" | 21.74 cy | - | - | - | 4.80 /bags | |
| | 4070.100 | Mortar: Grout Fill Conc | | | | | | |
| | | 3000 psi Fill Lintel Blocks | 11.56 CY | 20.00 /CY | 20.00 /CY | 231 | 57.60 /CY | |
| | | 3000 psi Core Fill Blocks | 15.04 CY | 20.00 /CY | 20.00 /CY | 301 | 57.60 /CY | |
| | 4080.100 | Reinforce: Horizontl Wall | | | | | | |
| | | Horiz Wall Reinf 8" Ladder | 7,341.75 LF | 20.00 /LF | 20.00 /LF | 146,835 | 1.14 /LF | |
| | 4101.100 | Control Joints | | | | | | |
| | | PVC Control Joints in 8" Wall | 336.00 LF | 20.00 /LF | 20.00 /LF | 6,720 | /LF | |
| | 4220.110 | Conc. Block: 8" | | | | | | |
| | | Blk 8" Standard Face Lt Wt | 11,013.00 ea | 2.00 /ea | 20.00 /hr | 22,026 | 1.20 /ea | |
| | 4220.170 | Conc. Block: 8" Lintel | | | | | | |
| | | Lintel 8" Stand Face Lt Wt | 1,115.00 ea | 2.00 /ea | 20.00 /hr | 2,230 | 1.20 /ea | |
| **7000.000** | | **THERMAL & MOISTURE PROT.** | | | | | | |
| | 7100.000 | Vapor Barrier | | | | | | |
| | | 6Mil Poly Vapor Barrier | 21,436.00 sf | 20.00 /sf | 20.00 /sf | 428,720 | /sf | |
| | 7210.050 | Insulation: Loose Fill | | | | | | |
| | | Vermiculte Insulation | 78.56 CY | 20.00 /CY | 20.00 /CY | 1,571 | 6.12 /bags | |

**Figure 6.6**   Spreadsheet

## SUMMARY

- Masonry work, which includes construction with clay bricks, concrete bricks and blocks, clay tiles, natural and artificial stone, is now typically performed by subcontractors.
- A thorough knowledge of masonry construction is required to prepare a detailed estimate of this trade.
- In the takeoff process, first the surface area of the masonry is measured in square feet or square meters of area, then a factor is applied to convert these measurements to number of pieces such as bricks or blocks required to complete the work.
- A number of other items associated with unit masonry are also measured in the takeoff, including mortar, wire "ladder" reinforcement to joints, metal ties used between walls, and extra work and materials involved in control joists, sills, lintels and arches.
- Estimators need to be familiar with the definition of the scope of work for the masonry trade in their geographical location.
- The main factors affecting the measurement of brick masonry are the size of the brick units, the size of the joints between bricks, the wall thickness and the pattern of brick bond utilized.
- Where bricks are required to be laid in a different pattern than the other bricks in a wall, such as a soldier course, all the wall is measured as regular bricks, then the extra cost of the nonstandard pattern is allowed for by measuring "extra over" the cost of the regular bricks.
- There are fewer variables to consider with concrete blocks than with clay bricks, but blocks do come in different sizes and the thickness of joints between blocks can vary.
- A block wall is measured as if it were constructed entirely of the same type of block; then bond beams, copings and any other nonstandard block are measured "extra over."
- Various publications provide information about conversion factors to calculate such things as the number of blocks per square foot or square meter of wall.
- Conversion factors are also available to enable the estimator to calculate the volume of masonry required per 1000 bricks or blocks of different sizes.
- Given the size of masonry units and the thickness of the mortar joints between them, estimators can calculate their own conversion factors.
- Quantities of masonry units are measured net in place.
- Bricks are measured in units of a thousand bricks.
- Concrete block masonry is enumerated.
- Masonry is classified in terms of the types of units and the use.
- Mortar is measured in cubic yards or cubic meters and any add mixtures are described.
- Masonry work takeoffs can be done manually or via computer

## REVIEW QUESTIONS

1. Which two variables, apart from brick or block size, have an appreciable effect on masonry material usage and mortar consumption?
2. Describe the usual method of calculating quantities of bricks or blocks in a masonry takeoff.

3. For each of the following work items indicate whether it is included or excluded from the masonry scope of work (consult Chapter 13):
   a. Cavity wall insulation
   b. Brick anchors
   c. Anchor slot
   d. Hoarding and heating a temporary enclosure for masonry work
   e. Hoisting masonry materials
   f. Scaffolding
   g. Wind-bracing of masonry walls
4. State the standard brick size adopted by the Common Brick Manufacturers Association.
5. List three sources that could be referred to for information about the number of bricks and amount of mortar required for different brick sizes.
6. How are special brick courses such as soldier or brick-on-edge courses dealt with in a masonry takeoff?
7. How many custom bricks of size $3\frac{1}{4} \times 3\frac{3}{4} \times 11\frac{7}{8}$ inches are there per square foot of wall area when the mortar joint is $\frac{1}{4}"$?
8. How much mortar is required per square foot of wall area with the custom bricks described in question 7?
9. Openings greater than what area are deducted from masonry wall areas?
10. How are scaffolding and hoisting operations accounted for in a masonry work estimate?
11. How is insulation to masonry work measured?

# 7

# MEASURING CARPENTRY AND MISCELLANEOUS ITEMS

## OBJECTIVES

*After reading this chapter and completing the review questions, you should be able to do the following:*

- Describe a system that will allow an estimator to accurately measure detailed carpentry work on a large project.
- Calculate quantities of lumber using board measure units.
- Describe how to measure lumber and sheet materials using metric units.
- Describe how framing lumber is classified in a takeoff.
- Describe how to measure the following items in a takeoff:
  - Rough carpentry work
  - Finish carpentry work
  - Doors, frames and windows
  - Miscellaneous metals items
  - Specialties items
  - Bathroom accessories
  - Finish hardware
  - Exterior and interior finishes
- Measure rough carpentry, finish carpentry and miscellaneous items from drawings and specifications.
- Complete a manual takeoff of carpentry work and miscellaneous items.

## Measuring Rough Carpentry

In order to prepare a detailed quantity takeoff of the carpentry work that is realistic, a comprehensive knowledge of carpentry details and practices is required of the estimator. This is particularly true for estimating housing and other buildings of wood frame construction because, on these types of projects, few specific details of framing requirements are shown on the design drawings. Instead, the builder is left to construct the framing in accordance with standard practices and code requirements.

As a consequence, in order to assess rough carpentry requirements adequately, the estimator has to be completely familiar with framing methods and also be alert to design requirements such as the need for extra joists or additional studs in certain locations.

It is not our intention to examine the particulars of carpentry construction and the detailing of carpentry work in this text; students who are not experienced in this work are advised to study one or more of the many reference books available on this subject to help them to better understand the essentials of a rough carpentry takeoff.

When estimating commercial and institutional buildings, different skills are required of the estimator measuring rough carpentry work. In contrast to the drawings of housing-type projects, design drawings for larger projects include many details that contain carpentry items. The task of the estimator on these projects is first to identify which parts of the building the details apply to, and second to establish the extent of the carpentry work involved in each of these details. In order to pursue these tasks effectively, the estimator needs a comprehensive knowledge of the use of such things as wood blocking, nailers, roof cants and systems of wood strapping.

Because of the large number of detail drawings and sketches found on the more complex projects such as schools and hospitals, the estimator needs to develop a systematic approach to the measurement of this work, otherwise it will become impossible to keep track of progress. One way to organize the rough carpentry takeoff process is to proceed through the large-scale detail drawings one by one, identifying the carpentry items contained in each detail. Then, by examining the small-scale drawings, determine where the detail applies and, therefore, the length or area of the carpentry items involved. This task is another that benefits from the use of highlighter pens to keep track of the details and carpentry items considered and to outline the extent of the different components as they are measured.

## Board Measure

The unit of measurement of lumber is generally the board measure (BM), which is sometimes referred to as board foot (BF); a 1000-board feet unit is written MBF. Board measure is a cubic measure in which one unit of BM is equivalent to a $1 \times 12$ board, 1 foot long. To calculate the BM of lumber, multiply the length in feet by the nominal width and thickness of the pieces in inches, then divide the product by 12. Quantities are rounded off to the nearest whole board foot. Pieces less than 1 inch thick are usually calculated using a thickness of one full inch. See Figure 7.1 for examples of BM calculations.

## Metric Units

When using the metric system, the cross-section of a piece of lumber will be stated in millimeters using the actual dimensions in preference to the nominal dimensions. The size of a 2-inch by 4-inch piece of lumber will remain the same, but it will be referred to as a $38 \times 89$ since the actual dimensions of the cross-section is 38 mm by 89 mm.

The thickness of sheathing does not change but is described in mm rather than inches. Thus a standard ½-inch sheet will become a 12.7-mm sheet in metric units. However, the size of a standard sheet of plywood or composite wood product will have to be adjusted on a metric project because the size of a metric module is 400

| Item | Calculation |
|------|-------------|
| 1. 14 pieces of 4 × 12 <br> 16 feet long | $\dfrac{14 \times 16 \times 4 \times 12 \times 1}{12} = 896 \text{ BM}$ |
| 2. 20 pieces of 3 × 10 <br> 22 feet long | $\dfrac{20 \times 22 \times 3 \times 10 \times 1}{12} = 1100 \text{ BM}$ |
| 3. 19 pieces of 2 × 14 <br> 20 feet long | $\dfrac{19 \times 20 \times 2 \times 4 \times 1}{12} = 887 \text{ BM}$ |
| 4. 120 pieces of 2 × 4 <br> 8 feet long | $\dfrac{120 \times 8 \times 2 \times 4 \times 1}{12} = 640 \text{ BM}$ |
| 5. 60 pieces of ¾ × 3 <br> 12 feet long | $\dfrac{60 \times 12 \times 1 \times 3 \times 1}{12} = 180 \text{ BM}$ |

**Figure 7.1**    Sample Board Measure Calculations

mm; for example, studs will be spaced at 400 mm on center rather than 16 inches. So the size of a 4-foot by 8-foot sheet will become 1200 mm by 2400 mm, which is 19.5 mm narrower and 38 mm shorter than the standard sheet.

In the takeoff process, items of lumber are measured in linear meters stating the cross-section size of the pieces in mm, and sheeting material is measured in square meters stating the thickness of the sheets in mm.

## Measuring Notes—Rough Carpentry

### Generally

1. Lumber is measured in board measure or linear meters as described previously.
2. Lumber generally shall be differentiated and measured separately on the basis of the following categories:
   a. Dimensions
   b. Dressing
   c. Grade
   d. Species
3. Lumber required to have a special treatment, that is, kiln dried, pressure treated and so on, shall be kept separate and described.
4. Wall boards shall be measured in square feet or square meters.
5. No deductions shall be made for openings less than 40 square feet (4 square meters).
6. Wall boards shall be classified and measured separately in the following categories:
   a. Type of material
   b. Thickness

### Framing Work

7. Lumber for framing shall be classified and measured separately in the following categories:

   a. Plates
   b. Studs
   c. Joists
   d. Bridging
   e. Lintels
   f. Solid beams
   g. Built-up beams
   h. Rafters
   i. Ridges
   j. Hip and valley rafters
   k. Lookouts and overhangs
   l. Gussets and scabs
   m. Purlins
   n. Other items of framing

### Trusses, Truss Joists and Truss Rafters

8. Prefabricated trusses, truss joists and truss rafters shall be enumerated and fully described.

### Manufactured Beams, Joists and Rafters

9. Manufactured beams, joists and rafters shall be measured in linear feet or linear meters and fully described.

### Sheathing

10. Sheathing shall be measured in square feet or square meters; wall, floor and roof sheathing shall be described and measured separately.
11. Diagonal work shall be kept separate.
12. Common boards, ship-lap, tongued and grooved, plywood and other types of sheathing shall be measured separately.
13. Work to sloping surfaces shall be described and measured separately.

### Copings, Cant Strips, Fascias and the Like

14. Copings, cant strips, fascias and so on shall be measured in linear feet or linear meters.

### Soffits

15. Soffits shall be measured in square feet or square meters; different materials shall be measured separately.

### Sidings

16. Sidings shall be measured in square feet or square meters, describing the type of material and stating whether vertical or horizontal.

### Vapor Barriers and Air Barriers

17. Vapor barriers and air barriers shall be measured in square feet or square meters, describing the type of material used.

### Underlay and Subfloors

18. Underlay and subfloors shall be measured in square feet or square meters, stating the type of material used.

### Blocking Furring and So On

19.  Blocking furring and suchlike shall be measured in board measure or linear meters and classified in the following categories:
     a. Blocking stating purpose and location
     b. Furring stating purpose and location
     c. Nailing trips
     d. Strapping
     e. Grounds
     f. Rough bucks
     g. Sleepers

### Rough Hardware

20.  An allowance for rough hardware shall be made based on the value of the carpentry material component.

## Measuring Finish Carpentry and Millwork

Materials for finish carpentry on a commercial project are usually supplied by a millwork subcontractor and installed by the general contractor. Also included with finish carpentry is architectural woodwork, which may be supplied by the same millwork contractor as the finish carpentry or, alternatively, could be supplied and installed by a separate custom woodwork subcontractor. While there are shortcut methods that make use of add-on factors to estimate an installation price as a percentage of the supply price for this work, when something more than an approximate price is required, a detailed takeoff of finish carpentry items is necessary.

Estimators should be familiar with the trade scope definitions of finish carpentry and architectural woodwork in the project location so that they can evaluate which items are being supplied by the millwork subtrade and, perhaps more importantly, which items are not being supplied. The millwork trade may consider some items to be outside of its scope of work and not include them in its supply quote; this leaves them for the general contractor to obtain if no other trade has picked them up. All such items that are outside of the scope of the subtrades have to be priced for both the costs of installation and supply by the general contractor.

## Measuring Notes—Finish Carpentry

### Generally

1.  Items generally shall be classified and measured separately according to materials, size and method of fixing.
2.  Grounds, rough bucks, backing and so on shall be measured under rough carpentry.
3.  Rough hardware shall be included as an allowance.
4.  All metal work such as counter legs, framing around counter tops, bins, stainless steel items and suchlike shall be measured under miscellaneous metals.
5.  Glazing shall be included in the glazing section unless forming an integral part of prefabricated cabinet work.

### Trim

6. Trim shall be measured in linear feet or linear meters, stating size.
7. Built-up items such as valance boxes, false beams and so on shall be kept separate and fully described.
8. Baseboard and carpet strip shall be measured through door openings.

### Shelving

9. Shelving shall be measured in linear feet or linear meters, stating width.

### Stairs

10. Prefabricated stairs shall be enumerated and fully described stating the width of stair and the number of risers.
11. Balusters and handrails shall be measured in linear feet or linear meters and fully described.

### Cabinets, Counters and Cupboards

12. Cabinets, counters and cupboards shall be enumerated and fully described. Alternatively, units may be measured in linear feet or linear meters, describing the type and size of the unit and stating whether it is floor mounted, wall mounted or ceiling hung.

### Paneling

13. Paneling shall be measured in square feet or square meters and fully described.

## Doors and Frames

Doors and door frames are usually obtained from subcontractors who quote prices to supply and deliver the specified goods to the site, so the general contractor needs only to estimate the cost of handling and installing the doors and frames. Estimating these installation costs can again be accomplished by using a percentage of the price of the materials, but, once again, a more accurate price will be obtained if the work is taken off in some detail. For this purpose doors and door frames shall be enumerated stating their size and type of material.

Different sections of the specifications deal with metal doors and frames, wood doors and frames, door assemblies and special doors. All of these specification sections should be examined by the estimator to determine the general contractor's work involved in each section. For example, some doors may arrive at the site "pre-hung" complete with finish hardware, while other doors require hanging at the site and being fitted with necessary hardware.

On most projects a door schedule will be included with the drawings or bound in the specifications providing information about the type and size of doors and frames, and often also details of the hardware specified for each door. The estimator will find these schedules invaluable for preparing door takeoffs as all that remains to be done is to count the number of doors required of each type listed on the schedule. Sometimes even this information is provided on the schedule.

## Windows

On most projects the general contractor will obtain prices from subtrades to supply and install complete windows, storefronts and so on for the entire project. In these cases there will be no work for the general contractor to measure in connection with windows other than blocking and trim previously mentioned.

However, on small jobs and especially renovation works, there may be a need for only a few windows and, perhaps, just one or two doors; in this case it may not be worthwhile hiring separate specialist subcontractors for this work. In this situation window and door components may be obtained directly from suppliers and be installed on site by the general contractor. To measure this work, individual windows shall be enumerated and fully described together with the rough opening sizes for the units.

## Miscellaneous Metals

Prices for the miscellaneous metals trade are received from subcontractors who specialize in this work. Some prices quoted by these subtrades will be for the supply and installation of certain parts of the work, while other prices will be for "supply only." This is another situation in which the general contractor has to estimate the cost of handling and installing items that are supplied but not installed by a trade contractor. Here again the general contractor will be able to prepare a more realistic estimate for his or her work from a takeoff of the items involved rather than by merely allowing a percentage against the supply price of these miscellaneous metals.

There is a second advantage to preparing a takeoff of the miscellaneous metals work. Mostly because miscellaneous metal items are scattered throughout a project and it is often uncertain whether an item is included in the miscellaneous metal section or another section, a subcontractor who quotes the supply of these goods usually does not give a blanket price for miscellaneous metals but, instead, offers a price for a specific list of items. The general contractor then has to determine that the subtrade's list includes all the items required on the job, so preparing his or her own takeoff enables the contractor to better check the item list obtained from the subtrade. This task is made easier by obtaining from the subtrade, as early as possible in the bid period, the list of items the subtrade proposes to include in the bid. The estimator can then discuss with this trade, and with other trades involved, what needs to be added to or deleted from the list of miscellaneous metal items.

The miscellaneous metals trade has a wide scope that in very general terms can be defined as metal items that are neither part of the structural steel work nor part of any other trade section. Some of the more common items contained in this section and their measuring methods are:

1.  Access doors and frames shall be enumerated and fully described.
2.  Miscellaneous angles and channels shall be measured in linear feet or linear meters, stating the size and location.
3.  Bolts required for anchoring miscellaneous steel items shall be enumerated, stating size and location.
4.  Bollards shall be enumerated and fully described.
5.  Foot scrapers and mud and foot grilles shall be enumerated and fully described.
6.  Grates, grilles, grillwork and louvers that are not part of the mechanical system shall be measured in square feet or square meters and fully described; frames for this work shall be measured in linear feet or linear meters, stating size and type of section.
7.  Metal handrails, railings and balusters shall be measured in linear feet or linear meters and fully described.
8.  Mat recess frames shall be enumerated and fully described.
9.  Steel ladders shall be measured in linear feet or linear meters and fully described.

10. Ladder rungs shall be enumerated and fully described.
11. Steel stairs and landings shall be enumerated and fully described.
12. Vanity and valance brackets shall be enumerated and fully described.
13. Metal corner guards shall be enumerated and fully described.

## Specialties

The specialties trade is another wide-ranging trade section of the specifications that includes items of work such as:

| | |
|---|---|
| a. Bulletin boards | h. Chalk boards |
| b. Signs | i. Lockers |
| c. Canopies | j. Mail boxes |
| d. Folding partitions | k. Toilet partitions |
| e. Storage shelving | l. Bathroom accessories |
| f. Coat racks | m. Security vaults |
| g. Theater and stage equipment | n. Loading dock equipment |

Different subcontractors will offer price quotations for each of the distinct parts of the specialties section. If there are fifteen parts to the section as listed, the general contractor will probably have to engage fifteen different subcontractors, each responsible for their own particular constituent of this section.

As with the miscellaneous metals section, prices from specialties subtrades will sometimes cover the full cost of their work and at other times cover only the cost of supplying the materials involved, leaving the general contractor to allow for the cost of handling and installing the goods. This will result in the need for a takeoff of the general contractor's work for the reasons discussed in the preceding Miscellaneous Metals section. Because the scope of this section is so large, only measurement of bathroom accessories, which is probably the most common specialties item, is considered here as an example.

### Bathroom Accessories

The following items shall be enumerated and fully described:

| | |
|---|---|
| a. Shower curtain rods | f. Toilet roll holders |
| b. Soap dispenser units | g. Towel dispensers |
| c. Grab bars and towel bars | h. Waste receptacles |
| d. Mirrors | i. Medicine cabinets |
| e. Napkin dispensers | j. Coat hooks |

## Finish Hardware

A **cash allowance** for the supply cost of finish hardware is often specified on larger projects (see the discussion of cash allowances in Chapter 15); otherwise a finish hardware price is obtained from a subcontractor. Whether a cash allowance or a subcontractor's price is included for finish hardware, the sum will invariably represent only the cost of supplying the finish hardware products so, yet again, associated handling and installation costs will have to be estimated by the general contractor. Estimating these costs calls for a quantity takeoff of all finish hardware requirements on the project.

Finish hardware includes the following items that shall be enumerated and fully described:

a. Hinges (in sets of two hinges)      h. Latch sets
b. Flush bolts                         i. Kick plates
c. Bumper plates                       j. Panic hardware
e. Deadlocks                           k. Push plates
f. Doorstops                           l. Pull bars
g. Lock sets                           m. Door closers

## Measuring Exterior and Interior Finishes

The general contractor's estimator does not usually measure finishes work for an estimate as this work is almost invariably subcontracted these days. But there are occasions when some finishing work on a small project may be performed by the prime contractor's work force, and the estimator may sometimes need to check the quantity of subtrade work. Because of this, we have provided an example of the measurement of finishes on the house example we are estimating in the text. See Figure 7.3 on page 167 and the comments on the exterior and interior finishes takeoff.

## Carpentry and Miscellaneous Work Takeoff—House Example

### Comments on the Takeoff Notes Shown as Figure 7.2a

*Rough Carpentry—Floor System*

1. The rough carpentry takeoff begins at the bottom with the supports for the floor beam. The beam is supported by the concrete walls at the ends and by adjustable steel posts in the middle. These posts are commonly known as "teleposts."
2. The floor beam consists of four pieces of 2 × 10 and the full width of the building is taken for the length of this beam even though the length would be a little less than this.
3. Floor joists are required at 16" on center, so the number of joists is obtained by dividing the width of the floor by 1'4", rounding up and adding the extra end joist.
4. The distance from the outer face of the north wall to the center of the beam is 14'0" so 14-foot joists will be used here. To the south of the beam, 14-foot joists would be used on the right and 12-foot joists on the left.
5. Double joists are required on each side of the stair opening and to the side of the jog in the south foundation wall, so three 14-foot joists are added. The joist system is completed with header joists, which are attached to the ends of the main joists.
6. Because the joists do not overlap above the beam but meet end-on, cleats have been measured. These cleats are attached to the joists over the beam to add support to the butt joint between joists.
7. Cross-bridging is required at mid span of the joists; as the cross-bridging is installed between joists, the number of sets of bridging is equal to the number of joists less one.

*Rough Carpentry—Wall System*

8. The perimeter of the outside face of the building is used to calculate the length of the exterior stud wall. This wall comprises three plates and studs at 16" on center.

**QUANTITY SHEET**                                           SHEET No.  | 1 of 13 |

JOB.... HOUSE EXAMPLE ........................................    DATE.........................

ESTIMATOR........ A B F ...................EXTENDED........................EXT. CHKD.......................

| DESCRIPTION | TIMES | DIMENSIONS | | | | |
|---|---|---|---|---|---|---|
| ROUGH CARPENTRY — FLOOR SYSTEM | | | | | | |
| | | | | | | |
| 3" DIA. STEEL TELEPOSTS | 3 | | | | 3 NR | |
| | | | | | | |
| | | | | | | |
| 2 × 10 IN BUILT-UP | 4 | 40.00 | | | 160 LF | |
| BEAM | | | | | 267 BM | |
| 40'-0" | | | | | | |
| 1'-4" | | | | | | |
| = 30 +1 | | | | | | |
| | | | | | | |
| 2 × 10  JOISTS | 31 | 14.00 | | | 434 | |
| | 16 | 14.00 | | | 224 | |
| 19'-10" | 15 | 12.00 | | | 180 | |
| 1'-4" | 3 | 14.00 | | | 42 | |
| = 15 +1 | 2 | 40.00 | | | 80 | |
| 31 - 16 = 15 | | | | | 960 | |
| | | | | | 1600 BM | |
| | | | | | | |
| | | | | | | |
| 2 × 4  CLEATS | 31 | 2.00 | | | 62 | |
| | | | | | 41 BM | |
| | | | | | | |
| | | | | | | |
| | | | | | | |
| 2 × 2  X-BRIDGING | 2 | 30 | | | 60 SETS | |
| | | | | | | |
| | | | | | | |
| | | | | | | |
| 3/4" T + G  PLY | | 40.00 | 28.00 | | 1120 | |
| FLOOR  SHEATHING.  ADT | | 20.17 | 2.00 | | ( 40 ) | |
| | | | | | 1080 SF | |

**Figure 7.2a**    House Example—Rough Carpentry Work

**QUANTITY SHEET**                                               SHEET No.  | 2 of 13 |

JOB.......... *House* ............................................................DATE................................................

ESTIMATOR......... *A B F* ......................EXTENDED..........................EXT. CHKD.........................

| DESCRIPTION | TIMES | DIMENSIONS | | | | | |
|---|---|---|---|---|---|---|---|
| _WALL SYSTEM_ | | | | | | | |
| MAIN FLOOR EXTR. | | | | | | | |
| | | | | | | | |
| 2 x 40'-0"  80'-0" | | | | | | | |
| 2 x 28'-0"  56'-0" | | | | | | | |
| 136'-0" | | | | | | | |
| | | | | | | | |
| 2 x 6 PLATES | 3 | 136.00 | | | 408 | | |
| | | | | | 408 BM | | |
| | | | | | | | |
| 136'-0" | | | | | | | |
| 1'-4" | | | | | | | |
| = 102 | | | | | | | |
| | | | | | | | |
| 2 x 6 STUDS | 102 | 8.00 | | | 816 | | |
| (CORNERS) | 6 x 2 | 8.00 | | | 96 | | |
| (OPENINGS) | 9 x 2 | 8.00 | | | 144 | | |
| (PARTITIONS) | 10 | 8.00 | | | 80 | | |
| | | | | | 1136 BM | | |
| | | | | | | | |
| | | | | | | | |
| ½" WALL SHEATHING | | 136.00 | — | 8.00 | 1088 SF | | |
| | | | | | | | |
| | | | | | | | |
| | | | | | | | |
| 2 x 10 LINTELS  WND. ① | 2 | 7.83 | | | 16 | | |
| ② ⑥ ⑦ | 2 x 3 | 4.33 | | | 26 | | |
| ③ ⑤ | 2 x 2 | 6.33 | | | 25 | | |
| ④ | 2 | 2.33 | | | 5 | | |
| ⑧ ⑨ ⑩ ⑪ | 2 x 4 | 3.33 | | | 27 | | |
| DOOR ① | 2 | 4.33 | | | 9 | | |
| ② | 2 | 3.50 | | | 7 | | |
| | | | | | 115 | | |
| | | | | | 192 BM | | |

**Figure 7.2a  continued**

**QUANTITY SHEET**

SHEET No. 3 of 13

JOB.......... *House* .........................................................DATE...................................

ESTIMATOR.............. *ABF* ...............EXTENDED.........................EXT. CHKD..........................

| DESCRIPTION | TIMES | DIMENSIONS | | | | | |
|---|---|---|---|---|---|---|---|
| Wall System (Cont'd) | | | | | | | |
| Interior — Basement | | | | | | | |
| | | | | | | | |
| Extr. Perim: 136'-0" | | | | | | | |
| LESS | | | | | | | |
| 4 x 2 x 8" 〈 5'-4" 〉 | | | | | | | |
| 130'-8" | | | | | | | |
| | | | | | | | |
| 2 x 4 Plates | 3 | 130.67 | | | 392 | | |
| | 3 x 2 | 15.67 | | | 94 | | |
| | 3 | 3.08 | | | 9 | | |
| 130'-8" | | | | | 495 | | |
| 1'-4" | | | | | 330 BM | | |
| = 98 | | | | | | | |
| | | | | | | | |
| 2 x 4 Studs | 98 | 8.00 | | | 784 | | |
| (Corners) | 6 x 2 | 8.00 | | | 96 | | |
| (Windows) | 4 x 2 | 8.00 | | | 64 | | |
| (Partitions) | 2 | 8.00 | | | 16 | | |
| | | | | | 960 | | |
| | | | | | 640 BM | | |
| | | | | | | | |
| | | | | | | | |
| 2 x 10 Lintels | 4 | 3.33 | | | 13 | | |
| | | | | | 22 BM | | |
| 3'-0¾" | | | | | | | |
| 2 x 1½"   + 3" | | | | | | | |
| 3'-3¾" | | | | | | | |
| | | | | | | | |
| 2 x 4 Lintels | 2 | 2.92 | | | 6 | | |
| | | | | | 4 BM | | |
| 2'-8" | | | | | | | |
| 2 x 1½"   + 3" | | | | | | | |
| 2'-11" | | | | | | | |

**Figure 7.2a  continued**

**QUANTITY SHEET**

SHEET No. | 4 of 13 |

JOB............ *House* ............................DATE.........................

ESTIMATOR.... *A.B.F* .........EXTENDED....................EXT. CHKD...................

| DESCRIPTION | TIMES | DIMENSIONS | | | | | |
|---|---|---|---|---|---|---|---|
| WALL SYSTEM (CONT'D) | | | | | | | |
| INTERIOR — MAIN FLOOR | | | | | | | |
| | | | | | | | |
| 2 × 6 PLATES | 3 | 5.33 | | | 16 BM | | |
| 5'–4" | | | | | | | |
| 1'–4" | | | | | | | |
| = 4 + 1 | | | | | | | |
| 2 × 6 STUDS | 5 | 8.00 | | | 40 BM | | |
| | | | | | | | |
| | | | | | | | |
| 2 × 4 PLATES    (L–R) | 3 | 2.42 | | | 7 | | |
| | | 18.50 | | | 19 | | |
| | | 7.25 | | | 7 | | |
| | | 24.42 | | | 24 | | |
| | | 2.42 | | | 2 | | |
| | | 13.00 | | | 13 | | |
| | | 5.08 | | | 5 | | |
| | | 3.67 | | | 4 | | |
| (T–B) | 2 | 1.50 | | | 3 | | |
| | | 3.50 | | | 4 | | |
| | | 2.75 | | | 3 | | |
| | | 9.42 | | | 9 | | |
| 161'–0" | 2 | 12.75 | | | 26 | | |
| 1'–4" | | 7.83 | | | 8 | | |
| = 121 | | 1.83 | | | 2 | | |
| CORNERS 10×2  20 | | 11.42 | | | 11 | | |
| OPENINGS 13×2  26 | | 6.83 | | | 7 | | |
| 167 | | 4.75 | | | 5 | | |
| | | 2.42 | | | 2 | | |
| | | | | | 3 × 161 | 483 | |
| | | | | | | 322 | BM |
| 2 × 4 STUDS | 167 | 8.00 | | | 1336 | | |
| | | | | | 891 | BM | |
| | | | | | | | |
| 2 × 4 LINTELS (DOORS) ③ | | 2.25 | | | 2 | | |
| ④⑤⑥⑦ | 4 | 2.75 | | | 11 | | |
| ⑧ | | 1.75 | | | 2 | | |
| ⑨ | | 2.92 | | | 3 | | |
| | | CARRIED FWD: | | | 18 | | |

**Figure 7.2a  continued**

**QUANTITY SHEET**                                           SHEET No. | 5 of 13

JOB.......... _House_ ....................................................DATE.................................

ESTIMATOR....... _ABF_ ................EXTENDED..........................EXT. CHKD.......................

| DESCRIPTION | TIMES | DIMENSIONS | | | | | |
|---|---|---|---|---|---|---|---|
| 2 x 4 LINTELS (CONT'D) | | BROUGHT | FWD: | | 18 | | |
| BIFOLDS  ⑩ | | 2.25 | | | 2 | | |
| ⑪ ⑬ | 2 | 5.25 | | | 11 | | |
| ⑫ ⑭ | 2 | 4.25 | | | 9 | | |
| ⑮ ⑯ | 2 | 3.25 | | | 7 | | |
| | | | | | 47 | | |
| | | | | | 31 BM | | |
| | | | | | | | |
| ROOF SYSTEM | | | | | | | |
| 40'-0" | | | | | | | |
| 2'-0" | | | | | | | |
| = 20 + 1 – 2 | | | | | | | |
| = 19 | | | | | | | |
| "W" TRUSSES  28'-0" SPAN | | 19 | | | 19 No. | | |
| | | | | | | | |
| GABLE ENDS | | 2 | | | 2 No | | |
| | | | | | | | |
| 28'-0" | | | | | | | |
| 2 x 2'-6"  5'-0" | | | | | | | |
| 33'-0" | | | | | | | |
| ½" PLY WALL SHEATHING | 2 x ½ | 33.00 | – | 5.33 | 176 SF | | |
| | | | | | | | |
| 1 x 3 RIBBONS | 5 | 40.00 | | | 200 LF | | |
| | | | | | 50 BM | | |
| 40'-0" | | | | | | | |
| 2 x 2'-6"  5'-0" | | | | | | | |
| 45'-0" | | | | | | | |
| 2 x 4 RIDGE BLOCKING | | 45.00 | | | 30 BM | | |

Figure 7.2a  continued

**QUANTITY SHEET**

SHEET No. | 6 of 13

JOB...... *House* ...............................................................  DATE......................................

ESTIMATOR........ *ABF* ............EXTENDED.........................EXT. CHKD......................

| DESCRIPTION | TIMES | DIMENSIONS | | | | | |
|---|---|---|---|---|---|---|---|
| $\frac{28'-0''}{2} = \frac{14'-0''}{+ 2'-6''}$ | | | | | | | |
| $\overline{16'-6''}$ | | | | | | | |
| $\frac{16'-6''}{3} = 5'-6''$ | | | | | | | |
| $5'-6'' \times \sqrt{10} = 17'-5''$ | | | | | | | |
| 2×4 BARGE RAFTERS | 4 | 17.42 | | | 70 LF | | |
| | | | | | 46 BM | | |
| $\frac{17'-5''}{2'-0''} = 9-1$ | | | | | | | |
| $= 8$ | | | | | | | |
| 2×4 LOOKOUTS | 4×8 | 4.00 | | | 128 | | |
| | | | | | 85 BM | | |
| 2×6 ROUGH FASCIA | 2 | 44.00 | | | 88 BM | | |
| 2×4 CEILING BLKS. | | 136.00 | | | 91 BM | | |
| ½" PLY ROOF SHEATHING | 2 | 44.00 | 17.42 | | 1533 SF | | |
| EAVES | | | | | | | |
| VENTED ALUM. SOFFIT | 2 | 40.00 | 2.00 | | 160 | | |
| | | 20.17 | 2.00 | | 40 | | |
| (SLOPED) | 4 | 17.42 | 2.00 | | 139 | | |
| | 4×½ | 2.00 | 0.67 | | 3 | | |
| | | | | | 342 SF | | |
| ALUM. "J" MOULD | 2 | 40.00 | | | 80 | | |
| | | 2.00 | | | 2 | | |
| | 4 | 17.42 | | | 70 | | |
| | | | | | 152 LF | | |
| ALUM. FASCIA 6" WIDE | 2 | 44.00 | | | 88 | | |
| | 4 | 17.42 | | | 70 | | |
| | | | | | 158 LF | | |

**Figure 7.2a  continued**

**QUANTITY SHEET**

SHEET No. 7 of 13

JOB...... *House* ......................................................DATE......................

ESTIMATOR...... *A.B.F* ............EXTENDED..................... EXT. CHKD........................

| DESCRIPTION | TIMES | DIMENSIONS | | | | | |
|---|---|---|---|---|---|---|---|
| *FINISH CARPENTRY* | | | | | | | |
| *STAIRS* | | | | | | | |
| | | | | | | | |
| *3'-0" WIDE STAIR* | 1 | | | | | 1 Nº | |
| *W. 13- RISERS* | | | | | | | |
| | | | | | | | |
| | | | | | | | |
| *2 x 2 HANDRAIL* | | 12.08 | | | | 12 LF | |
| | | | | | | | |
| *13 x 7 7/8" = 102.375 RISE* | | | | | | | |
| *12 x 8 1/2" = 102.000 RUN* | | | | | | | |
| *√102.375² + 102² = 144.5 SLOPE* | | | | | | | |
| *= 12'-1"* | | | | | | | |
| | | | | | | | |
| | | | | | | | |
| *RAILINGS COMPS 2 x 6* | | | | | | | |
| *H/RAIL & 2 x 2 BALUSTERS* | | 11.67 | | | | 12 LF | |
| *8'-4"* | | | | | | | |
| *3'-4"* | | | | | | | |
| *11-8"* | | | | | | | |
| | | | | | | | |
| | | | | | | | |
| *DOORS* | | | | | | | |
| *2'-8" x 6'-8" x 1 3/4" SOLID* | 1 | | | | | 1 Nº | |
| *CORE EXTR. DR & FRM* | | | | | | | |
| *C/W ALUM STORM DOOR* | | | | | | | |
| *①* | | | | | | | |
| | | | | | | | |
| | | | | | | | |
| | | | | | | | |
| *3'-0" x 6'-8" x 1 3/4" DITTO* | 1 | | | | | 1 Nº | |
| *③* | | | | | | | |

**Figure 7.2a continued**

**QUANTITY SHEET**

SHEET No.   8 of 13

JOB.......... *Horse*

DATE..................

ESTIMATOR.......... *ABF*

EXTENDED..................    EXT. CHKD..................

| DESCRIPTION | TIMES | DIMENSIONS | | | | | |
|---|---|---|---|---|---|---|---|
| 2'-0" x 6'-8" x 1⅜" HOLLOW CORE INTR. DR + FRM. ③ | 1 | | | | | 1 N⍛ | |
| 2'-6" x 6'-8" x 1⅜" DITTO ④ ⑤ ⑦ ⑧ | 4 | | | | | 4 N⍛ | |
| 1'-6" x 6'-8" x 1⅜" DITTO ⑥ | 1 | | | | | 1 N⍛ | |
| 2'-8" x 6'-8" x 1⅜" DITTO ⑨ | 1 | | | | | 1 N⍛ | |
| 2'-0" x 6'-8" BIFOLD ⑩ | 1 | | | | | 1 N⍛ | |
| 5'-0" x 6'-8" DITTO ⑪ ⑫ | 2 | | | | | 2 N⍛ | |
| 4'-0" x 6'-8" DITTO ⑫ ⑭ | 2 | | | | | 2 N⍛ | |
| 3'-0" x 6'-8" DITTO ⑮ ⑯ | 2 | | | | | 2 N⍛ | |
| ATTIC ACCESS HATCH 20" x 30" | 1 | | | | | 1 N⍛ | |

**Figure 7.2a  continued**

| DESCRIPTION | TIMES | DIMENSIONS | | | | | | |
|---|---|---|---|---|---|---|---|---|
| *DOOR HARDWARE* | | | | | | | | |
| 4" BUTT HINGES | 2 | 1½ | | | | 3 | PR. | |
| 3½" BUTT HINGES | | 7 | | | | 7 | PR. | |
| KEY-IN-KNOB LOCK/LATCH SETS | | 2 | | | | 2 | № | |
| DEAD BOLTS | | 2 | | | | 2 | № | |
| PASSAGE SETS | | 4 | | | | 4 | № | |
| PRIVACY SETS | | 3 | | | | 3 | № | |

**QUANTITY SHEET**

SHEET No. [ 9 of 13 ]

JOB .......... House .......... DATE ..................

ESTIMATOR .......... ABF .......... EXTENDED .......... EXT. CHKD ..........

**Figure 7.2a continued**

**QUANTITY SHEET**    SHEET No. | 10 of 13 |

JOB.......... _House_

ESTIMATOR.......... _ABF_ ..........EXTENDED..........EXT. CHKD..........

| DESCRIPTION | TIMES | DIMENSIONS | | | | | |
|---|---|---|---|---|---|---|---|
| WINDOWS | | | | | | | |
| 36"×12" WNW IN CONC. | | | | | | | |
| WALLS  ⑧ ⑨ ⑩ ⑪ | | 4 | | | | 4  N⁰ | |
| | | | | | | | |
| WNDW COMPS: 2 × 18" | | | | | | | |
| CASEMTS & 54" FIXED | | | | | | | |
| × 48" HIGH  ① | | 1 | | | | 1  N⁰ | |
| | | | | | | | |
| DITTO COMPS: 24" CASEMT | | | | | | | |
| & 24" FIXED × 42" HI | | | | | | | |
| ② | | 1 | | | | 1  N⁰ | |
| DITTO COMPS: 24" CASEMT | | | | | | | |
| & 2× 24" FIXED × 48" ③ | | 1 | | | | 1  N⁰ | |
| | | | | | | | |
| DITTO COMPS: 24" CASEMT | | | | | | | |
| × 42"  ④ | | 1 | | | | 1  N⁰ | |
| | | | | | | | |
| DITTO COMPS: 24" CASEMT | | | | | | | |
| & 2× 24" FIXED × 42" ⑤ | | 1 | | | | 1  N⁰ | |
| | | | | | | | |
| DITTO COMPS: 24" CASEMT | | | | | | | |
| & 24" FIXED × 42"  ⑥ ⑦ | | 2 | | | | 2  N⁰ | |
| | | | | | | | |
| ¾" × 1⅝" WNW TRIM  ① | 2 | 7.50 | | | | 15 | |
| | 2 | 4.00 | | | | 8 | |
| ② | 2 | 4.00 | | | | 8 | |
| | 2 | 3.50 | | | | 7 | |
| ③ | 2 | 6.00 | | | | 12 | |
| | 2 | 4.00 | | | | 8 | |
| ④ | 2 | 2.00 | | | | 4 | |
| | 2 | 3.50 | | | | 7 | |
| ⑤ | 2 | 6.00 | | | | 12 | |
| | 2 | 3.50 | | | | 7 | |
| ⑥ & ⑦ | 2×2 | 4.00 | | | | 16 | |
| | 2×2 | 3.50 | | | | 14 | |
| | | | | | | 118  LF | |

Figure 7.2a  continued

| DESCRIPTION | TIMES | DIMENSIONS | | | | | |
|---|---|---|---|---|---|---|---|
| QUANTITY SHEET | | | | | SHEET No. | 11 of 13 | |
| JOB _House_ | | | | DATE | | | |
| ESTIMATOR _ABF_ | | EXTENDED | | EXT. CHKD | | | |
| _CABINETS_ | | | | | | | |
| 2'-0" x 3'-0" FLOOR MOUNTED | | 5.50 | | | 6 | | |
| CABS. c/w COUNTER TOP | | 5.00 | | | 5 | | |
| | 2 | 4.00 | | | 8 | | |
| | | 3.75 | | | 4 | | |
| | | | | | 23 LF | | |
| 1'-0" x 2'-8" WALL | 2 | 2.00 | | | 4 | | |
| MOUNTED CABS | | 4.00 | | | 4 | | |
| | | 3.83 | | | 4 | | |
| | | 5.00 | | | 5 | | |
| | | | | | 17 LF | | |
| 1'-0" x 1'-10" DITTO | | | | | | | |
| (ABOVE RANGE) | | 2.50 | | | | | |
| (ABOVE FRIG.) | | 2.50 | | | | | |
| | | | | | 5 LF | | |
| 1'-0" x 2'-8" CEILING | | | | | | | |
| HUNG DITTO | | 4.50 | | | 5 LF | | |
| 2'-0" x 2'-6" BATHROOM | | 5.25 | | | | | |
| VANITY | | 4.00 | | | | | |
| | | | | | 9 LF | | |
| 12" WIDE CLOSET SHELVES | 5 | 3.50 | | | 18 LF | | |
| (PANTRY) | | | | | | | |

**Figure 7.2a  continued**

**QUANTITY SHEET**                                                SHEET No. | 12 of 13 |

JOB...... *House* .................................................................DATE.........................

ESTIMATOR...... *A B F* ......................EXTENDED.........................EXT. CHKD.........................

| DESCRIPTION | TIMES | DIMENSIONS | | | | | |
|---|---|---|---|---|---|---|---|
| 1'-4" WIDE SHELVES ⑥ | 4 | 2.33 | | | 9 | | |
| ⑩ | | 2.33 | | | 2 | | |
| ⑪ | | 7.00 | | | 7 | | |
| ⑫ | | 5.00 | | | 5 | | |
| ⑬ | | 6.33 | | | 6 | | |
| ⑭ | | 5.00 | | | 5 | | |
| ⑯ | | 4.75 | | | 5 | | |
| | | | | | 39 LF | | |
| | | | | | | | |
| | | | | | | | |
| ADJUSTABLE CLOSET RODS | | 5 | | | 5 No | | |
| | | | | | | | |
| | | | | | | | |
| BATHROOM ACCESSORIES | | | | | | | |
| | | | | | | | |
| T.R. HOLDER | | 2 | | | 2 No | | |
| | | | | | | | |
| 2'-0" x 3'-0" MIRROR | | 2 | | | 2 No | | |
| | | | | | | | |
| MED. CABINET | | 2 | | | 2 No | | |
| | | | | | | | |
| SHOWER CURTAIN RAIL | | 1 | | | 1 No | | |

**Figure 7.2a  continued**

| DESCRIPTION | TIMES | DIMENSIONS | | | | | |
|---|---|---|---|---|---|---|---|
| 3/4" x 2 1/2" BASEBOARD | 2 | 26.00 | | | 52 | | |
| | 2 | 31.75 | | | 64 | | |
| (LR. KIT. HALL) | 2 | 13.00 | | | 26 | | |
| | 2 | 1.50 | | | 3 | | |
| (PANTRY) | 2 | 3.17 | | | 6 | | |
| | 2 | 1.00 | | | 2 | | |
| VESTIBULE | 2 | 3.50 | | | 7 | | |
| + CLOSET) | 2 | 4.75 | | | 10 | | |
| | 2 | 2.00 | | | 4 | | |
| (CLOSET) | 2 | 2.17 | | | 4 | | |
| | 2 | 4.50 | | | 9 | | |
| (BATH) ① | 2 | 7.67 | | | 15 | | |
| | 2 | 7.50 | | | 15 | | |
| (MASTER B.R.) | 2 | 10.83 | | | 22 | | |
| | 2 | 12.75 | | | 26 | | |
| (ENSUITE) | 2 | 5.25 | | | 11 | | |
| | 2 | 5.08 | | | 10 | | |
| (CLOSET) | 2 | 6.67 | | | 13 | | |
| | 2 | 2.33 | | | 5 | | |
| (LINEN) | 2 | 2.33 | | | 5 | | |
| | 2 | 1.75 | | | 4 | | |
| (B.R. ①) | 2 | 10.50 | | | 21 | | |
| | 2 | 11.25 | | | 23 | | |
| (CLOSET) | 2 | 5.50 | | | 11 | | |
| | 2 | 2.33 | | | 5 | | |
| (B.R. ②) | 2 | 11.42 | | | 23 | | |
| | 2 | 8.58 | | | 17 | | |
| (CLOSET) | 2 | 6.83 | | | 14 | | |
| | 2 | 2.42 | | | 5 | | |
| | | | | | 432 LF. | | |

**QUANTITY SHEET**

JOB HOUSE

ESTIMATOR A8F

SHEET No. 13 of 13

Figure 7.2a  continued

9. Two extra studs are allowed at each corner in the wall and at each opening. Also, an extra stud is added at each location where a partition meets the exterior wall. Note that the full height of the wall is used for the stud length even though the studs will be cut to fit between the plates.

10. 2 × 10 lintels are specified for openings in the exterior walls. The length of these lintels is equal to the width of the rough opening plus 3".

11. Non-loadbearing interior partitions may have just two plates, but three plates have been allowed because many builders like partitions to match the exterior walls to allow them to use the same standard stud length for all walls.

12. Plates for partitions are measured through door openings.

13. There are so many interior walls that a systematic approach is needed to ensure they are all included in the takeoff. The technique used here is to consider first the walls that run left to right (L–R) starting at the top of the drawing, then the walls that run from top to bottom (T–B) starting at the left of the drawing. It is a good idea to highlight the lengths of the walls as they are measured to keep track of progress and ensure that all walls have been accounted for.

14. To avoid having to examine in detail the complex layout of interior partitions to determine the need for extra studs, a quick method of calculating the approximate number of studs required is to allow one stud every 12" along the length of the wall. This should provide sufficient extra studs to allow for the places where double studs are required.

    While this method is perhaps not as accurate as calculating studs at 16" on center and adding the necessary extra studs, it does generally provide a number that is quite close and in a lot less time, so many estimators consider this shortcut to be justified.

*Rough Carpentry—Roof System*

15. The number of trusses required is calculated in a manner similar to joists: the length of the building is divided by the truss spacing plus one for the end truss, but two trusses are deducted as gable ends will be substituted for these.

16. The 1 × 3 ribbons are not shown on the drawings; they are attached along the length of the roof to the truss members that exceed 6'0" long to preform the same function as bridging.

17. Ridge blocking is measured along the ridge of the trusses plus the width of the overhang outside of the gables.

18. The length of the barge rafter is equal to the slope length of the roof. This length can be determined using the slope ratio of the roof. With a 1:3 slope, the slope length (hypotenuse) is the square root of 10 times the roof height; see sketch on Figure 7.2b.

19. See Figures 7.2b and 7.2c for details of the ridge blocking, barge rafters, lookouts and ceiling blocking.

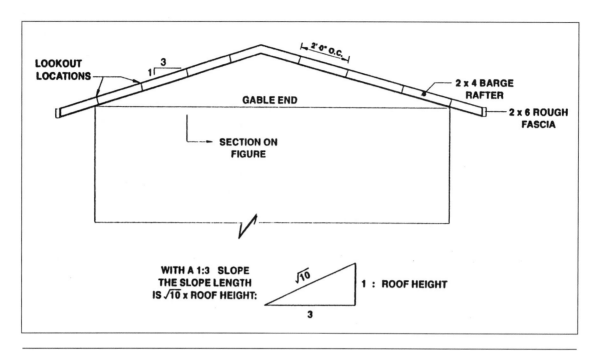

**Figure 7.2b**  Elevation of Roof Gable

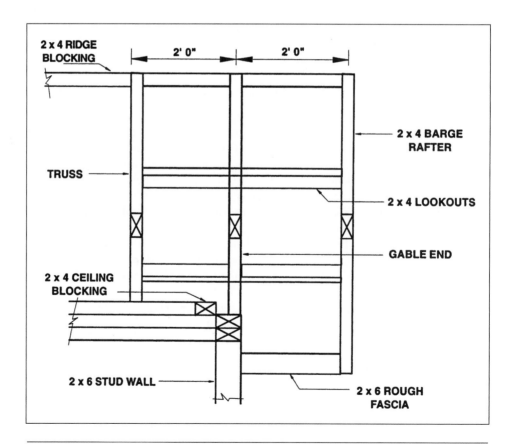

**Figure 7.2c**  Section Through Roof Gable

20. The rough fascia is attached to the tails of the trusses and the bottom ends of the barge rafters.

21. See Figure 7.2d for details of the eaves.

*Finish Carpentry*

22. To calculate the slope length of the balustrade, the total rise and run are determined and then, using the Pythagorean theorem, the length of hypotenuse is obtained.

23. If landings are required in connection with stairs, they would be measured here even though the work is actually *rough* carpentry.

    The components of a landing would be measured like a floor system and may include posts, beams, joists and sheathing.

24. Minimal details are included in the takeoff descriptions of doors, hardware and windows; this allows the takeoff to proceed without having to constantly check on the different item specifications. At this stage there is no need, for instance, to ascertain the precise type of interior doors required when they are all to be of the same type. Specific requirements can be verified when these items are priced.

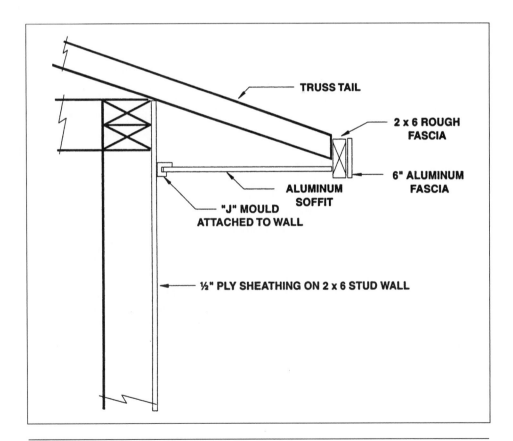

**Figure 7.2d**   Detail of Eaves

25. Doors will be priced as complete units so it is not necessary to takeoff door frames, trim and door stops as the full door price will include for all of these items.

26. Note that the quantity of hinges is given in pairs while the other items of hardware are enumerated.

27. The length of baseboard is determined by taking 2 times the overall length plus 2 times the overall width of all rooms regardless of shape. Note that closets are considered to be separate rooms and no deductions are made for door openings.

28. In the baseboard length calculation, the dimensions indicated on the drawings are used without adjustments for wall thicknesses and so on. It is not necessary to calculate the precise length of each piece of baseboard because the time it takes to do this is not justified by the cost of the item.

## Exterior and Interior Finishes Takeoff—House Example

### Comments on the Takeoff Shown as Figures 7.3 and 7.4

*Exterior Finishes*

1. The measurement of exterior finishes begins at the foundations and proceeds up to the roof finish.

2. On this house the finishes consist of asphalt damp proofing on the foundations below grade, parging on the exposed foundations and aluminum siding on the framed walls, all of which are measured in square feet. No deductions are made for openings less than 40 square feet of damp proofing and parging, nor for openings less than 1 square foot of siding.

3. The width of the exposed top of the foundation wall varies from 8½" to 1'9½", with an average width of 1'3". This leaves 6'9" as the depth of the foundation below grade that is to receive asphalt damp proofing.

4. Cement parging is to be applied to the exposed area of foundation; this is 1'3" plus 2" overlap with the damp proofing.

5. The siding is applied over the joist headers but only up to the eaves soffit, which gives a height of 8'7".

6. The areas of all openings (over 1 square foot) are deducted from the siding but not from the building paper as only openings over 40 square feet are deducted from building papers, vapor barriers and the like.

7. A 4-inch wide flashing is allowed for over all window and door openings although this is not clearly detailed on the drawings.

8. Eaves, gutters, and down spouts are not shown on the drawings but are allowed for here.

9. The quantity of roof shingles is left in square feet here because this is a small area; roofing is usually measured in squares that are areas of 100 square feet with no deduction for openings of less than 40 square feet.

**QUANTITY SHEET**

JOB........*HOUSE EXAMPLE*........................................................  DATE.............................

ESTIMATOR.........*ABF*.................EXTENDED..............................EXT. CHKD...................

SHEET No.  [ *1 of 3* ]

| DESCRIPTION | TIMES | DIMENSIONS | | | | | |
|---|---|---|---|---|---|---|---|
| EXTERIOR FINISHES | | | | | | | |
| 8½" | | | | | | | |
| 1'- 9½" | | | | | | | |
| 2'- 6" | | | | | | | |
| AVE:   1'- 3" | | | | | | | |
| DEPTH | | | | | | | |
| OF D/PROOFING | | | | | | | |
| 8'- 0" | | | | | | | |
| LESS 1'- 3" | | | | | | | |
| 6'- 9" | | | | | | | |
| ASPH. D/PROOFING | | 136.00 | — | 6.75 | 918 SF | | |
| 1'- 3" | | | | | | | |
| OVERLAP +  - 2" | | | | | | | |
| 1'- 5" | | | | | | | |
| ½" PARGING | | 136.00 | — | 1.42 | 193 SF | | |
| 8'- 1" | | | | | | | |
| +  1'- 0" | | | | | | | |
| -  0- 6" | | | | | | | |
| 8'- 7" | | | | | SIDING | BLDG. PAPER | |
| HORIZ. ALUM SIDING | | 136.00 | — | 8.58 | 1167 | 1167 | |
| FRONT { (WNW) DDT | 2 | 4.08 | — | 3.58 | ⟨ 29⟩ | — | |
|     " " | | 7.58 | — | 4.08 | ⟨ 31⟩ | — | |
|     (DOOR) " | | 3.75 | — | 6.67 | ⟨ 25⟩ | — | |
|     (WNW) " | | 4.08 | — | 3.58 | ⟨ 15⟩ | — | |
|     " " | | 6.08 | — | 4.08 | ⟨ 25⟩ | — | |
| BACK {  " " | | 2.08 | — | 3.58 | ⟨ 7⟩ | — | |
|     " " | | 6.08 | — | 3.58 | ⟨ 22⟩ | — | |
|     (DOOR) " | | 3.08 | — | 6.67 | ⟨ 20 ⟩ | — | |
| 28'- 0"<br>5'- 0"  ADD GABLES | 2×½ | 33.00 | — | 5.50 | 182 | 182 | |
| 33'- 0" | | | | | 1175 SF | 1349 | |
| 6 | | | | | | + 135 | 10% LAPS |
| = 5'- 6" GABLE HEIGHT. | | | | | | 1484 SF | |

**Figure 7.3**    House Example—Exterior Finishes

**QUANTITY SHEET**

JOB _HOUSE EXAMPLE_  DATE

ESTIMATOR _ABF_  EXTENDED  EXT. CHKD.

SHEET No. 2 of 3

| DESCRIPTION | TIMES | DIMENSIONS | | | | | |
|---|---|---|---|---|---|---|---|
| 4" WIDE GALV. FLASH G. | 2 | 4.58 | | | 9 | | |
| | | 8.08 | | | 8 | | |
| | | 4.58 | | | 5 | | |
| | | 6.58 | | | 7 | | |
| | | 2.58 | | | 3 | | |
| | | 6.58 | | | 7 | | |
| | | 3.58 | | | 4 | | |
| 40'-0" | | | | | 45 LF | | |
| 2 × 2'-6"   5'-0" | | | | | | | |
| 45'-0" | | | | | | | |
| | | | | | | | |
| EAVES GUTTER | 2 | 45.00 | | | 90 LF | | |
| | | | | | | | |
| 2'-6" | | | | | | | |
| 9'-0" | | | | | | | |
| EXTENSION  3'-0" | | | | | | | |
| 14'-6" | | | | | | | |
| | | | | | | | |
| 3" DOWNSPOUTS | 2 | 14.50 | | | 29 LF | | |
| | | | | | | | |
| | | | | | | | |
| 21016 ASPH. SHINGLES | 1.054 | 45.00 | 33.00 | | 1565 | | |
| (STARTER) | 2 | 45.00 | 1.00 | | 90 | | |
| | | | | | 1655 SF | | |
| 40'-0" × 28'-0" | | | | | | | |
| ADD  5'-0"   5'-0" | | | | | | | |
| 45'-0" × 33'-0" | | | | | | | |
| | | | | | | | |
| RIDGE CAP | | 45.00 | | | 45 LF | | |
| | | | | | | | |
| | | | | | | | |
| 4" WIDE DRIP-EDGE | | | | | | | |
| FLASHING | 2 | 45.00 | | | 90 LF | | |

**Figure 7.3  continued**

**QUANTITY SHEET**

JOB.......... *House Example* ................................................ DATE...............................

ESTIMATOR......... *ABF* .........EXTENDED........................... EXT. CHKD.......................

SHEET No. | 3 of 3

| DESCRIPTION | TIMES | DIMENSIONS | | | | | |
|---|---|---|---|---|---|---|---|
| 6 MIL POLY. EAVES PROTECTION | 2 | 45.00 | 4.00 | | | 360 SF | |
| | | | | | | | |
| 48" × 42" P.C. CONC. STEPS W. 4 - RISERS | 1 | | | | | 1 N° | |
| | | | | | | | |
| DITTO W. 2 - RISERS | 1 | | | | | 1 N° | |
| | | | | | | | |
| W.I. RAILING 2'-6" HI. | | 7.00 | | | | 7 LF | |
| | | | | | | | |
| PAINT W.I. RAILING | | DITTO | | | | 7 LF | |

**Figure 7.3  continued**

**QUANTITY SHEET**

SHEET No. | 1 of 3

JOB.... *HOUSE EXAMPLE* ........ DATE.........

ESTIMATOR.... *ABF* ....EXTENDED............EXT. CHKD............

| DESCRIPTION | TIMES | DIMENSIONS | | | | |
|---|---|---|---|---|---|---|
| *INTERIOR FINISHES* | | | | | | |
| | | | | | | |
| 32 oz. *CARPET* | | 40.00 | 28.00 | | 1120 | |
| DDT. | | 20.17 | 2.00 | | ⟨ 40 ⟩ | |
| (KIT & DNG. RM) .. | | 21.50 | 12.00 | | ⟨ 258 ⟩ | |
| (FRONT ENT.) .. | | 9.42 | 5.08 | | ⟨ 48 ⟩ | |
| (BATHS + CLOS.) ~ | | 7.67 | 12.75 | | ⟨ 98 ⟩ | |
| | | | | | 676 SF | |
| | | | | | | |
| *SHEET VINYL FLOORING* | | | | | | |
| (KIT. & DNG. RM.) | | 21.50 | 12.00 | | 258 | |
| (FRONT ENT.) | | 9.42 | 5.08 | | 48 | |
| (BATHS + CLOS.) | | 7.67 | 12.75 | | 98 | |
| | | | | | 404 SF | |
| | | | | | | |
| $\frac{1}{2}''$ D/WALL *CEILING* | | 40.00 | 28.00 | | 1120 | |
| DDT | | 20.17 | 2.00 | | ⟨ 40 ⟩ | |
| | | | | | 1080 SF | |
| | | | | | | |
| *TEXTURED FINISH* | | DITTO | | | 1080 SF | |
| | | | | | | |
| R 35 *LOOSE INSULN.* | | DITTO | | | 1080 SF | |
| | | | | | | |
| 6 MIL *POLY. V.B.* | | DITTO + 10% | | | 1188 SF | |
| | | | | | | |
| *INSULN. STOPS* | 2 | 23 | | | 46 N° | |
| 45'-0" | | | | | | |
| 2'-0" | | | | | | |
| = 23 | | | | | | |

**Figure 7.4**    House Example—Interior Finishes

**QUANTITY SHEET**                                                    SHEET No. | 2 of 3 |

JOB............ *House* ...............................................    DATE...............................

ESTIMATOR.......... *ABF* ...............EXTENDED...........................EXT. CHKD.....................

| DESCRIPTION | TIMES | DIMENSIONS | | | | | |
|---|---|---|---|---|---|---|---|
| — EXTERIOR WALLS | | | | | | | |
| R 20 BATT INSULN. | | 136.00 | — | 8.00 | 1088 | SF | |
| | | | | | | | |
| V.B. a.b. | | DITTO + 10% | | | 1197 | SF | |
| | | | | | | | |
| R 12 BATT INSULN. | | 133.33 | — | 7.67 | 1023 | SF | |
| (BASEMENT) | | | | | | | |
| 136'-0" | | | | | | | |
| LESS 4 × 8" 2'-8" | | | | | | | |
| 133'-4" | | | | | | | |
| V.B. a.b. | | DITTO + 10% | | | 1125 | SF | |
| | | | | | | | |
| | | | | | | | |
| — ALL WALLS | | | | | | | |
| ½" D/WALL WALLS (MAIN) | | 432.00 | — | 8.00 | 3456 | | |
| (BSMT.) | | 133.33 | — | 7.67 | 1023 | | |
| | 2 | 34.42 | — | 7.67 | 528 | | |
| 2 × 15'-8" 31'-4" | | | | | 5007 | SF | |
| 3'-1" | | | | | | | |
| 34'-5" | | | | | | | |
| PAINT WALLS | | FROM ABOVE | | | 3456 | SF | |
| (MAIN FLOOR ONLY) | | | | | | | |
| | | | | | | | |
| PAINT EXTR. DOORS | 2 | 2.83 | 6.67 | | 38 | | |
| | 2 | 3.17 | 6.67 | | 42 | | |
| 2'-8" | | | | | 80 | SF | |
| EDGE + -1¾" | | | | | | | |
| 2'-9¾" | | | | | | | |

**Figure 7.4  continued**

| DESCRIPTION | TIMES | DIMENSIONS | | | | | |
|---|---|---|---|---|---|---|---|
| STAIN INTR. DOORS ③ | 2 | 2.17 | — | 6.67 | 29 | | |
| ④ ⑤ ⑦ ⑧ | 2×4 | 2.67 | — | 6.67 | 142 | | |
| ⑥ | 2 | 1.67 | — | | 22 | | |
| ⑨ | 2 | 2.83 | — | | 38 | | |
| 2'-0" BIFOLDS ⑩ | 2 | 2.25 | — | | 30 | | |
| +2×1⅜" 2¾" ⑪ ⑬ | 2×2 | 5.50 | — | | 147 | | |
| 2'-2¾" ⑮ ⑯ | 2×2 | 3.50 | — | 6.67 | 93 | | |
| 5'-0" ⑫ ⑭ | 2×2 | 4.50 | — | 6.67 | 120 | | |
| +4×1⅜" 5½" | | | | | 621 | SF | |
| 5'-5½" | | | | | | | |
| | | | | | | | |
| | | | | | | | |
| PAINT ATTIC ACCESS | | | | | | | |
| HATCH | | | — | | 1 | N° | |
| | | | | | | | |
| | | | | | | | |
| | | | | | | | |
| PAINT HANDRAIL | | 12.08 | | | 12 | LF | |
| | | | | | | | |
| | | | | | | | |
| | | | | | | | |
| PAINT WOOD RAILINGS | | 11.67 | | | 12 | LF | |
| 3'-6" HIGH | | | | | | | |
| | | | | | | | |
| | | | | | | | |
| PAINT SHELVES | 2×5 | 3.50 | 1.00 | | 35 | | |
| | 2 | 39.00 | 1.33 | | 104 | | |
| | | | | | 139 | SF | |

**Figure 7.4  continued**

10. Precast concrete steps and exterior railings are taken off here although they are not strictly exterior finishes.
11. Windows, eaves, soffits, fascias and trim are all prefinished—otherwise painting would have to be measured here.
12. Painting the exterior doors is measured with the interior finishes.

*Interior Finishes*

13. Flooring, drywall wallboards, insulation, vapor barriers, paint and other finishes are measured in square feet with no deductions for openings less than 40 square feet.
14. The main floor of the house is carpeted except for the kitchen, dining area, front entrance, bathrooms and the closet off the kitchen that are to be finished with sheet vinyl flooring.
15. Insulation stops are installed under the roof between the tails of the trusses to maintain a space for ventilation above the insulation over the eaves.
16. The exterior wall of the basement is insulated and drywalled but left unpainted.
17. The length of the drywall, 432 feet on the main floor, is obtained from the previous takeoff of the baseboard.
18. Painting railings and balustrades is measured in linear feet stating the height of the railing to be painted.

## SUMMARY

- A thorough knowledge of carpentry construction is required to prepare a detailed estimate of this trade. This is particularly true for estimating wood frame structures where carpentry details are often lacking.
- The carpentry requirements of commercial and institutional projects are usually well detailed. With this type of job the estimator has to identify carpentry items and determine their extent.
- The estimator needs to apply a systematic approach to measuring carpentry on large projects to ensure that all items are accounted for.
- The unit of measure of lumber is generally the board measure (BM).
- To calculate the BM of a piece of lumber, multiply the length of the piece in feet by the nominal width and thickness of the piece in inches, and then divide by 12.
- Items of lumber are classified in terms of:
  a. Dimension
  b. Dressing
  c. Grade
  d. Species
- Framing lumber is further classified in terms of use. For example, plates, joists, lintels and suchlike.
- Trusses and rafters are described and enumerated.

- Sheathing and siding is measured in square feet or square meters.
- Materials for finish carpentry are usually supplied by a millwork subcontractor and installed by the general contractor.
- Items of finish carpentry are generally classified and measured separately according to materials, size and method of fixing.
- Subcontractors usually supply doors, frames and windows. General contractors generally have to estimate the cost of installing doors and door frames, but subcontractors generally quote a price for windows that includes supply and installation.
- Items of miscellaneous metal are usually obtained from subcontractors who quote a price to supply these items and may or may not also offer an installation price. When the general contractor needs to determine an installation price, it is most accurately obtained from a detailed takeoff of the items of miscellaneous metals.
- The specialties trade that includes items such as bulletin boards, canopies, coat racks, lockers and so on is usually estimated by obtaining subtrade prices for each of the items required on a project. Some of the subtrades will price only the supply of the required items, in which case the general contractor will have to estimate the price of installing the items.
- Bathroom accessories are enumerated and fully described in the takeoff process.
- Finish hardware is often covered by a cash allowance but where it is not, it is enumerated and fully described in the takeoff.
- Exterior and interior finishes are usually subcontracted, but the general contractor's estimator should be able to take off and price this work if required.

## REVIEW QUESTIONS

1. Why does the estimator of carpentry work, especially on housing projects, require a comprehensive knowledge of carpentry details and practices?
2. Describe a strategy an estimator could adopt to cope with the large number of carpentry details found on some commercial and institutional projects.
3. Define the term "board foot."
4. What is meant by the abbreviation MBF?
5. Describe how lumber generally is measured in a quantity takeoff.
6. How would trusses as part of the construction of a house be measured?
7. How is rough hardware accounted for in a carpentry estimate?
8. How are the following items measured in a finish carpentry takeoff:
   a. Stairs
   b. Shelves
   c. Cedar paneling
   d. Custom cabinets
   e. Counter tops
9. On larger projects, where can the estimator usually find details of the type and size of doors and windows required on the project?
10. Why should the general contractor's estimator prepare a detailed takeoff of miscellaneous metal items when a subtrade will be supplying these materials?
11. Make a list of specialty items that can be expected in a school project.
12. Contractors are often required to include a cash allowance for finish hardware in their bid price. What, precisely, is this cash allowance to be expended on?

# 8

# Pricing Generally

## OBJECTIVES

*After reading this chapter and completing the review questions, you should be able to do the following:*

- Describe the general process of pricing a construction estimate.
- List the five price categories considered in an estimate.
- Explain why project costs may exceed estimated prices for the work.
- Describe the two main components considered when pricing labor and equipment.
- Explain the influence of job factors and labor and management factors on productivity.
- Describe a strategy for dealing with factors that are difficult to assess on a project.
- Describe the use of cost reports as a source of information for pricing estimates.
- Explain how cost reports should be compiled for use with future estimates.
- Describe how materials are classified and priced in an estimate.
- Explain why there is less risk of material cost overruns than there is for labor and equipment.
- Convert material prices per package unit to prices per takeoff unit.
- Explain why pricing subcontractors' work can be far more difficult than it first appears.

## Introduction

The process of pricing an estimate can be divided into two stages. The first stage consists of preparing the takeoff for pricing by sorting and listing all takeoff items under the trade section breakdown, and the second stage consists of pricing this sorted list of items, which is referred to as the *recap*. The reason for preparing the recap is to bring together all work of a similar nature so that the estimator can concentrate on the needs of one trade at a time. Also, when the recap is complete, prices of each trade will be presented in summary form, which facilitates easier evaluation of the estimate. If items were not recapped into trades but, instead, the takeoff priced directly, it would be far more difficult to answer questions such as: what is the total amount in the estimate for excavation equipment? or, what is the total amount of concrete required to be placed on this project?

The nature of the recap breakdown depends on the type of work involved in the project and the type of contract the contractor will be entering into with the owner. For a general contractor working on nonhousing projects that are to be performed under lump-sum contracts, the recap will usually comprise the following trade breakdown:

1. Excavation and backfill work
2. Concrete work
3. Formwork
4. Concrete finishes
5. Carpentry work
6. Miscellaneous work

Other trade headings could be added to the breakdown, such as Reinforcing Steel and Masonry Work, when these trades are included in the contractor's usual scope of work or where the contractor wishes to estimate the price of these trades to compare with subtrade quotations.

When considering the takeoff process earlier, we mentioned that some estimators takeoff each of the recap trades separately, which makes the preparation of the recap much easier, but this approach has few devotees for the reasons we discussed in Chapter 3. So, when the work is not taken off trade-by-trade and especially on larger projects, processing the recap can be extremely time-consuming as the entire takeoff has to be scanned for each of the trades listed on the breakdown. The estimator will first scan for all the items of Excavating and Fill and list these in the recap, then he or she will scan for all the Concrete items and so on, and the process will be repeated until the takeoff is fully recapped. This procedure has to be performed with great care and a system of checking off items as they are carried to the recap is essential to confirm that all items that were in the takeoff have indeed been processed and are now included in the recap.

In computer estimating systems, the process of recapping takeoff items is handled automatically. In most systems a numerical code is applied to takeoff items that enables the computer to classify items and list them under the appropriate trade section for pricing. Depending on the system used, takeoff items can also be priced automatically as part of this same process. This demonstrates a significant advantage of using a computer in the estimating process because not only does the computerized system eliminate the time an estimator otherwise would have to spend recapping items, but it also ensures all takeoff items will be accounted for. There is very little chance of a takeoff item being overlooked and not priced once it has been entered into the computer.

Returning to the manual process, once the recap is prepared, the pricing procedure can commence. In pricing a construction estimate there are five price categories that need to be considered:

1. Labor
2. Equipment
3. Materials
4. Subcontractors
5. Job overheads

We will consider the particular factors that affect the level of prices in each of these categories but there is one factor that impacts on all of those categories: the contractor's risk.

## Contractor's Risk

In Chapter 15, we suggest that the estimator, when considering the amount of profit to add to a bid, should assess the amount of risk the contractor will be assuming if awarded a contract to construct the project. In essence, the risk we are contemplating here is the risk of the contractor losing money in the process of a building project, which happens where the actual cost of constructing the project exceeds the estimated cost. The precise nature and value of this risk on any given project is not a simple thing to evaluate. First we have to consider the nature of the contract that the contractor enters into with the owner, for it is from the terms of this contract that much of the contractor's risk emerges.

For example, the contract could state that the contractor will be reimbursed all project costs by the owner plus an added amount for profit; this is known as a *cost-plus contract* (see Chapter 1 for a discussion of various contract types). With a cost-plus contract, the contractor's risk will be reduced to a minimum. A negligent contractor who fails to accurately track all costs could still lose money under these terms, but the majority of contractors will find cost-plus contracts profitable since most of the risk is borne by the owner rather than the contractor.

If the contract is a stipulated *lump-sum* type, which is the type of contract used on the vast majority of construction projects, the contractor agrees to perform the work of the project for the bid price. In this situation, if the work costs more to put in place than the price offered, the contractor will lose money. This is because the terms of these contracts make it clear that the amount of money forthcoming from the owner to pay for the work is limited to the contractor's bid price. Only where the owner changes the contract will the contract price change. Otherwise, if the work costs more than the amount of the contractor's bid price, the extra costs will have to be paid by the contractor.

When a contractor is working under the terms of a lump-sum contract there are three general reasons why actual costs may exceed estimates:

1.  Takeoff quantities are too low.
2.  Actual productivity does not meet anticipated productivity.
3.  Subcontractors or material suppliers fail to meet obligations.

The possibility of inaccuracy in the measurement of takeoff quantities is a risk that is fundamental to the estimating process. The principal objective of following the systematic methodology of preparing a quantity takeoff outlined in preceding chapters is to minimize the possibility of error in the results. But even the most careful estimator can make a mistake in the takeoff—a fact that every experienced estimator will admit—and although computer systems can help reduce this risk, they cannot eliminate the risk. So an important task of senior estimators is still the review of their colleague's takeoffs to try to identify anything that may have been done in error.

Time and cost restraints usually make it impossible for a second estimator to check every single step that was taken in the preparation of a takeoff, but veteran estimators develop an ability to make broad assessments of takeoffs that can often enable them to detect deficiencies. This assessment, nevertheless, is not perfect; inevitably some errors will still go undetected. Consequently, there is always some takeoff risk that needs to be addressed when the project fee is considered and the bid price is finalized.

A discussion of the risk associated with productivities, subcontractors and suppliers follows.

## Pricing Labor and Equipment

There are two basic methods of pricing labor and equipment. In the first method the estimator uses productivity rates to convert the takeoff quantities into labor-hours and equipment-hours and then applies wage rates and equipment rates to the total hours to obtain the estimated labor and equipment prices for the project. The second method of pricing labor and equipment makes use of unit prices that are applied to listed takeoff quantities to calculate the total labor or equipment price for the work involved.

While there are some advantages attached to the first method of pricing, particularly when assessing productivities of the work involved, it can get complicated when there are a number of different trades involved, each with their own set of wage rates. The second method of pricing is adopted here because it is a more straightforward approach to pricing and it is the method most used by contractors in the commercial construction sector.

Whichever method of pricing is used, it is clear that there is a high risk factor in pricing labor and equipment because each involves two components that are subject to variations that can be difficult to predict. The two factors are:

1.  The hourly wage rate of labor or hourly cost of equipment
2.  The productivity of the labor or equipment

### Wage Rates

All of the labor unit prices used in the estimates that follow are derived from the base wages of the labor craft involved. The wage rates used in this text represent average rates, but it must be noted that rates vary from state to state and even from place to place within a state. Before the labor costs of an estimate can be priced, the estimator has to ascertain the craft wage rates that apply in the location of the project. Where union labor is used, the wage rate will be the rate that is found in the local union agreement for the craft involved. Nonunionized labor is paid the open-shop wage rate that is agreed to between the contractor and its individual workers. On projects funded by the state or federal government money, crafts are usually paid the prevailing wage for the vicinity of the work site. The prevailing wage is calculated from a survey of wages in a geographical area.

In addition to the base wage rates, employers also have to consider such items as social security tax, unemployment tax, workers' compensation insurance, liability insurance and other fringe benefits. The amount of these additional contributions, referred to as *payroll additive* in the following, is calculated on the estimate Summary Sheet as a percentage of the total labor content of the bid price. See Chapter 15 for a discussion of payroll additive.

While a single wage rate can be used for each craft over the course of a project of short duration, the estimator has to predict how wage rates will change over the term of longer projects for each of the different crafts used in the work. This task is perhaps easier when dealing with unionized labor when definite wage agreements are in place, but the assessment can never be done with total certainty. Even with apparently firm union agreements, events can render the terms of these agreements invalid, leaving the contractor with wage bills that differ greatly from those anticipated.

Faced with the prospect of a construction project of long duration, historical trends can be studied to help predict the level of future wages. The usual procedure is to examine the shape of the graph of actual wage rates plotted over a number of years and from this graph determine the average rate of wage increment. From this the analyst can proceed to extrapolate along the graph line to arrive at the projected wage rates for the term of the project. Yet, no matter how scientifically this process is accomplished, in times when economic stability cannot be guaranteed, estimating future wage rates from previous trends amounts to little more than guesswork.

### Equipment Rates

Accurate predictions of future hourly costs of construction equipment can be just as difficult to evaluate as future wage rates. Ownership costs of a contractor's own equipment can be analyzed with some certainty (see Chapter 9 for examples), but even these calculations are based on assumptions about such things as life expectancy, maintenance costs and salvage values, all of which can vary from expectations.

Rental rates of equipment can also be difficult to tie down because they fluctuate in response to a number of variables including the level of economic activity in the construction industry, interest rates, taxes, import duties and even construction labor wage rates. (Steeply rising wage rates cause managers to substitute equipment for labor wherever possible; this increases the demand for equipment, which can raise rental rates.)

The rental rate at the time the work is executed, therefore, can be quite different from the rate that is forecast at the time of the estimate. The risk of rental rate increases can be managed to some extent by procuring guaranteed price quotations from rental companies for major equipment on some projects, but these firm price agreements are not always obtainable.

In addition to the risk that rental rates at the time of the work may be higher than those quoted at the time of the bid, there are a number of other factors regarding equipment rental rates that the estimator should take into account. Whenever rental rates are solicited, the estimator should obtain the following rental information:

1. What is the age and condition of the equipment being offered?
2. Is the cost of an operator included in the rates quoted?
3. Do the rates include the cost of necessary maintenance of the equipment (on longer rentals)?
4. Is the cost of fuel included in the rental?
5. Is there a minimum rental fee or a minimum rental duration that translates into a minimum rental fee?
6. Will there be additional charges for delivery?
7. Are there any accessories needed to operate the equipment; if so, at what price are they available?
8. What taxes or license fees over and above the rental fee are applicable?
9. What insurance coverage is included in the rental agreement?
10. Are there any special qualifications needed before the contractor can legally operate the equipment? In some jurisdictions, some proof of operator training is required before certain equipment can be legally used.

### Productivity of Labor and Equipment

The productivity of labor and equipment is influenced by a large number of factors that can differ greatly according to the time and place of a project. These factors can be classified into two main groups: job factors and labor and management factors.

Examples of job factors include:

1. Weather conditions expected at the site
2. Access to and around site
3. Site storage space
4. Nature of the project, its size and complexity
5. Distance from materials and equipment sources
6. Wage and price levels at the job location

Examples of labor and management factors include:

1.  Quality of job supervision
2.  Quality of job labor
3.  Motivation and morale of workers
4.  Type and quality of tools and equipment
5.  Experience and records of similar projects in the past

*Job Factors*

The contractor has very little influence, if any, over the job factors that prevail on an individual project, but the nature of these factors should be carefully investigated as they can have a major impact on the rates of productivity and thus the cost of work done. Each of the job factors listed can vary over a wide range of possible conditions. The estimator has to learn to determine the implications of the particular conditions expected on the project with regard to their impact on job factors. Following are some observations on the method of investigating job factors and examples of some situations that may be encountered on different projects.

Weather Conditions    If the estimator is unfamiliar with the prevailing weather conditions at the location of the job, weather reports may be obtainable from a weather-monitoring agency in the vicinity of the site. Particular attention should be paid to data about the extreme and average summer and winter temperatures, together with information regarding expected rainfall and snowfall amounts at the site location. Many contractors have discovered to their detriment that weather conditions that are normal at the site of the work make it impossible to attain the productivity levels they experienced on previous projects in other locations. A little investigative work at the time of the estimate can often help to avoid overoptimistic production forecasts.

Site Access    The quality of site access can affect productivity in many ways. Poor access to some sites may restrict the size of equipment that can be used for the work so that smaller, less efficient items of plant have to be used rather than items of a more appropriate size. Because large trucks are not able to get close to a site due to restricted access, they may have to unload some distance away and then move the material, some time later, to the actual work area. This double handling of materials can significantly raise the cost of work.

Storage Space    Cramped storage space at a site can also necessitate double handling of materials where materials have to be temporarily stored at one location and then moved on site some time later. "Just-in-time" delivery of materials in accordance with a precise schedule can be used to avoid this kind of double handling, but working to this kind of schedule requires extensive and careful management of operations that is not always available.

Nature of the Project and Other Job Factors    Clearly, certain types of projects will have a different cost structure from other types of projects. Even though a school building and a parking garage may both have concrete walls and slabs, the cost of these components will probably be different because they perform a different function on each job and, thus, their design will not be the same in each case.

Large jobs are generally less cost per unit than small jobs. Where there are large quantities of repetitive work to be performed on a project, the contractor may obtain increased productivity from work improvements attained because operations are repeated. Studies have shown that when a relatively complex work activity is repeated, there is an opportunity to improve production rates as those involved in the work learn to perform it more efficiently. But situations in which certain operation is performed just once or only a few times exclude the contractor from the benefits of a learning rate achieved in repetitious operations. Even on large projects, the work may be too diverse to enable the contractor's crews to achieve a significant learning rate.

*Labor and Management Factors*

Like job factors, some of the labor and management factors are also beyond the control of the contractor. Over the long term, a contractor may be able to develop effective supervisors and workers inside the company but, because of the cyclical nature of the construction business, contractors often find it difficult to maintain enough employment to hold on to these good workers. Consequently, the staffing of most jobs requires the contractor to hire many workers from outside the company. Therefore, the quality of supervision and the labor available for the project often depends most on the economic climate at the time of the work. In a strong economy in which there is plenty of construction work underway and contractors are busy, the better supervisors and workers will not be available, they will already be fully employed, so the contractor may have to make do with lower quality personnel for the next job. In a weak economy, good workers are far more easy to find.

Some labor and management factors can, however, be influenced by the contractor. Efforts can be made to cultivate high morale and motivation of workers at the job site; a contractor can try to ensure that projects are equipped with high-quality, well-maintained tools and plant to pursue the work. The concepts of "total quality management" have been embraced by a number of construction companies that endeavor, through a process of constant improvement, to develop its personnel into a highly productive work force. These contractors are convinced that these efforts will improve their company's competitive edge, which will ensure their future success in the industry.

## Use of Cost Reports

If an estimator had a range of accurate historic prices for the labor or equipment requirements of each work item on a project, then by accounting for all the factors and variables associated with that project, the estimator would be able to determine the price of each item with some confidence. But how does an estimator come by such a range of prices? What is required is a data base of information on the precise cost of work items over a variety of past jobs that were completed under carefully monitored conditions. From this painstakingly gathered historic data, a range of realistic prices for each work item of an estimate can be established. This then allows the estimator to assess the cost of any particular work item on the range and arrive at a reasonable prediction of the cost of that item.

In order to be effective, a historic cost database for use in estimating has to be assembled systematically; a haphazard collection of previous project costs gathered using different systems of measurement and costing will be more of a hindrance to good estimating than a help. A satisfactory cost database can only be developed from an accurate and consistent cost reporting system, which is pursued in four key processes:

1. The priced takeoff items on the estimate are coded to produce the project budget. Some of the estimate takeoff items may be combined to provide a simpler breakdown that is easier to use than a detailed estimate in the cost reporting system.

2.  As the work is performed on the project, labor-hours and equipment-hours expended are coded to the applicable work items of the budget. This coding function is best performed by the site supervisors who are responsible for the work in progress as they are in the best position to determine exactly which work item their work crews are working on.

3.  The quantity of work done in each of the work items underway is measured in accordance with the same rules of measurement followed by the estimators. Measurements that are most consistent with the estimate quantities will be obtained if work completed is measured by estimators, but this is generally not possible, so site personnel usually have to perform the task.

4.  Company accountants apply wage rates and equipment prices to the hours coded. This determines the labor cost and equipment cost for each work item. Then these costs are combined with the work quantities to generate the actual unit costs of the project work items.

See Figure 8.1 for an example of a cost report prepared in this way. The cost report shows the budget amounts and budget unit prices in the form of an estimate summary. Next, the amounts that have been spent on each of the items in the current month are shown together with the unit costs calculated from the quantity of work done on each item. The amount spent to date and the unit costs to date are then shown and, finally, based on the rates of productivity realized to date, projections of the final costs are indicated.

## Project Information with Cost Reports

A carefully and systematically prepared cost report is necessary for obtaining accurate data for estimating purposes but does not provide sufficient information to accurately price the work of a new project. Before the data from cost reports can be used effectively two basic questions must be answered:

1.  Why did the estimator choose the original unit price for an item?
2.  Why did the actual unit price for the item vary from the estimated unit price?

While it is not the objective of the estimating process to predetermine the exact cost of every item of the work, it is useful for several reasons to learn why the actual cost of the work did vary from the estimated cost. If the quality of estimating is to improve, some feedback of information about actual results is necessary. The unit prices used in the estimate could have been inappropriate because of an estimating problem, because of a production problem or simply due to unfortunate circumstances. We need to know which.

As an example, consider the amounts shown on Figure 8.1 to be a record of the costs of Job #1. As you see, the final costs for stripping topsoil on this project were $1.24 and $2.46 per cubic yard for labor and for materials, respectively. This compares with the amounts $0.80 and $2.00 that were originally estimated for this work. An estimator who is faced with the task of estimating the cost of stripping topsoil on a similar project (Job #2) has to decide whether to use $0.80 and $2.00 again or raise the prices to reflect the higher costs encountered on Job #1. There could be many different reasons why the costs of Job #1 were different from the estimate, but essentially what the estimator needs to know is were the estimate prices wrong or did something happen in the execution of the work that caused the costs to be higher than expected?

If the estimator of Job #1 is the same person as the estimator now considering Job #2, and Job #1 was constructed quite recently, then this estimator will be aware of the factors determining the unit prices of Job #1 and will probably have a good

**ESTIMATE SUMMARY**

\*\*These items are now completed.

| ITEM | QUANT | | LABOR $ | EQUIP. $ | MATL. $ | TOTAL $ |
|------|-------|--|---------|----------|---------|---------|
| STRIP TOPSOIL\*\* | 500 | CY | 400.00 | 1,000.00 | 0.00 | 1,400.00 |
| unit price | | | 0.80 | 2.00 | 0.00 | 2.80 |
| EXCAV. TRENCH | 1,200 | CY | 1,500.00 | 2,136.00 | 0.00 | 3,636.00 |
| unit price | | | 1.25 | 1.78 | 0.00 | 3.03 |
| EXCAV. PITS | 950 | CY | 3,182.50 | 4,560.00 | 0.00 | 7,742.50 |
| unit price | | | 3.35 | 4.80 | 0.00 | 8.15 |
| TRANS. EQUIPMENT | 4 | Mo | 0.00 | 10,000.00 | 0.00 | 10,000.00 |
| unit price | | | 0.00 | 2,500.00 | 0.00 | 2,500.00 |
| BACKFILL TRENCH | 872 | CY | 2,180.00 | 6,060.40 | 12,208.00 | 20,448.40 |
| unit price | | | 2.50 | 6.95 | 14.00 | 23.45 |
| BACKFILL PITS | 795 | CY | 2,504.25 | 5,604.75 | 11,130.00 | 19,239.00 |
| unit price | | | 3.15 | 7.05 | 14.00 | 24.20 |

**CURRENT MONTH**

| ITEM | QUANT | | LABOR $ | EQUIP. $ | MATL. $ | TOTAL $ | UNDER/(OVER) $ |
|------|-------|--|---------|----------|---------|---------|----------------|
| STRIP TOPSOIL | 85 | CY | 89.25 | 204.85 | 0.00 | 294.10 | (56.10) |
| unit price | | | 1.05 | 2.41 | 0.00 | 3.46 | (0.66) |
| EXCAV. TRENCH | 890 | CY | 1,076.90 | 1,335.00 | 0.00 | 2,411.90 | 284.80 |
| unit price | | | 1.21 | 1.50 | 0.00 | 2.71 | 0.32 |
| EXCAV. PITS | 702 | CY | 2,730.78 | 3,524.04 | 0.00 | 6,254.82 | (533.52) |
| unit price | | | 3.89 | 5.02 | 0.00 | 8.91 | (0.76) |
| TRANS. EQUIPMENT | 1 | Mo | 0.00 | 2,500.00 | 0.00 | 2,500.00 | 0.00 |
| unit price | | | 0.00 | 2,500.00 | 0.00 | 2,500.00 | 0.00 |
| BACKFILL TRENCH | 644 | CY | 1,635.76 | 4,520.88 | 9,048.20 | 15,204.84 | (103.04) |
| unit price | | | 2.54 | 7.02 | 14.05 | 23.61 | (0.16) |
| BACKFILL PITS | 0 | CY | 0.00 | 0.00 | 0.00 | 0.00 | 0.00 |
| unit price | | | 0.00 | 0.00 | 0.00 | 0.00 | 0.00 |

**TOTAL TO DATE**

| ITEM | QUANT | | LABOR $ | EQUIP. $ | MATL. $ | TOTAL $ | UNDER/(OVER) $ | PROJECTED TOTAL | UNDER/(OVER) |
|------|-------|--|---------|----------|---------|---------|----------------|-----------------|--------------|
| STRIP TOPSOIL | 499 | CY | 618.76 | 1,227.54 | 0.00 | 1,846.30 | (446.30) | 1,846.30 | (446.30) |
| unit price | | | 1.24 | 2.46 | 0.00 | 3.70 | (0.90) | | |
| EXCAV. TRENCH | 1,134 | CY | 1,008.00 | 1,791.72 | 0.00 | 2,799.72 | 636.30 | 3,492.00 | 144.00 |
| unit price | | | 1.33 | 1.58 | 0.00 | 2.91 | 0.12 | | |
| EXCAV. PITS | 702 | CY | 2,730.78 | 3,524.04 | 0.00 | 6,254.82 | (533.52) | 8,464.50 | (722.00) |
| unit price | | | 3.89 | 5.02 | 0.00 | 8.91 | (0.76) | | |
| TRANS. EQUIPMENT | 2 | Mo | 0.00 | 5,000.00 | 0.00 | 5,000.00 | 0.00 | 10,000.00 | 0.00 |
| unit price | | | 0.00 | 2,500.00 | 0.00 | 2,500.00 | 0.00 | | |
| BACKFILL TRENCH | 861 | CY | 2,221.38 | 6,121.71 | 12,200.37 | 20,543.46 | (353.01) | 20,805.92 | (357.52) |
| unit price | | | 2.58 | 7.11 | 14.17 | 23.86 | (0.41) | | |
| BACKFILL PITS | 0 | CY | 0.00 | 0.00 | 0.00 | 0.00 | 0.00 | 0.00 | 0.00 |
| unit price | | | 0.00 | 0.00 | 0.00 | 0.00 | 0.00 | | |

**Figure 8.1**  Cost Report

idea of the reasons why the final unit costs varied from the unit prices. That being the case, good use can be made of the cost information contained in the cost report for Job #1. But if the estimator of Job #1 is no longer around, or if Job #1 took place some time ago, the information necessary to properly evaluate the Job #1 cost report may be lacking. Also, if an estimator works on more than 30 bids per year and numerous projects are underway all generating cost reports, it can be difficult to keep track of all the information involved with the result that, when the cost report of a particular project is studied, the estimator may be uncertain of the actual circumstances surrounding that project.

To ensure that the needed project information is available to future estimators, the final cost report on a project should be accompanied by two commentaries: one from the project estimator describing the basis of the unit prices, at least for the substantial items of the estimate. A second report should come from the project supervisor, or the estimator if he or she has been closely observing the progress of the job, describing the actual conditions that precipitated the unit costs realized. Armed with this information, the estimator is far more likely to obtain appropriate unit prices when estimating subsequent projects.

We can add that there are additional benefits that may be obtained from having estimators write a commentary on the unit prices they chose for a project. By performing this task, the estimator is disciplined to seriously think about the factors affecting unit prices rather than just picking a price for vague, if any, reasons. Additionally, the mere process of writing out their justifications often contributes to clearer thinking in estimators. Certainly, an estimator who has prepared detailed notes can provide a far more coherent explanation when questioned about unit prices at the management review of the estimate.

### Strategy for Pricing Labor and Equipment

One pricing strategy used by some contractors, especially when producing a competitive bid, consists of adjusting unit prices for job factors and management factors that are relatively certain and then being optimistic with the other factors. For example: the estimator may know that access to the site is good, the size of the job is small and supervision and labor at the site will be good quality. item prices are thus adjusted to reflect the effect of these factors. But for weather conditions, worker morale and other factors of which there is uncertainty, the estimator makes optimistic assumptions that keep the prices inexpensive. The total dollar value of these "optimistic" assessments is calculated and, when the fee for the job is determined, the appraised risk factor is accounted for.

There is certainly some danger involved in this strategy, but whether or not the strategy is adopted, estimators should always be encouraged to take time to try to identify the estimating risks because this information is needed to properly complete a bid. Quantifying and subsequently dealing with the project risk is studied later when we consider the process of completing the bid in Chapter 15.

### Pricing Materials

The materials used in a construction project fall into two broad categories: materials that form part of the finished structure such as concrete, and materials that are consumed in the construction operations such as formwork and fuel oil. Items in the first category are priced using unit prices against the recapped work quantities in the same way as labor and equipment are priced. Some of the consumable materials used on a project are included in the price of the permanent work. For instance, the cost of form materials is added to the concrete work price for the project. The cost of other consumables such as fuel and lube oil can be found in equipment prices (see Chapter 9), and consumables are priced in a number of general expense items (see Chapter 14).

There should be less risk involved in the pricing of materials than there is with labor and equipment pricing. Once the work item quantities are established in the take-off process, the estimator should be able to obtain firm prices from material suppliers for the supply and delivery of materials, which will minimize the contractor's risk. Simply using a catalog price for materials may be acceptable for pricing minor items, but where significant quantities are involved, the estimator should communicate quantities and specifications to suppliers and ask them to return firm prices for the supply of materials that meet the project requirements. By following this procedure, volume discounts may be obtained and firm prices for the project can be secured.

Although the unit prices used to price materials are based on quotations from the suppliers of these materials, seldom are the *unit* prices used in the estimate obtained directly from the suppliers. The packaging units of materials obtained from suppliers are different from the units measured in the estimate. Even where suppliers do quote prices for the same unit of measurement as the estimate, like concrete, the suppliers' prices need to be adjusted to account for wastage of materials. But, when pricing grout materials, for example, the estimator needs a price per cubic foot, which is the unit of measurement, but the supplier's quote might be on the basis of a price per cartridge of grout. In this case the estimator will have to determine the size of the grout cartridge in order to calculate the price per cubic foot. These types of adjustments will be examined in more detail in the sections concerning the pricing of specific trades found in the following chapters.

When pricing any materials, there are a number of questions that should be answered before the prices are used in an estimate:

1. *Do the materials offered by the supplier comply with the specifications?*
   If gravels are specified to meet certain grading requirements, for instance, the estimator should ensure that the prices quoted by the supplier are for gravels that satisfy these requirements.
2. *Do the prices quoted include delivery of the materials to the site?*
   Wherever possible suppliers should be required to quote the price of materials F.O.B. (free on board) the site, which, in the absence of a definition to the contrary, is interpreted to mean the price includes delivery to the site but the contractor is responsible for the cost of any unloading requirements at the site.
3. *Can the contractor rely on the supplier's prices to remain firm until the owner awards the contract?*
   If the contractor is required to give the owner 30 days to award the contract, consent should be obtained from the material suppliers to hold their prices firm for this time period. When the owner awards the contract, the contractor can ensure the prices continue to remain firm by issuing purchase orders to the suppliers.
4. *Does the supplier's price include state or city sales taxes?*
   Materials are usually quoted exclusive of taxes, then state and city taxes are added, where they apply, to the item prices for materials used in the estimate.
5. *Will there be any storage or warehousing requirements for the materials?*
   On-site storage is often required for construction materials before they are incorporated into the structure. The estimator should ensure that storage

space requirements can be met at the site; if they cannot, temporary storage of materials off-site may be necessary, which can have significant costs associated with it.

6. *What are the vendor's terms of offer?*

Almost all price quotations received from material suppliers will have terms and conditions attached to them. In the chapters that follow, we will consider many of the common terms and conditions of sale that attach to specific types of materials such as concrete and lumber, but there are some terms that are found on quotes for all kinds of materials. Payment terms fall into this category.

### Owner-Supplied Materials

Most project specifications provide detailed descriptions of the materials to be supplied by the contractor. The contractor is often told in no uncertain terms what materials are required for the project, what brands are acceptable and, sometimes, where to procure the goods required. However, when some of the project materials or equipment are to be supplied by the owner, specifications of exactly what is to be provided may be quite skimpy. The estimator has to be careful in these situations particularly where the owner-supplied materials or equipment are to be installed by the contractor or even if they are simply to be handled in some way by the contractor.

Details of size, weight and packaging are usually necessary to estimate handling costs, but additional, more detailed information can also be required. On some projects an owner may supply concrete that is to be placed and finished by the contractor. If the contractor's price for placing the concrete is based on the use of concrete pumps, the estimator is wise to ascertain details of the proposed concrete mix because some mixes are easier to pump than others and some mixes are virtually impossible to pump into place.

### Pricing Subcontractors' Work

On the face of it, the process of pricing subcontractors' work in the general contractor's office would appear to be quite simple. Subtrades prepare takeoffs and price the work of their own trade, then they submit lump-sum prices to the general contractors who are preparing bids to go to the owner. The general contractors examine the prices received from subtrades and customarily select the lowest price for each trade, which, in turn, is incorporated into the general contractor's bid price. By following this procedure, the risk of cost overruns is shifted to the subtrade who agrees to perform the work of their trade for a firm price they quote. However, the process is not quite as straightforward as it appears. Some of the problems associated with pricing subcontractors' work are:

1. Not all subtrades offer lump-sum bids; some trades bid unit prices for work. For instance, a subcontractor may quote a price of so much per foot to supply and install concrete curb and gutter. In this case the contractor will need to takeoff the length of curbs required for the project then apply the unit price to the quantity obtained.

   Other subtrades offer prices on an hourly basis. Excavation trades may present a price per hour for the use of operated excavation equipment; in this case, not only will work quantities have to be taken off, but an assessment of probable productivity rate of this crew will also be needed before a complete price for the work can be calculated.

2.  The subtrade's interpretation of what is and what is not part of the work of its trade may differ from the general contractor's interpretation. This problem is reduced where there are definitions of the scope of work in a trade available from trade associations or local bid depository rules. Even when these trade scope definitions are available, the general contractor's estimator has to ensure that all the work he or she expects the trade to perform is, in fact, covered in the subtrade's bid.

3.  The subcontractor whose price is used in the general contractor's bid to the owner and who is awarded the subtrade work may be unable to perform the work. The most common reason for failure to perform is the financial collapse of the subtrade. While legal action against the defaulting subtrade is possible, suing a bankrupt outfit is a fruitless endeavor. This risk can be reduced by calling for performance bonds from subtrades and by making a habit of checking the financial health of major subtrades.

These points will be examined in more detail when the pricing of individual subtrade's work is dealt with in Chapter 13.

## SUMMARY

■ Pricing an estimate consists of first sorting and listing all takeoff items according to the required trade breakdown (the recap) and then applying prices to this sorted list of items.

■ If computer estimating is used, the process of sorting into a recap is automated using a numerical item coding system in the takeoff.

■ There are five price categories considered in the pricing of an estimate:
  1. Labor
  2. Equipment
  3. Materials
  4. Subcontractors
  5. Job overheads

■ There are three general reasons why project costs may exceed estimated prices:
  1. Takeoff quantities are too low.
  2. Actual productivity does not meet anticipated productivity.
  3. Subcontractors or material suppliers fail to meet obligations.

■ The estimator's work is usually reviewed in order to reduce the risk of takeoff errors.

■ There are two main components to consider when pricing labor and equipment, both of which may be difficult to predict:
  1. The hourly wage rate of labor or hourly cost of equipment
  2. The productivity of labor or equipment

■ In order to price labor, the estimator has to determine the wage rates that apply to the project and predict how these rates will change over the life of the project.

■ Equipment rates depend on ownership costs or rental rates, both of which can be difficult to predict for a project.

■ The productivity of labor and equipment is governed by two groups of factors:
  1. Job factors that have to do with the nature and location of the particular project under consideration
  2. Labor and management factors that mainly relate to the quality of supervision and the skills of workers on the project.

■ The contractor will have little if any influence over job factors, but must carefully examine these factors before pricing the work, including probable weather conditions, site access, and repetitive work.

■ Because a contractor will often need to hire supervisors and workers on a project-by-project basis, the quality of project workers mostly depends on the economic conditions at the time and in the vicinity of the project. It is difficult to find good workers when busy economic conditions prevail.

■ A database of accurate historic labor and equipment costs is a useful source of information when pricing an estimate. This is generally obtained from previous project cost reports.

■ It is also beneficial to have information about why the final cost of work shown on a cost report varied from the estimated costs.

■ Estimators are usually optimistic about conditions that are difficult to predict. The financial risk of this optimism is then accounted for when assessing the fee to be added to the estimate.

■ Construction materials are classified as those that are part of the finished structure such as concrete and steel, and those that are consumed in the construction process such as formwork and fuel.

■ There is less risk of cost overruns with material estimates since firm prices can be obtained from suppliers for the supply of materials.

■ The price of materials entered into an estimate has to be calculated because the units that materials are packaged in often differ from takeoff units.

■ On most projects the contractor is required to supply and install materials that comply with specifications, but some materials may be owner-supplied and then installed by the contractor. In such a case, the contractor needs to obtain detailed information about the nature of what is to be installed.

■ While pricing subcontract work may appear to be straightforward—the subtrade estimates the price of its work and submits the price to the general contractor—in practice there can be many problems associated with pricing this work. Such problems range from concern about the subtrade's ability to perform the work to overlap or gaps in the coverage of subtrade work.

## REVIEW QUESTIONS

1. Describe the two stages involved in pricing an estimate.
2. Describe how computers can help with the first stage in pricing an estimate.
3. List the five price categories that need to be considered when pricing a construction estimate.
4. What is the "contractor's risk" and how does it relate to the type of construction contract to be used on a project?
5. List three general reasons why actual costs may exceed estimated costs on a lump-sum contract.
6. Describe the two basic methods of pricing labor and equipment in an estimate.
7. Research the wage rates of the following crafts in your area:
   a. Laborer
   b. Carpenter
   c. Cement finisher
   d. Brick mason
   e. Equipment operator
   f. Teamster
8. Why is the accurate pricing of labor and equipment very difficult on some projects?
9. What 10 questions should be considered regarding equipment rental rates?
10. List five job factors and five labor and management factors that can affect the productivity of labor and equipment on a construction project.

11. Why should an estimator be on the lookout for situations in which construction operations on a project are repetitive?
12. How does site access affect the price of work?
13. Which labor and management factors can a contractor have some influence over?
14. How are cost reports used by an estimator?
15. What two questions have to be answered before the data from a cost report can be used effectively by an estimator?
16. Why is the process of pricing materials usually easier than pricing labor and equipment on a project?
17. How can owner-supplied materials present a problem to estimators?

# 9

# PRICING CONSTRUCTION EQUIPMENT

## OBJECTIVES

*After reading this chapter and completing the review questions, you should be able to do the following:*

- Identify the three main equipment categories and describe how each is priced in an estimate.
- List and describe the advantages of renting rather than owning construction equipment.
- Explain in what circumstances it can be financially beneficial to own rather than rent construction equipment.
- List and describe the types of expenses that should be accounted for when calculating the ownership cost of equipment.
- Use each of the following methods to calculate depreciation allowances:
  a. Straight-line
  b. Declining balance
  c. Production or use
- Explain why tire depreciation has to be calculated separately from vehicle depreciation.
- Calculate maintenance and repair costs on an item of equipment.
- Calculate financing costs on an item of equipment.
- Calculate taxes, insurance and storage costs on an item of equipment.
- Calculate fuel and lube oil costs on an item of equipment.
- Determine the complete hourly ownership cost of an item of equipment.
- Calculate the rental rate for an item of equipment.

## Introduction

Plant, equipment and tools used in construction operations are priced in three different categories in the estimate. First, hand tools up to a certain value together with blades, drill bits and other consumables used in the work are priced as a percentage of the total labor price of the estimate. This percentage, referred to as the "Small Tools Allowance," is included with the "add-ons" that are found toward the end of the estimate Summary Sheet. Pricing small tools is discussed in Chapter 14.

The second category consists of larger items of equipment that are usually shared by a number of work activities; the kind of equipment items that are kept at the site over a period of time and used only intermittently on the work in progress. This equipment is priced on the estimate General Expenses Sheet under the "Site Equipment" heading. See Chapter 13 in which the process of pricing general expenses is described.

The third category of equipment comprises items that are used for specific tasks on the project such as digging a trench or hoisting materials into place. This equipment is priced directly against the takeoff quantities for that work it is to be used on. The equipment is not kept on-site for extended periods like those in the previous classification, but is shipped to the site, used for its particular task and then immediately shipped back to its source. Excavation equipment, cranes, hoisting equipment and costly, highly specialized items such as concrete saws all fall into this category.

Pricing this equipment calls for unit prices expressed in terms of hourly, weekly or monthly rates for each item of equipment. In the situation in which the contractor's equipment is rented, the pricing process is straightforward as the rental rates obtained from rental companies can be used directly against the time the equipment is required for. Where the equipment is owned by the contractor, ownership costs have to be calculated as detailed in the following, but this raises the question: should a contractor rent or own construction equipment?

## Renting versus Purchasing Equipment

For the contractor who is new in the construction business, the decision whether to rent or purchase equipment is usually quite easy to make because, lacking surplus cash and without a well-established credit rating, the only viable alternative is renting. For the older, more mature construction business, the decision may be a great deal more difficult. This contractor, who is more likely to be in a position in which funds and credit sources are available for equipment investments, has to determine if such investments are justified. Buying construction equipment is justified only where the investment promises net benefits in comparison with the alternative of renting equipment and investing the cash elsewhere.

A contractor does not necessarily have to own any construction equipment in order to carry on business. In most parts of the country there are many companies in the construction equipment rental business offering competitive rental rates on a large selection of equipment. There can be distinct advantages to renting equipment, including:

1. The contractor does not have to maintain a large inventory of specialized plant and equipment where individual items are used infrequently.
2. The contractor has continuous access to the newest and most efficient items of equipment available.
3. There is little or no need for equipment warehouse and storage facilities.
4. There is a reduced need for the contractor to employ maintenance staff and operate facilities for their use.
5. Accounting for equipment costs can be simpler when equipment is rented.
6. There may be significant savings on company insurance premiums when a contractor is not maintaining a inventory of plant and equipment.

However, when the construction operations of a contractor generate a steady demand for the use of certain items of equipment or plant, there can be distinct financial benefits gained by owning equipment. There can also be a marketing advantage to the contractors who own their own equipment due to the perception that these contractors are more financially stable and committed than others who own no equipment. In fact, some owners require contractors who bid on their projects to list on the bid the company-owned equipment they propose to use in the work. This information is utilized in the owner's assessment of the bidder.

Where a comparison of equipment ownership with the rental alternative strictly on the basis of cost is needed, the full cost per unit of time of owning an item of equipment has to be determined. To estimate the full ownership cost, the following aspects of equipment ownership have to be considered:

1. Depreciation expense
2. Maintenance and repair costs
3. Financing expenses
4. Taxes
5. Insurance costs
6. Storage costs
7. Fuel and lubrication costs

## Depreciation

In everyday usage the term *depreciation* refers to the decline in market value of an asset. To accountants the term has a more narrow meaning having to do with allocating the acquisition cost of an item of plant over the useful life of that asset. The way this allocation of cost is calculated may or may not reflect the loss of market value; more often than not it does not. Also, the allocation of depreciation costs considered here is not related in any way to tax considerations. For tax purposes a completely different depreciation schedule may be adopted.

In our appraisal of depreciation, some factors are explicit while other factors have to be estimated. Generally what we know is that the asset costs a certain amount to acquire (the initial cost); the asset will be used for a number of years (the useful life); and the asset will be sold at the end of this period for a sum of money (the salvage value). There is, however, some uncertainty about the exact length of the useful life of the asset and about the precise amount of salvage value that will be realized when the asset is disposed of. Any assessment of depreciation, therefore, requires these values to be estimated.

The process of allocating the cost of the item over its useful life is known as amortization, and there are several depreciation methods available to calculate amortization of an asset. Here we will consider three methods:

1. The straight-line method
2. The declining-balance method
3. The production or use method

### Straight-Line Depreciation

The straight-line method is the most commonly used method of calculating depreciation and is the method that we will use in the examples that follow. Depreciation on a straight-line basis is allocated equally per year over the useful life of the asset, thus the annual depreciation amount is constant and is equal to the cost of the asset minus any salvage value divided by the years of life of the asset:

$$\text{Annual Depreciation} = \frac{\text{Initial Cost} - \text{Estimated Salvage Value}}{\text{Estimated Useful Life (years)}}$$

| End of Year | Depreciation for the Year | Book Value |
|---|---|---|
| | $ | $ |
| 0 | 0 | 150,000 |
| 1 | 20,000 | 130,000 |
| 2 | 20,000 | 110,000 |
| 3 | 20,000 | 90,000 |
| 4 | 20,000 | 70,000 |
| 5 | 20,000 | 50.000 |

**Figure 9.1**   Straight-Line Depreciation Schedule

## EXAMPLE 1

The total initial cost of an excavator is $150,000; the useful life is expected to be 5 years; and the estimated salvage value at the end of this period is $50,000:

$$\text{Annual Depreciation} = \frac{\$\,150,000 - \$50,000}{5}$$

$$= \$20,000$$

Figure 9.1 shows the depreciation schedule based on straight-line depreciation over the life of this excavator.

At least two valid criticisms can made of the straight-line method of depreciation: first, that it does not reflect the fact that depreciation usually occurs at an accelerated rate in the early years of the life of an asset and, second, that this method of calculating depreciation does not account for the intensity of use of the asset. The two alternative methods of depreciation that follow try to address the deficiencies of the straight-line method.

### Declining-Balance Depreciation

With the declining-balance method of depreciation, the annual depreciation amounts decline as the asset gets older. Using this method of calculation, the depreciation in any given year is calculated by applying average rate of depreciation to the book value of the asset at the beginning of the year where:

$$\text{Average Rate of Depreciation} = \frac{1 \ (100\%)}{\text{Estimated Useful Life (years)}}$$

Book Value = Initial Cost − Accumulated Depreciation

Depreciation in Year$_n$ = Book Value (beginning of the year)

$\times$ Average Rate of Depreciation

| End of Year | Depreciation Rate | Depreciation for the Year | Book Value |
|---|---|---|---|
| | | $ | $ |
| 0 | 0.00% | 0 | 150,000 |
| 1 | 20.00% | 30,000 | 120,000 |
| 2 | 20.00% | 24,000 | 96,000 |
| 3 | 20.00% | 19,200 | 76,800 |
| 4 | 20.00% | 15,360 | 61,440 |
| 5 | 20.00% | 11,440* | 50,000 |

*This amount has been adjusted to give a book value of exactly $50,000 to reflect the true salvage value.

**Figure 9.2**   Declining-Balance Depreciation Schedule

## EXAMPLE 2

If we consider the same item as described in example 1:

$$\text{Average Rate of Depreciation} = \frac{1}{5}$$

$$= 20\%$$

$$\text{Depreciation in Year}_1 = \$150,000 \times 20\%$$

$$= \$30,000$$

Figure 9.2 shows the depreciation schedule based on declining balance depreciation over the life of this excavator.

### Production or Use Depreciation

Unlike the previous two methods, the production or use method of calculating depreciation is not a function of the age of the asset. Instead, the depreciation value in a specific year depends on the amount the asset is used in that year. First the number of units of work the asset will produce over its useful life is estimated, then the depreciation in any given year can be calculated by multiplying the number of units produced in that year by the depreciation amount per unit:

$$\text{Estimated Number Production Units} = \text{the number of units produced by the asset over its useful life}$$

$$\text{Depreciation Amount per Unit} = \frac{\text{Initial Cost} - \text{Salvage Value}}{\text{Estimated Number Production Units}}$$

$$\text{Depreciation in Year}_n = \text{Depreciation amount per unit} \times \text{the number of units produced that year}$$

| End of Year | Units Produced | Depreciation for the Year | Book Value |
|:---:|:---:|:---:|:---:|
| | CY | $ | $ |
| 0 | 0 | 0 | 150,000 |
| 1 | 2,800 | 28,000 | 122,000 |
| 2 | 1,600 | 16,000 | 106,000 |
| 3 | 2,000 | 20,000 | 86,000 |
| 4 | 1,500 | 15,000 | 71,000 |
| 5 | 2,100 | 21,000 | 50,000 |

**Figure 9.3**    Production Depreciation Schedule

### EXAMPLE 3

Considering once again the excavator described in example 1. If it is estimated that this unit will excavate a total of 10,000 cu. yd. over its useful life of 5 years and the unit excavates 2,800 cu. yd. in its first year of operation:

$$\text{Depreciation Amount per cu. yd.} = \frac{\$\,150,000 - \$50,000}{10,000}$$

$$= \$10.00 \text{ per cu. yd.}$$

$$\text{Depreciation in Year}_1 = 2,800 \times \$10.00$$

$$= \$28,000$$

Figure 9.3 shows the depreciation schedule based on production depreciation over the life of this excavator.

### Rubber-Tired Equipment

Because the life expectancy of rubber tires is generally far less than the life of the equipment they are used on, the depreciation rate of tires will be quite different from the depreciation rate on the rest of the vehicle. The repair and maintenance cost of tires as a percentage of their depreciation will also be different from the percentage associated with the repair and maintenance of the vehicle.

Consequently, when considering the depreciation of a rubber-tired vehicle, the cost of a set of replacement tires should be deducted from the initial cost of the vehicle and the depreciation on the tires and the depreciation on the vehicle without the tires should each be calculated separately. The repair and maintenance of the vehicle and of the tires can then also be calculated separately.

## Maintenance and Repair Costs

The costs of maintenance and repairs of plant and equipment is a factor that cannot be ignored when considering ownership costs. Equipment owners will agree that good maintenance, including periodic wear measurement, timely attention to recommended service and daily cleaning when conditions warrant it, can extend the life of equipment and actually reduce the operating costs by minimizing the effects of adverse conditions. All items of plant and equipment used by a construction contractor will require maintenance and probably also some repairs during the course of their useful life. The contractor who owns equipment usually sets up facilities with workers qualified to perform the necessary maintenance operations on equipment. It is the cost of operating this setup that we have to consider and include in the total ownership charges applied to items of plant and equipment.

Construction operations can subject equipment to considerable wear and tear but the amount of wear varies enormously between different items of equipment used and between different job conditions. The rates used in the following examples are based on the average costs of maintenance and repair, but since these costs can vary so much, the contractor formulating equipment operating prices should adjust the rates for maintenance and repairs according to the conditions the equipment is to work under. Again, as in many places in estimating, good records of previous costs in this area will much improve the quality of the estimator's assessment of probable maintenance costs.

Maintenance and repair costs are calculated as a percentage of the annual depreciation costs for each item of equipment. When depreciation is calculated using the straight-line method, as in the examples 6 and 7 that follow, the result is a constant amount being charged yearly for depreciation and then a second constant amount is allowed for maintenance and repairs. Realistically, depreciation will be high in the early years of ownership, while actual maintenance and repair costs in these years should be low. The relative values of yearly depreciation and maintenance costs will gradually reverse until, in the later years, low depreciation will be accompanied by high maintenance and repair bills. Using a constant amount yearly for these two expenses, therefore, would seem reasonable as the variance of one factor is offset by the countervariance of the other factor.

## Financing Expenses

Whether the owner of construction equipment purchases the equipment using cash or whether the purchase is financed by a loan from a lending institution, there is going to be an interest expense involved. The interest expense is the cost of using capital; where cash is used, it is the amount that would have been earned had the money been invested elsewhere, that is, the forgone interest revenue. Where the purchase is financed by a loan, the interest expense is the interest charged on the loan. In both cases the interest expense can be calculated by applying an interest rate to the owner's average annual investment in the unit. The average annual investment is approximately midway between the total initial cost of the unit and its salvage value. Thus:

$$\text{Average Annual Investment} = \frac{\text{Total Initial Cost} + \text{Salvage Value}}{2}$$

The interest rate used to calculate the financing expense will vary from time to time, from place to place and also from one company to another depending mostly on its credit rating and how good a deal it can get from the lending institution. In the examples that follow we will use a rate of 6%.

## Taxes, Insurance and Storage Costs

Just as with investment expenses, significant variations can be expected in the cost of the annual taxes, insurance premiums, storage costs together with fees for licenses required and other fees expended on an item of equipment. Where these expenses are known, they should be added into the calculation of the annual ownership costs of the equipment. In the case where information on these costs is not available, they may be calculated as a percentage of the average annual investment cost of the piece of equipment. The interest expense rate and the rate for taxes, insurance and storage costs are often combined to give a total *equipment overhead rate*. Below we will use an equipment overhead rate of 11%, which comprises 6% for the investment rate and 5% to cover taxes, insurance and storage costs.

## Fuel and Lubrication Costs

Fuel consumption and the consumption of lubrication oil can be closely monitored in the field. Data from these field observations will enable the estimator to quite accurately predict future rates of consumption under similar working conditions. However, if there is no access to this information, consumption can be predicted where the size and type of engine is known and the likely engine operating factor is estimated. This operating factor is an assessment of the load under which the engine is operating. An engine continually producing full-rated horsepower is operating at a factor of 100%. Construction equipment never operates at this level for extended periods, so the operating factor used in calculating overall fuel consumption is always a value less than 100%. The operating factor is yet another variable with a wide range of possible values responding to the many different conditions that might be encountered when the equipment under consideration is used. In the examples that follow the specific operating factors used can be no more than averages reflecting normal work conditions. Again, there is no good substitute for hard data carefully obtained in the observation of actual operations in progress.

When operating under normal conditions, namely at a barometric pressure of 29.9 in. of mercury and at a temperature of 60°F, a gasoline engine will consume approximately 0.06 gallons of fuel for each horsepower-hour developed. A diesel engine is slightly more efficient at 0.04 gal. of fuel for each horsepower-hour developed.*

## EXAMPLE 4

Regarding the excavator mentioned in example 1, if the unit was equipped with a diesel engine rated at 120 hp. operating at a factor of 50%, the fuel consumption can be determined thus:

$$\text{Fuel Consumption} = \frac{120 \times 0.04 \times 50}{100}$$

$$= 2.4 \text{ gals. per hour}$$

The amount of lubricating oil consumed will vary with the size of the engine, the capacity of the crankcase, the condition of the engine components, the frequency of oil changes and the general level of maintenance. An allowance in the order of 10% of the fuel costs is used as an average value in the following examples.

---

*Robert L. Peurifoy and G. D. Oberlender, *Estimating Construction Costs*, Fifth Edition, McGraw-Hill, New York, 2001.

## Equipment Operator Costs

Whether a contractor decides to rent or own the equipment used on its projects, the cost of operating the equipment has to be considered. In some situations rentals may be available that include an operating engineer as part of the rental agreement. This variety of rental agreement is sometimes available for excavation equipment and it can be a preferred alternative when the rental company offers a high-caliber equipment operator who is familiar with the particular excavation unit and is capable of high productivity.

More often than not, however, equipment is rented without an operator. So, just as in the case in which the contractor is using company-owned equipment, the labor costs for operating the equipment have to be calculated and added to the estimate. The usual way to price these costs is to apply an operating engineer's hourly wage alongside the equipment hourly rate and then use the expected productivity of the equipment to determine a price per measured unit for labor and a price per measured unit for equipment. Example 5 illustrates this method of pricing equipment and operator's costs. Note that the unit prices for labor and for equipment should always be considered separately as the labor prices have to be included in the total labor content of the estimate so that "add-ons" can be applied to this amount at the close of the bid.

### EXAMPLE 5

Where the hourly cost of an excavator is $72.00, the wage of an operator for this equipment is $30.00 per hour and the expected productivity of the excavator is 50 cu. yd. per hour, the unit prices for labor and equipment would be calculated thus:

| *Equipment* | *Labor* |
|---|---|
| $72.00/50 cu. yd. | $30.00/50 cu. yd. |
| = $ 1.44 per cu. yd. | = $ 0.60 per cu. yd. |

These unit prices can now be applied to the total quantity of excavation that this equipment is expected to perform in accordance with the takeoff.

## Company Overhead Costs

Where the equipment ownership costs calculated in accordance with this chapter are to be used as a basis of rental rates charged by the contractor to others for the use of the contractor's equipment, the full rental rates should include an amount for company overhead costs and amount for profit. Company overhead costs are basically the fixed costs associated with running a business. They may include the cost of maintaining a furnished office, office equipment and personnel together with all the other costs of business operation.

Since the rental rate quoted by a contractor to another party for the use of the contractor's equipment is, in a sense, a kind of bid, the same considerations should be applied to the markup on the rental rate as are applied to markup on any of contractor's bids. Markup, comprising an amount for company overhead costs and an amount for profit is dealt with in some detail in Chapter 15 where the markup included in a bid price is discussed.

**EXAMPLE 6A**

Calculate the ownership cost per hour for a crawler-type excavator powered by a 250 hp. diesel engine based on the following data:

| | |
|---|---|
| ENGINE: | 250 hp. diesel |
| OPERATING FACTOR: | 50% |
| PURCHASE PRICE: | $320,000 |
| FREIGHT CHARGES: | $2,000 |
| ESTIMATED SALVAGE VALUE: | $150,000 |
| USEFUL LIFE: | 6 years |
| HOURS USED PER YEAR: | 2,000 |
| MAINTENANCE & REPAIRS: | 110% of annual depreciation |
| EQUIPMENT OVERHEAD RATE: | 11% |
| DIESEL FUEL PRICE: | $1.80 per gallon |

First, preliminary calculations are made to determine the *average annual investment* and the fuel consumption rate:

Average Annual Investment = (Total Cost + Salvage Value)/2

= ($320,000 + $2,000 + $150,000)/2

= $236,000

Fuel Consumption = 250 × 0.04 × 50% gals. per hour

= 5 gals. per hour

The annual cost of depreciation, maintenance and repairs, and equipment overheads can now be calculated:

*Annual Costs*

Depreciation $= \dfrac{\text{Initial Cost} - \text{Estimated Salvage Value}}{\text{Estimated Useful Life (years)}}$

$= \dfrac{\$322,000 - \$150,000}{6}$ = $28,667

Maintenance and Repairs = 110% of Annual Depreciation

= 1.1($28,666.67) = $31,533

Equipment Overheads = 11% × Average Annual Investment

= 0.11($236,000) = $25,960

*Total Annual Costs—Vehicle:* $86,160

Now the hourly costs can be calculated including the cost of fuel and lube oil required:

*Hourly Costs*

Vehicle Cost $= \dfrac{\text{Total Annual Cost}}{\text{Hours Used per Year}}$

$$= \frac{\$86,160}{2,000} \qquad = \$43.08$$

| | | |
|---|---|---|
| Fuel Cost | = Fuel Consumption $\times$ Cost of Fuel | |
| | = 5 gals. per hour $\times$ \$1.80 per gal. = | \$ 9.00 |
| Lube Oil | = 10% of Fuel Cost | |
| | = 0.1 $\times$ \$9.00 | $= \underline{\$\ 0.90}$ |
| | *Excavator Cost per Hour:* | $\underline{\underline{\$52.98}}$ |

## EXAMPLE 6B

What would be the estimated unit prices for excavating basements using the excavator described in example 6A together with an operating engineer at a wage of $30.00 per hour when the expected productivity of this unit is 36 cu. yd. per hour?

| | *Equipment* | *Labor* |
|---|---|---|
| Unit Prices = | \$52.98/36 cu. yd. | \$30.00/36 cu. yd. |
| | = \$  1.47 per cu. yd. | \$  0.83 per cu. yd. |

## EXAMPLE 6C

What would be the hourly charge-out rates (rental rates) for the previously described excavator and operator based on the following overheads and profit requirements?

**1.** The company overheads on equipment is 20%
**2.** The company overhead on labor is 50%
**3.** The required profit margin is 10%

*Charge-Out Rate for Excavator*

| | | |
|---|---|---|
| Excavator Cost per Hour | | = \$52.98 |
| Company Overhead at 20% | | = $\underline{\$10.60}$ |
| | Subtotal: | \$63.58 |
| Profit Margin at 10% | | = $\underline{\$\ 6.36}$ |
| | *Total Charge-Out Rate—Excavator:* | $\underline{\underline{\$69.94}}$ |

*Charge-Out Rate for Operator*

| | | |
|---|---|---|
| Operator Cost per Hour | | = \$30.00 |
| Company Overhead at 50% | | = $\underline{\$15.00}$ |
| | Subtotal: | \$45.00 |
| Profit Margin at 10% | | = $\underline{\$\ 4.50}$ |
| | *Total Charge-Out Rate—Operator:* | $\underline{\underline{\$49.50}}$ |

In the next example, the item of equipment considered is a rubber-tired vehicle, so the depreciation cost of the tires is calculated separately for the depreciation cost of the rest of the vehicle.

### EXAMPLE 7

Calculate the ownership cost per hour for a dump truck powered by a 120 hp. gasoline engine based on the following data:

| | |
|---|---|
| ENGINE: | 120 hp. gasoline |
| OPERATING FACTOR: | 40% |
| PURCHASE PRICE: | $75,000 (including tires) |
| FREIGHT CHARGES: | $1,000 |
| ESTIMATED SALVAGE VALUE: | $7,500 |
| USEFUL LIFE: | 5 years |
| HOURS USED PER YEAR: | 1,800 |
| MAINTENANCE & REPAIRS: | 130% of annual depreciation |
| TIRE COST: | $5,000 |
| TIRE LIFE: | 4,000 hours |
| MAINTENANCE & REPAIRS (TIRES): | 15% of tire depreciation |
| EQUIPMENT OVERHEAD RATE: | 11% |
| GASOLINE FUEL PRICE: | $2.20 per gallon |

Preliminary calculations are made to determine the *average annual investment* and the fuel consumption rate:

$$\text{Average Annual Investment} = \text{(Total Cost + Salvage Value)}/2$$
$$= (\$75,000 + \$1,000 + \$7,500)/2$$
$$= \$41,750$$

$$\text{Fuel Consumption} = 120 \times 0.06 \times 40\% \text{ gals. per hour}$$
$$= 2.88 \text{ gals. per hour}$$

The annual cost of depreciation, maintenance and repairs, and equipment overheads can now be calculated:

*Annual Costs*

Depreciation

$$\text{(Vehicle)} = \frac{\text{Initial Cost (less tires) – Estimated Salvage Value}}{\text{Estimated Useful Life (years)}}$$

$$= \frac{(\$76,000 - \$5,000) - \$7,500}{5} \qquad = \$12,700$$

| | | |
|---|---|---|
| Maintenance and Repairs | = 130% of Annual Depreciation | |
| | = 1.30($12,700) | = $16,510 |
| Equipment Overheads | = 11% × Average Annual Investment | |
| | = 0.11($41,750) | = $ 4,593 |
| | *Total Annual Costs—Vehicle:* | $33,803 |

*Hourly Costs*

Vehicle Cost:      $= \dfrac{\text{Total Annual Cost}}{\text{Hours Used per Year}}$

$= \dfrac{\$33{,}803}{1{,}800}$      = $18.78

Tire Depreciation:      $= \dfrac{\text{Tire Cost}}{\text{Life of Tires}}$

$= \dfrac{\$5{,}000}{4{,}000}$      = $ 1.25

Maintenance and Repairs   = 15% × $1.25      = $ 0.19
   on tires

Fuel Cost:      = Fuel Consumption × Cost of Fuel

      = 2.88 gals. per hour × $2.20 per gal.   = $ 6.34

Lube Oil:      = 10% of Fuel Cost

      = 0.1 × $6.00      = $ 0.63

      *Dump Truck Cost per Hour:*      $27.19

## Use of Spreadsheets

Because of the repetitive nature of the calculations involved, a computer spreadsheet program is very useful for calculating equipment ownership costs. Use of computer spreadsheet applications is particularly befitting where it is necessary to calculate the ownership costs of large numbers of equipment items. Also, the process of updating the data on which the ownership cost calculations are based can be accomplished far more conveniently when the original calculations are stored in the form of spreadsheet calculations.

A spreadsheet template can be readily set up to provide the basic format of the calculation process; then the data applicable to the specific item of equipment can be inserted to generate the ownership cost of that item. This process can be repeated for any number of equipment units enabling the estimator to calculate their ownership costs in a matter of minutes. Figure 9.4 shows the format of a spreadsheet template for this use. Figure 9.5 and Figure 9.6 show the computer calculations of the ownership costs of the items of equipment considered in examples 6 and 7.

## SUMMARY

- Categories of equipment and how they are priced:
  a. Hand tools, priced as a percentage of labor price (add-ons)
  b. On-site equipment used intermittently, priced as a general expense item
  c. Equipment used for specific tasks, priced directly against the takeoff items
- Advantages of renting equipment:
  a. The contractor does not have to maintain a large inventory specialized plan and equipment where individual items are used infrequently.

**EQUIPMENT OWNERSHIP COSTS**

VEHICLE:

OPERATING FACTOR:

PURCHASE PRICE:

FREIGHT:

SALVAGE VALUE:

LIFE EXPECTANCY:

HOURS PER YEAR:

MAINTENANCE & REPAIRS:

TIRE COST:

TIRE LIFE:

MAINTENANCE & REPAIRS (TIRES):

EQUIPMENT OVERHEAD:

FUEL COST:

AVERAGE ANNUAL INVESTMENT:

FUEL CONSUMPTION:

**ANNUAL COSTS**

DEPRECIATION

VEHICLE:                                                =

MAINTENANCE & REPAIRS:    % OF DEPRECIATION    =

EQUIPMENT OVERHEAD:    % OF                   =  _____

    VEHICLE FIXED COST:                         $  _____

**HOURLY COST**

VEHICLE COST:                                        =

TIRE DEPRECIATION:

MAINTENANCE & REPAIRS ON TIRES:

FUEL:  Gal ×                                        =

OIL: 10% OF FUEL COST                               =  _____

    **COST PER HOUR:**                               $  _____

**Figure 9.4**    Spreadsheet Template for Equipment Ownership Costs

**EQUIPMENT OWNERSHIP COSTS**

|                                          |              |                            |
|------------------------------------------|-------------:|----------------------------|
| VEHICLE:                                 | EXCAVATOR    | 250 hp. DIESEL             |
| OPERATING FACTOR:                        | 50.00%       |                            |
| PURCHASE PRICE:                          | $320,000     |                            |
| FREIGHT:                                 | $2,000       |                            |
| SALVAGE VALUE:                           | $150,000     |                            |
| LIFE EXPECTANCY:                         | 6            | YEARS                      |
| HOURS PER YEAR:                          | 2,000        | HOURS                      |
| MAINTENANCE & REPAIRS:                   | 110.00%      | DEPRECIATION               |
| TIRE COST:                               | $0           | (TRACK VEHICLE)            |
| TIRE LIFE:                               | N/A          |                            |
| MAINTENANCE & REPAIRS (TIRES):           | N/A          |                            |
| EQUIPMENT OVERHEAD:                      | 11.00%       | AVERAGE ANNUAL INVESTMENT  |
| FUEL COST:                               | $1.80        | PER GAL. (DIESEL)          |

AVERAGE ANNUAL INVESTMENT:

$$(\$322,000 + \$150,000)/2 = \$236,000.00$$

FUEL CONSUMPTION:

$$250 \text{ hp.} \times 50\% \times 0.04 \text{ gal./hp. hour} = 5 \text{ gal. per hour}$$

**ANNUAL COSTS**

DEPRECIATION:

| | | |
|---|---|---:|
| VEHICLE: (320,000 + 2,000 − 150,000)/6 | = | 28,666.67 |
| MAINTENANCE & REPAIRS: 110% OF DEPRECIATION | = | 31,533.33 |
| EQUIPMENT OVERHEAD: 11% OF $236,000 | = | 25,960.00 |
| VEHICLE FIXED COST: | | $86,160.00 |

**HOURLY COST**

| | | |
|---|---|---:|
| VEHICLE COST: 80,861.67/ 2,000 | = | 43.08 |
| TIRE DEPRECIATION:      N/A | = | 0.00 |
| MAINTENANCE & REPAIRS ON TIRES:  N/A | = | 0.00 |
| FUEL: 5.0 Gal × $1.80 | = | 9.00 |
| OIL: 10% OF FUEL COST | = | 0.90 |
| **EXCAVATOR COST PER HOUR:** | | $      52.98 |

**Figure 9.5**   Example 6—Computer Calculation of Excavator Ownership Costs

## EQUIPMENT OWNERSHIP COSTS

|  |  |  |
|---|---|---|
| VEHICLE: | DUMP TRUCK | 120 hp. GASOLINE |
| OPERATING FACTOR: | 40.00% | |
| PURCHASE PRICE: | $75,000 | |
| FREIGHT: | $1,000 | |
| SALVAGE VALUE: | $7,500 | |
| LIFE EXPECTANCY: | 5 | YEARS |
| HOURS PER YEAR: | 1,800 | HOURS |
| MAINTENANCE & REPAIRS: | 130.00% | DEPRECIATION |
| TIRE COST: | $5,000 | |
| TIRE LIFE: | 4,000 | HOURS |
| MAINTENANCE & REPAIRS (TIRES): | 15.00% | TIRE DEPRECIATION |
| EQUIPMENT OVERHEAD: | 11.00% | AVERAGE ANNUAL INVESTMENT |
| FUEL COST: | $2.20 | PER GAL. (GASOLINE) |

AVERAGE ANNUAL INVESTMENT:

$$($76,000 + $7,500)/2 \qquad = $41,750.00$$

FUEL CONSUMPTION:

$$120 \text{ hp.} \times 40\% \times 0.06 \text{ gal./hp. hour } = 2.88 \text{ gal. per hour}$$

**ANNUAL COSTS**

DEPRECIATION:

| | | |
|---|---|---|
| VEHICLE: (76,000 − 5,000 − 7,500)/5 | = | 12,700.00 |
| MAINTENANCE & REPAIRS: 130% OF DEPRECIATION | = | 16,510.00 |
| EQUIPMENT OVERHEAD: 11% OF $41,750 | = | 4,592.50 |
| VEHICLE FIXED COST: | | $33,802.50 |

**HOURLY COST**

| | | |
|---|---|---|
| VEHICLE COST: 34,226/1,800 | = | 18.78 |
| TIRE DEPRECIATION: $5,000/4,000 Hours | = | 1.25 |
| MAINTENANCE & REPAIRS ON TIRES: 15% of Tire Depreciation | = | 0.19 |
| FUEL: 2.88 Gal × $2.20 | = | 6.34 |
| OIL: 10% OF FUEL COST | = | 0.63 |
| **DUMP TRUCK COST PER HOUR:** | | $ 27.19 |

**Figure 9.6** Example 7—Computer Calculation of Dump Truck Ownership Costs

b. The contractor has continuous access to the newest and most efficient items of equipment available.

c. There is little or no need for equipment warehouse and storage facilities.

d. There is a reduced need for the contractor to employ maintenance staff and operate facilities for their use.

e. Accounting for equipment costs can be simpler when equipment is rented.

f. There may be significant savings on company insurance premiums when a contractor is not maintaining an inventory of plant and equipment.

■ There are financial advantages to owning equipment when a contractor can provide steady use of the equipment.

■ The following expenses should be considered when determining the full ownership cost of material:

a. Depreciation

b. Maintenance and repair

c. Financing

d. Taxes

e. Insurance

f. Storage

g. Fuel and lubrication

■ When straight-line depreciation is used, the amount of depreciation from initial cost to salvage value is distributed equally each year over the life of the asset.

■ Depreciation can also be calculated using the declining-balance method or the production or use method. Using these methods results in depreciation amounts that are high to begin with but decline as the asset gets older.

■ When considering the depreciation of a rubber-tired vehicle, the tire depreciation is calculated separately from the vehicle depreciation.

■ Maintenance and repair costs on construction equipment vary considerably depending on the type of equipment and the job conditions encountered.

■ Whether equipment purchase is financed by loans or by use of the contractor's cash, interest charges will apply. These are calculated as a percentage of the average annual investment amount.

■ Taxes, insurance charges and storage costs vary over a wide range of values depending on particular circumstances. An allowance for these costs is also calculated as a percentage of the average annual investment amount.

■ Fuel costs depend on the type of engine and are proportional to the engine's horsepower rating. An operating factor that is always less than 100% is also introduced to account for the fact the engine does not operate continuously at full throttle.

■ An allowance for company overheads and profit should be added to hourly ownership costs of an item of equipment when quoting a price for the rental of the item.

## REVIEW QUESTIONS

1. Describe the three different ways tools and equipment are priced in an estimate.

2. What factors should be considered by the contractor who is deciding whether to rent or buy construction equipment?

3. What are the advantages and disadvantages to the contractor of renting rather than buying equipment?

4. Suggest how an accountant might explain the depreciation of an asset.

5. Why is the "straight-line method" not a satisfactory way to calculate depreciation?

6. Calculate the ownership cost per hour for a crawler-type hydraulic crane powered by a 350 hp. diesel engine based on the following data:

| | |
|---|---|
| ENGINE: | 350 hp. diesel |
| OPERATING FACTOR: | 60% |
| PURCHASE PRICE: | $570,000 |
| FREIGHT: | $2,500 |
| ESTIMATED SALVAGE VALUE: | $350,000 |
| LIFE EXPECTANCY: | 5 years |
| HOURS PER YEAR: | 2,000 |
| MAINTENANCE & REPAIRS: | 120% of annual depreciation |
| EQUIPMENT OVERHEAD: | 11% |
| DIESEL FUEL PRICE: | $1.80 per gallon |

7. What would be the hourly charge-out rates (rental rates) for the crane and operator described in question 6 based on the following overheads and profit requirements?

   —The operator's wage is $20.00 per hour.
   —The company overheads on equipment is 20%.
   —The company overhead on labor is 50%.
   —The required profit margin is 10%.

8. Calculate the ownership cost per hour for a motor grader powered by a 190 hp. gasoline engine based on the following data:

| | |
|---|---|
| ENGINE: | 190 hp. gasoline |
| OPERATING FACTOR: | 45% |
| PURCHASE PRICE: | $95,000 (including tires) |
| FREIGHT: | $1,500 |
| SALVAGE VALUE: | $17,500 |
| LIFE EXPECTANCY: | 7 years |
| HOURS USED PER YEAR: | 1,700 |
| MAINTENANCE & REPAIRS: | 125% of annual depreciation |
| TIRE COST: | $9,000 |
| TIRE LIFE: | 5,000 hours |
| MAINTENANCE & REPAIRS (TIRES): | 15% of tire depreciation |
| EQUIPMENT OVERHEAD: | 11% |
| GASOLINE FUEL PRICE: | $2.20 per gallon |

# 10

# PRICING EXCAVATION AND BACKFILL

## OBJECTIVES

*After reading this chapter and completing the review questions, you should be able to do the following:*

- Explain why specialized subcontractors usually perform excavation and backfilling operations.
- Calculate equipment transportation expenses.
- Identify different types of excavation including:
  a. Site cut and fill operations
  b. Basement-type excavations
  c. Trench excavations
  d. Pit and sump excavations
- Describe the job factors and the labor and management factors that influence the price of excavation and backfill operations.
- Describe the four factors that have an effect on the labor price for excavation work.
- Explain how excavation and backfilling productivity rates are obtained.
- Use historic productivity rates to calculate the unit price of labor and equipment for excavation and backfill work.
- Calculate the price of gravels and other backfill materials.
- Explain how swell factors and compaction factors impact the price of excavation and backfill work.
- Calculate trucking requirements for hauling excavated material and gravels.
- Complete the recap and pricing of excavation and backfill work using manual methods.
- Use the Precision Estimating software to price excavation and backfill work.

## Excavation Equipment and Methods

Like much construction work today, the excavation trade is usually performed by contractors who specialize in this work. Further specialization is then found among excavation contractors who tend to concentrate on a single specific type of construction. For instance, one company's business may involve operations that relate only to site preparation, while another company's work may be restricted to small pit and trench excavations. The main reason for this specialization can be found in the vast array of excavation equipment that is available to pursue the many different possible types of excavation activity. No contractor, even a large-size excavation company, can hope to be equipped to perform effectively all types of excavation work—the scope is just too wide. Despite this specialization, some general contractors do not subcontract excavation work on every project. They overcome the need to maintain a large inventory of specialized equipment by hiring items that are appropriate to the work to be done and owning only the kind of units that can be used on a wide variety of projects.

In this chapter our objective is to develop an understanding of the principles of estimating the cost of excavation work generally. We will not be dealing with highly specialized excavation undertakings that call for special techniques and equipment. Instead, we will limit our scope by considering only four of the more common types of excavation and the backfill activities associated with them:

1. Site cut and fill operations
2. Basement-type excavations
3. Trench excavations
4. Pit and sump excavations

Within each of these excavation categories there are many ways to proceed as well as a large selection of different types of equipment that could be used to execute the work. Again we will restrict our analysis to the more usual methods and types of equipment employed in the four common types of excavation activities by contractors engaged in general building construction work.

The particular items of equipment used in excavation operations may be owned or rented by the contractor performing the work. See Chapter 9 for a discussion of rental or ownership of construction equipment generally. Figure 10.1 shows a list

| ITEM | PRICE per DAY |
|---|---|
| 105-HP DOZER | $650.00 |
| 165-HP DOZER | $875.00 |
| MOTOR GRADER | $785.00 |
| 10-T ROLLER | $600.00 |
| 3-T ROLLER | $350.00 |
| WATER TRUCK | $300.00 |
| 1.5-CY TRACK LOADER | $425.00 |
| 12-CY DUMP TRUCK | $275.00 |
| 0.75-CY TRACK BACKHOE | $670.00 |
| 0.5-CY WHEEL BACKHOE | $385.00 |
| BOBCAT LOADER | $280.00 |

**Figure 10.1** Excavation Equipment

of excavation equipment prices that are used in the calculations that follow. These prices reflect the cost of equipment employed in a contractor's work, whether the equipment is owned or rented, and the prices include fuel and maintenance expenses but do not include the cost of an operator or any allowance for transportation of the equipment to and from the construction site.

### Equipment Transportation Expenses

The cost of transporting minor items of equipment such as plate compactors, small rollers, bobcats and small tractors to and from the site is priced on the General Expense Sheet under the item "Trucking and Material Handling." This item covers the cost of trucking associated with carrying incidental construction goods and equipment where the supplier of these items does not deliver them to the site.

Transportation of major items of equipment—backhoes, bulldozers, loaders, power shovels and the like—is better priced on an item-by-item basis taking into account the cost of loading, unloading, trucking and any rental charges levied while the unit is in transit. When pricing the trucking component of these costs, the estimator has to be aware of the additional requirements such as "scout trucks" when loads are oversize and also the possible need to take special routes when road or bridge weight restrictions make some routes unavailable.

### Site Cut and Fill Operations

The equipment used in cut and fill operations depends mostly on the scale of the project. Extensive operations call for large size earth-moving scrapers and bulldozers often working in teams excavating and depositing large quantities of material, which is then leveled and compacted by a second team of rollers, water trucks and graders. Small cut and fill operations can be undertaken with no more than a small dozer and grader or even with just a bobcat and a "walk behind" roller.

### Basement Excavations

Basements are often excavated using a track backhoe of between ½- and 1-cubic yard capacity filling dump trucks that dispose of surplus material. The excavation of large-size basements may be undertaken using one or more track loaders of up to 2- or 3-cubic yard capacity loading fleets of trucks that access the bottom of the basement by means of ramps down the sides of the excavation. Large-scale excavations of this type may call for the pricing of grading and maintenance of these access ramps.

### Trench Excavations

For smaller trenches such as those required for shallow pipes and cables, trenching machines are available that attach to rubber-tired tractors and provide a highly maneuverable and efficient trench excavating setup.

Larger trenches are more often dug using backhoes. Small backhoes are available as attachments to tractors; larger units of up to one 1- cubic yard or more capacity are usually mounted on crawler tracks. Large backhoes are used for excavating deep trenches and typically they side cast excavated material (spoil) to either side of the trench where it remains until required for backfilling. Where it is not feasible to store spoil in heaps next to the trench due to lack of space, backhoes may have to load the excavated material onto trucks that remove it from the site. Material for backfilling will then have to be trucked back to the site when it is required.

### Pit and Sump Excavations

Pit and sump excavation is mostly performed using backhoes. The particular type and size of backhoe used in any situation depends mostly on the size of the holes to be excavated. Where there are many small pits to be dug over the area of the site, backhoes mounted on rubber-tired tractors are generally favored because of their superior maneuverability over track mounted machines.

### Hand Excavation

While it is almost always more economical to use appropriate items of equipment to excavate rather than to perform the work by hand, there are circumstances in which hand excavation methods offer a more effective means of obtaining the results required. One particular situation that calls for hand excavation is the situation in which the work is to be done in confined spaces, limiting equipment access. The introduction in recent years of items of very small equipment that are nimble enough to work in these small spaces has reduced the need for hand excavation, but for excavation around existing structures like footings, pipes and cables that are not to be disturbed, digging by hand may still be necessary.

A second common requirement for hand excavation arises where the bottom of trenches or pits for concrete footings has to be trimmed. This is necessary when contract specifications call for footing concrete to bear on undisturbed soil. Because the soil left at the bottom of excavations when mechanical excavators are used is usually loosened by the action of the teeth on excavator buckets, hand trimming is needed to remove this loose soil.

## Excavation Productivity

As we noted in Chapter 8, the cost of labor and equipment used in completing the work depends on the productivity of the project labor crews and the equipment they use. The amount of work excavation and backfilling crews can perform per hour is highly variable because there are so many factors that can affect the efficiency of the labor and equipment involved. Job factors and labor factors that particularly influence excavation and backfill operations include the following:

*Job Factors*

1. Type of material excavated or backfilled
2. Moisture condition of materials
3. Weather conditions expected
4. Access to and around site
5. Project size and complexity
6. Distance to haul materials for disposal
7. Availability of gravels and fill materials
8. Wage and price levels at the job location

*Labor and Management Factors*

1. Quality of job supervision
2. Quality of job labor
3. Motivation and morale of workers
4. Type and quality of tools and equipment
5. Experience and records of similar past projects

### Type of Material Excavated or Backfilled

Different soils and rocks will certainly provide a wide variation in possible excavation productivity rates on a project, but it can often be difficult to determine exactly what is the nature and quality of the materials to be excavated. Even when soils reports are available, the information they contain can be inaccurate because data is obtained from sampling procedures that often generate misleading results. So there is a large component of risk involved when the estimator makes assumptions about the kind of material that is to be excavated on a project. Because a faulty assumption can translate into a significant dollar amount, the estimator is advised to proceed with much care in this area.

We recommend that notes are kept in the estimate stating all the underlying assumptions made about the material to be excavated and explaining the basis of these assumptions. Good clear notes detailing the evidence obtained in site investigations and information gleaned from soils reports together with the conclusions drawn therefrom can provide useful support to possible future claims and to negotiations for additional contract reimbursement. The evidence of soils reports and investigations should be thoroughly discussed by the estimator and his or her colleagues during the estimating period; the opinion and advice of persons experienced in excavation operations can be most valuable in assessing likely soil conditions.

### Moisture Conditions of Materials

As can be seen in Figure 10.2, wet materials are heavier than dry materials, so excavation productivity will generally be slower when the material being excavated is

| MATERIAL | LOOSE | | BANK | | COMPACTION FACTOR |
|---|---|---|---|---|---|
| | lb./cu. yd. | kg/m³ | lb./cu. yd. | kg/m³ | |
| 1. CLAY—Dry | 2500 | 1483 | 3100 | 1839 | 24.00% |
| —Wet | 2800 | 1661 | 3500 | 2076 | 25.00% |
| 2. CLAY and GRAVEL—Dry | 2400 | 1424 | 2800 | 1661 | 16.67% |
| —Wet | 2600 | 1543 | 3100 | 1839 | 19.23% |
| 3. EARTH—Dry | 2500 | 1483 | 3200 | 1898 | 28.00% |
| —Wet | 2700 | 1602 | 3400 | 2017 | 25.93% |
| 4. LOAM | 2100 | 1246 | 2600 | 1543 | 23.81% |
| 5. GRAVEL—Pit-run, dry | 2550 | 1513 | 2850 | 1691 | 11.76% |
| —Put-run, wet | 3250 | 1928 | 3650 | 2165 | 12.31% |
| —Dry ¼ to 2" | 2850 | 1691 | 3200 | 1889 | 12.28% |
| —Wet ¼ to 2" | 3400 | 2017 | 3800 | 2254 | 11.76% |
| 6. SAND—Dry | 2400 | 1424 | 2700 | 1602 | 12.50% |
| —Damp | 2850 | 1691 | 3200 | 1898 | 12.28% |
| —Wet | 3100 | 1839 | 3500 | 2076 | 12.90% |
| 7. SAND and GRAVEL—Dry | 2900 | 1721 | 3250 | 1928 | 12.07% |
| —Wet | 3400 | 2017 | 3750 | 2225 | 10.29% |

**Figure 10.2**   Material Weights and Compaction Factors

wet. Not only does the added weight contribute to a decrease in productivity but also, especially in clay soils, the presence of moisture can produce slippery conditions that slow down operations considerably. Also, some soils can become sticky in wet conditions, which, again, reduces productivity.

While moisture in clay soil can be a hindrance to backfilling operations, adding water to dry backfill material is often desirable to assist in attaining the compaction requirements. Here close control of moisture content is required since too much or too little moisture can hamper backfilling productivity. The objective is to provide the optimum moisture content of soil so that the required density of the material can be obtained in the compaction operations.

### Weather Conditions Expected

The moisture content of the materials handled in excavation and backfilling operations is mostly dependent on the prevailing weather conditions at the site. In order to ascertain what the moisture content of excavation material is likely to be, an assessment of weather conditions expected at the site during excavation and fill operations is necessary.

Expected weather conditions also have to be considered in order to forecast extremes of cold, heat and wind conditions, all of which will adversely affect productivity at the site and entail additional costs in which specific provisions have to be made to allow work to proceed in these weather conditions.

Consequently it is very important for the estimator to anticipate at what time of the year particular site operations will take place and obtain information about the likely weather conditions in those periods. This is another place in the estimating process in which estimators tend to take an optimistic approach. Knowing that extreme weather conditions are possible at certain times of the year, the prices used in the estimate are based on the assumption that the weather will be good. This approach is acceptable only where risk is identified and taken into account when overall bid package and fee are assembled later in the process.

### Access to and Around Site

The condition of site access influences productivity in a number of ways. Restricted access can prohibit the use of large equipment on some sites. If a large backhoe would normally be used for the work involved in a project but lack of a suitable access road prevents a machine of the desired size from getting to the site, then prices have to be based on smaller, less productive equipment or the cost of improving the access has to be considered.

Productivity can also be reduced when the site is located in a congested city center or where a number of concurrent operations cause congestion on the site itself, forcing the excavator to work around other site activities. Here again the estimator has to anticipate probable site conditions and make allowances in the pricing for them.

### Project Size and Complexity

It is, perhaps, axiomatic to state that productivity will decline when projects are of a more complex nature. However, in order to make appropriate adjustments to productivity rates to reflect the difficulty of executing the work, an estimator has to develop the skill of assessing project complexity on a defined scale of difficulty. Exactly what makes one project more difficult than another has to be learned from experience. The junior estimator who has not yet developed this skill is advised to fully discuss the assessed difficulty of a project with more experienced colleagues so that necessary adjustments to rates and prices can be made.

Generally a large project allows crews and equipment to attain high rates of productivity as, through repetition of operations, this enables them to make method improvements that enhance production rates. Also, when fixed costs such as transportation expenses are spread over large quantities of work, unit prices per unit of work done are reduced as demonstrated in the following example:

1.  Calculate the equipment and labor prices per cu. yd. to excavate 3000 cu. yd. of trench using a ¾ cu. yd. backhoe costing $670.00 per day (8 hours) plus $4000 for transportation and setup charges. Expected output is 60 cu. yd. per hour with an operator and 0.5 of a laborer at wages of $30.00 and $21.00 respectively.

|  |  | *Labor* | *Equipment* |
|---|---|---|---|
| Unit Costs |  | $ | $ |
| Operator |  | 30.00 | — |
| Laborer | 0.5 × $21.00 | 10.50 | — |
| Backhoe | 670.00/8 | — | 83.75 |
| | Per hr. | 40.50 | 83.75 |
| | Per cu. yd. (/60) | 0.68 | 1.40 |
| Project Costs |  |  |  |
| Transportation: |  | — | 4,000.00 |
| Excavation |  |  |  |
| Labor | 3,000 cu. yd. × 0.68 | 2,040.00 | — |
| Equip | 3,000 cu. yd. × 1.40 | — | 4,200.00 |
| | | 2,040.00 | 8,200.00 |
| | Per cu. yd. (/3000) | 0.68 | 2.73 |

2.  Calculate the equipment and labor prices per cu. yd. to excavate 300 cu. yd. of trenching with the same crew as listed in item 1. Because the amount of work involved is so much less, overall equipment productivity would probably be lower, but even if it remained the same the price per unit of work increases:

|  |  | *Labor* | *Equipment* |
|---|---|---|---|
| Unit Costs (unchanged) |  | $ | $ |
| | Per cu. yd. (/60) | 1.40 | 0.68 |
| Project Costs |  |  |  |
| Transportation |  | — | 4,000.00 |
| Excavation |  |  |  |
| Labor | 300 cu. yd. × 0.68 | 204.00 | — |
| Equip | 300 cu. yd. × 1.40 | — | 420.00 |
| | | 204.00 | 4,420.00 |
| | Per cu. yd. (/300) | 0.68 | 14.73 |

### Other Job Factors

As we see later under Excavation Materials, the principal cost of the materials involved in excavation and fill operations is the cost of transportation, so, clearly, the availability near the site of suitable materials and disposal fields will lower the cost of the work significantly.

Perhaps the most important job factors that must be established before an estimate of any trade can be prepared are the trade wage rates and equipment prices prevailing at the location of the project under consideration. All subsequent pricing is based on the rates established by the estimator at the start of the pricing operations, so it goes without saying that great care should be taken in obtaining accurate information about the wage rates and prices that will be in force for the project duration.

### Labor and Management Factors

The cost of the work of the excavation trade, as all other trades, is much influenced by the quality of the labor and management engaged in the operations, but, because much of this work is accomplished by use of major items of equipment, the ability and skills of competent equipment operators are key to high productivity in excavation and fill operations. An estimator would benefit from maintaining a separate set of productivity rates for each of the operators of major equipment employed by the company so that, when the estimate is priced and operator "X" is to be employed, the estimator will be able to ascertain the productivity rate this individual operator is capable of for the work involved.

## Excavation Work Crews

In order to determine the labor price per unit of excavation work four elements have to be considered:

1. The number of workers that comprise the work crew and their trades
2. The wage rates of the trades involved
3. The probable productivity rate of the crew or the equipment they are operating
4. The productivity time factor (per hour, day or week)

### EXAMPLE

A crew comprising 3 laborers and 0.5 of a labor foreman is observed to take 2 days (16 hours) to excavate by hand 36 cu. yd of heavy soil. (Note that 0.5 of a foreman is included because the foreman is attached to this crew for only 50% of the time.) If the wage rate of laborers is $21.00 per hour and of foremen $24.00 per hour, the unit cost of hand excavation on this project can be obtained thus:

Crew Cost = 3 × $21.00 + 0.5 × $24.00
= $75.00 per crew-hr.

Productivity Rate = 16 hours/36 cu. yd.
= 0.44 crew-hr. per cu. yd.

Unit Cost = 0.44 × $75.00
= $33.00 per cu. yd.

If the same wage rates and productivity rate are expected to prevail on a future project, then hand excavation on this future project could be priced at the same $33.00. But because wage rates and productivity rates are liable to vary on different jobs, the price will probably have to be modified before it can be used for pricing a future job. The unit cost in this example is calculated in two steps so that changes in wage rates can be assessed separately from productivity variations.

If, for example, laborers' wages rise to $22.68 and a foreman's to $25.92, the crew cost would become:

Crew Cost        = 3 × $22.68 + 0.5 × $25.92
                 = $81.00 per crew-hr.

This represents an 8.0% increase in the crew cost. Then if the productivity is expected to decline by 10% due to some changed conditions the new productivity rate would become:

Productivity Rate  = 0.44 × 1.10
                   = 0.484 crew-hr. per cu. yd.

And, finally, the revised unit price would be:

Unit Price        = 0.484 × $81.00
                  = $39.20 per cu. yd.

This new unit price of $39.20 represents an overall increase of 18.8% over the original unit cost ($33.00) calculated at the beginning of this example.

---

Note that a different unit price would be obtained if simply 18.0% was added to the original unit rate (8.0% for the wage increase plus 10% for the productivity decline). Because these rate changes apply to different components of the unit price, changes in productivity should be dealt with separately from wage rate changes for most accuracy.

Tracking productivity rates separately from prices also has the advantage of "transportability"—the same productivity rates can be used in different localities, only the wage rates have to be adjusted to reflect pay rates that differ from place to place.

### Productivity Rates

Where contractors do not have their own productivity records available to use in estimating future projects, they can turn to publications such as *Means Building Construction Cost Data* that provide crew productivity rates.* Some production rate information is also available from equipment manufacturers. Figure 10.3 shows the productivity of hydraulic excavators from information published by Caterpillar Inc. All published productivity rates should be used with caution because they generally reflect production in ideal conditions using new equipment. But, when there is no alternative, the rates given do provide a basis for making an assessment of probable production rates.

However, there is no substitute for a good historic database of the costs established on a contractor's own projects, as described in Chapter 8. This database can be compiled from observations of excavation work made by site supervisors, project coordinators or estimators. An example of the data recorded and how the data is processed is shown on Figure 10.4. On this example the backhoe is taken to be owned by the contractor, and an equipment item number is used to identify the specific unit of equipment used on the work. The estimator could use this identification number to determine the make, age and condition of the equipment if necessary. Alternatively, this equipment information could be recorded on the data record.

A summary chart of historic productivity rates can be prepared from these data records. This chart lists excavation operations, shows the typical crew constituents for each operation and indicates its historic productivity range. A sample chart is shown on Figure 10.5 (page 219). The productivity rates on this chart are used in pricing the examples that follow.

---

*RSMeans (Kingston, MA) publishes a variety of cost data media.

| Estimated Bucket Payload Loose Cubic Yards | Hourly Production | |
|---|---|---|
| | Loose cu. yd./hr. (100% Efficiency) | Loose m³/hr. (100% Efficiency) |
| 0.25  (0.19 m³) | 37 to 67 | 28 to 51 |
| 0.50  (0.38 m³) | 75 to 135 | 57 to 103 |
| 0.75  (0.57 m³) | 90 to 202 | 69 to 154 |
| 1.00  (0.76 m³) | 120 to 270 | 92 to 206 |
| 1.25  (0.96 m³) | 150 to 300 | 115 to 229 |
| 1.50  (1.15 m³) | 154 to 360 | 118 to 275 |
| 1.75  (1.34 m³) | 180 to 420 | 138 to 321 |

*Source:* Caterpillar Inc.

**Figure 10.3**   Productivity of Hydraulic Excavators

---

**PRODUCTIVITY RECORD**

DATE: 22-Jul-03

| | |
|---|---|
| JOB NUMBER: | #94128 |
| PROJECT: | Acme Office Building |
| JOB SUPERINTENDENT: | J. J. Doe |
| LOCATION: | 123 XYZ Street |
| | Seattle, Washington |
| DATE OF WORK: | 20-22 Jul 03        (4 days) |
| WEATHER CONDS: | Mostly cloudy, sunny periods; rain one day |
| AVE. TEMP: | 75°F |

|  | NAME | TRADE |
|---|---|---|
| CREW: | John A. Jones | Operator |
| | William B. Smith | Laborer |

| | |
|---|---|
| EQUIPMENT: | 3/4 CY Backhoe     (Owned) |
| UNIT No: | B9133 |
| WORK ITEM: | Excav. Utility Trench |
| WORK QUANTITY: | 1562.00 cu. yd. |
| WORK DURATION: | 30.00 hr. |
| PRODUCTIVITY RATE: | 52.07 cu. yd./hr. |
| OTHER COMMENTS: | |

BY: J. J. Doe

**Figure 10.4**   Sample Data Record

| OPERATION | EQUIPMENT | CREW | OUTPUT | |
|---|---|---|---|---|
| 1. Hand Excavation<br>—Light to Medium Soil<br>—Heavy Soil to Rock | Site tools only<br>Site tools only | 0.5 Foreman<br>3.0 Laborers<br>0.5 Foreman<br>3.0 Laborers | 3–5 cu. yd./hr.<br><br>0.5–4 cu. yd./hr. | 2–4 m³/hr.<br><br>0.4–3 m³/hr. |
| 2. Hand Trimming | Site tools only | 0.5 Foreman<br>3.0 Laborers | 70–200 sq. ft./hr. | 6.5–18.5 m²/hr. |
| 3. Strip Topsoil<br>—Lot Size Areas<br>—Small Areas | Bobcat<br>105-HP dozer | 1.0 Operators<br>1.0 Operators | 20–50 cu. yd./hr.<br>40–70 cu. yd./hr. | 15–38 m³/hr.<br>31–54 m³/hr. |
| 4. Excavate to Reduce Levels—Small Areas | 105-HP dozer | 1.0 Operators | 40–70 cu. yd./hr. | 31–54 m³/hr. |
| 5. Site Grading | Motor Grader | 1.0 Operators | 100–250 cu. yd./hr. | 9–23 m²/hr. |
| 6. Basement Excavations | 1.5-CY loader<br>*Trucks | 1.0 Operators<br>* Drivers<br>0.5 Laborer | 50–150 cu. yd./hr. | 38–115 m³/hr. |
| 7. Trench Excavations | 0.75-CY backhoe | 1.0 Operators<br>0.5 Laborer | 15–75 cu. yd./hr. | 11–57 m³/hr. |
| 8. Excavate Pits and Sumps | 0.50-CY backhoe | 1.0 Operators<br>0.5 Laborers | 12–30 cu. yd./hr. | 9–23 m³/hr. |
| 9. Spread and Compact Backfill—Basements | 105-HP dozer<br>Roller | 1.0 F/M Operator<br>1.0 Operators<br>3.0 Laborers | 30–100 cu. yd./hr. | 23–76 m³/hr. |
| —Footings | Bobcat<br>Roller | 1.0 F/M Operator<br>1.0 Operators<br>3.0 Laborers | 20–60 cu. yd./hr. | 15–46 m³/hr. |
| 10. Spread and Compact Gravel 6" Thick | Bobcat<br>Roller | 1.0 F/M Operator<br>1.0 Operators<br>3.0 Laborers | 10–30 cu. yd./hr. | 8–23 m³/hr. |
| 11. Remove Surplus Excavated Material from site | 1.5-CY loader<br>*Trucks | 1.0 Operators<br>* Drivers<br>1.0 Laborer | 100–200 cu. yd./hr. | 76–153 m³/hr. |
| 12. Import Fill from Pit to Site | 1.5-CY loader<br>*Trucks | 1.0 Operators<br>* Drivers<br>1.0 Laborer | 50–100 cu. yd./hr. | 38–76 m³/hr. |
| 13. 6" Diameter Drain Tile | Site tools only | 0.3 Foreman<br>1.0 Skilled worker<br>1.0 Laborer | 25–35 ft./hr. | 7–11 m/hr. |
| 14. Place Drain Gravel, No Compaction | Bobcat | 1.0 Operators<br>1.0 Laborer | 20–50 cu. yd./hr. | 15–38 m³/hr. |

*Calculate number of trucks required.

**Figure 10.5**  Excavation Work—Productivity Rates

## Excavation Materials

Materials required in the excavation and backfill operations mostly comprise gravels and fill materials that are required to meet the designer's specifications for use in backfilling and providing base courses under slabs and paving. Material prices for backfill have two components: the price of supplying material to the site and the price of spreading and compacting the material. The estimator's task is to establish the "supply price" of these materials, which is the full price per takeoff unit, usually bank measure, to supply the materials to the site and unload them. The cost of spreading and compacting these materials after they have been unloaded is estimated separately on the excavation and backfill recap.

Whenever it is possible, the contractor will use the material excavated at the site for backfill requirements and, thus, eliminate the need to pay for importing outside materials for this purpose. For instance, if a basement is to be excavated and backfilled, the excavator will leave sufficient excavated material next to the basement so that, when the basement is constructed, this material can be used to backfill around the finished basement as required. In this case there will be just two items to price:

1. Excavation of the basement (including the cost of removing surplus material)
2. Backfill around the basement using the excavated material

However, where it is not possible to use the excavated material for backfill because it is not suitable or it does not meet specification requirements, a third item has to be priced on the estimate:

3. Import basement backfill material

This item may be priced in two ways: a quote may be obtained from a supplier who offers a price to deliver to the site material that meets specifications or the estimator can calculate the cost of obtaining suitable material from a pit or quarry and transporting it to the site including pit royalty charges and any additional expenses for grading or processing the material to meet specifications. An example of pricing the supply of gravel by this second method follows.

### Swell and Compaction Factors

Before we can calculate a price for supplying gravel, we must consider the effect of material compaction because, no matter which method of pricing the supply of fill material is used, the estimator has to account for the swell or compaction factors applicable to that material. Swell and compaction factors are discussed in Chapter 4 where the measurement of excavation work was addressed. However, these factors are not applied at the time of the takeoff since bank measure excavation and backfill quantities are not modified when they are measured. So now the necessary adjustments have to be made for swell and compaction.

Recall that when soil or gravel is excavated, the resulting material is less dense than it was in the ground before it was disturbed. This same material now occupies more volume, a concept called the *swell factor*. Similarly, backfill material that is compacted will be more dense than before it is compacted so, to fill a certain volume with compacted material, a larger volume of loose material will be required. The *compaction factor* is an allowance for the extra material required to fill volumes with compacted material rather than loose material.

A list of gravels and materials commonly used in backfill operations together with their compaction factors is shown as Figure 10.2.

When a supplier quotes a price of $12.00 per cu. yd. (loose) for wet sand, either the price or the bank measure takeoff quantity has to be increased by a compaction factor that, according to Figure 10.2, is approximately 13% in this case. In addition to the extra material required for compaction, there will also be some wastage of material in the operations, so it is convenient to also add a wastage factor to the compaction factor. In this example wastage might amount to a further 10%, which gives a combined total of 23% to be added for wastage and compaction of the wet sand. Where a quantity of 1000 cu. yd. (bank measure) of sand bedding had been taken off, the supply of material could be priced by either of the following two methods:

a.  Supply Sand—1000 cu. yd. @ $14.76 = $14,760.

b.  Supply Sand—1230 cu. yd. @ $12.00 = $14,760.

Some estimators prefer to use method (a) because it leaves the takeoff quantity unchanged. Others prefer method (b) because it reflects the actual quantity of sand that needs to be ordered; either method is acceptable.

Where the supply of fill material is quoted as a price per ton, the estimator has to use the unit weight of the compacted material to determine the price per cu. yd. bank measure. For instance, if a price of $10.00 per ton is offered for the supply of pit run gravel, using the weight of compacted pit run from Figure 10.2, the price per cu. yd. can be calculated thus:

$$\text{Price per cu. yd.} = \text{Price per ton} \times \frac{\text{weight of compacted cu. yd. (lbs)}}{2000 \text{ lbs per ton}}$$

So, in this case:

$$\text{Price per cu. yd. (bank measure)} = \$10.00 \times \frac{3650}{2000}$$

$$= \$18.25$$

If a price per cu. yd. of loose material is required, then, of course, the weight of a loose cubic yard is used in the preceding formula.

### Calculating Trucking Requirements

In the examples that follow, the estimator has to determine the optimum number of trucks required to transport excavated materials and gravels. A simple formula can be used for this calculation based on the premise that it is desirable to have sufficient trucking capacity to ensure that the excavation equipment is able to operate continuously and not have to waste time waiting for trucks. Clearly, three trucks will be required if it takes 10 minutes to load a truck and 20 minutes for that truck to unload and return for another load, because, while the first truck is away, two other trucks can be loaded. From this intuitive analysis, the following expression is obtained:

$$\text{Number of Trucks Required} = \frac{\text{Unloading Time}}{\text{Loading Time}} + 1$$

where

Unloading Time = Round-trip Travel Time + Time to Off-load the Truck

and

$$\text{Loading Time} = \frac{\text{Truck Capacity}}{\text{Loader Output}}$$

and

$$\text{Truck Capacity (bank measure)} = \frac{\text{Capacity (Loose)}}{1 + \text{Swell Factor}}$$

Note that the number of trucks obtained from this calculation should always be rounded up no matter how small the decimal. This is because to round down, even a small fraction may result in a shortage of truck capacity. Most estimators consider it better to have more rather than less capacity so that the excavator is kept occupied.

Also, this method of calculating the number trucks required is based on there being a sufficient amount of material to justify the use of a fleet of trucks. Where only small quantities of material are to be transported, one or two trucks may be all that is needed to move the entire quantity of excavated material regardless of the output of the excavator and the other variables. For example, if there is a total of just 20 cu. yd (bank measure) of excavated material to be removed from a site, and the capacity of trucks is 10.5 cu. yd. (bank measure), then, clearly, two trucks are all that is needed. The estimator merely has to calculate the time the trucks are required for given the output of the excavator, the speed of the trucks and the distance to be traveled by the trucks.

### EXAMPLE: GRAVEL SUPPLY PRICING

Calculate a price of obtaining gravel from a pit located 16 miles from the site where there is a pit royalty of $2.50 per cu. yd.; using a 1.5 cu. yd. track loader at the pit to load the gravel at a rate of 50 cu. yd. (bank-measure) per hour and 12 cu. yd. dump trucks to transport the gravel to the site. The loader and dump trucks are priced at the rates stated in Figure 10.1, and the labor crew for this operation consists of one equipment operator at $30.00 per hour, two laborers at $21.00 per hour and truck drivers at $22.00 per hour. In this case the dump trucks travel at an average speed of 20 miles per hour, the gravel has a swell factor of 12% and 5 minutes is required to off-load the truck.

Given this data, the first step in the calculation is to determine how many trucks are required for the operation. So, making use of the formula described above:

$$\text{Number of Trucks Required} = \frac{\text{Unloading Time}}{\text{Loading Time}} + 1$$

where, in this case:

$$\text{Unloading Time} = \frac{(2 \times 16 \text{ miles})}{20 \text{ miles per hour}} \times \frac{60 \text{ mins}}{1\text{-hour}} + 5 \text{ mins.}$$

$$= 96 + 5$$

$$= 101 \text{ minutes}$$

$$\text{Truck Capacity} = \frac{12 \text{ cu. yd. of loose material}}{1.12}$$

$$= 10.71 \text{ cu. yd. (bank measure)}$$

$$\text{Loading Time} = \frac{(10.71) \times 60 \text{ minutes}}{50 \text{ cu. yd./hr.}}$$

$$= 12.85 \text{ minutes}$$

$$\text{Number of Trucks} = \frac{101}{12.85} + 1$$

$$= 7.86 + 1$$

$$= 8.86$$

Therefore, nine trucks are required.

|  |  | Labor | Equipment |
|---|---|---|---|
| Gravel Supply Price |  | $ | $ |
| Track loader | ($425.00/8 hr.) | — | 53.13 |
| Operator |  | 30.00 | — |
| Trucks | 9 × ($275.00/8) | — | 309.38 |
| Drivers | 9 × $22.00 | 198.00 | — |
| Laborers | 2 × $21.00 | 42.00 | — |
|  |  | 270.00/hr. | 362.50/hr. |
| Price per cu. yd. |  |  |  |
| (Divide by output of 50 cu. yd./hr.) | = | 5.40/cu. yd. | 7.25/cu. yd. |

Therefore, the total price for gravel = $5.40 (labor + $7.25 (equipment) + $2.50 (royalty)

= $15.15 per cu. yd. (bank measure)

## Excavation and Backfill Recap and Pricing Notes Example 1—House

The page number of Figure 10.6, "Excavation and Backfill Recap," is designated as 1 of 14 because this is the first sheet of the entire recap for the house. Sheets 2, 3 and 4 of 14 for the concrete work are found as Figure 11.9; sheets 5 through 14 of 14 are found as Figures 12.5 through 12.9 where the rest of the work is priced.

1. Excavation and backfill items are listed on the recap generally in the same order they appear in the takeoff, but note that the item "Gravel under slab-on-grade," which was included in the Concrete Work section of the takeoff, is listed on this recap. This is because gravel is part of the excavation and backfill trade. Large takeoffs will often have items of different trades mixed together throughout their length, so, each time a trade is recapped, the estimator has to comb the entire takeoff for all the items of that particular trade.

2. A precise check-off system is required to indicate which items have been recapped and thus make it clear which items are yet to be considered. The accuracy of a takeoff can be completely undermined if items that were measured do not get onto the recap for pricing. Most estimators check off the takeoff items as they transfer them to the recap, then check them a second time to ensure that all items have been recapped.

3. For each of the items listed on the recap, there can be an equipment price, a labor price and, in some cases, also a material price. It is useful to be able to obtain separate values for the total labor, total material, total equipment and combined totals included in an estimate; the use of multiple column stationery can achieve this.

4. Labor, materials and equipment prices are entered in separate columns to provide information for the analysis of estimates. This also allows the estimator to calculate "add-ons" for each of these cost categories separately. For example, payroll additive is calculated as a percentage of labor, whereas waste factors and some taxes are often calculated as a percentage of material

**PRICING SHEET**

JOB: HOUSE EXAMPLE  DATE: ....................
ESTIMATED: A.B.F.

Page No. 1 of 14

| No. | DESCRIPTION | QUANTITY | UNIT | UNIT PRICE | LABOR | UNIT PRICE | MATERIALS | UNIT PRICE | EQUIPMENT | TOTAL |
|---|---|---|---|---|---|---|---|---|---|---|
| | EXCAVATION & BACKFILL | | | | | | | | | |
| 1. | STRIP TOPSOIL | 93 | CY | 1.50 | 140 | — | — | 1.75 | 163 | 303 |
| 2. | EXCAV. BASEMENT | 448 | CY | 2.14 | 959 | — | — | 3.18 | 1425 | 2384 |
| 3. | HAND EXCAV. | 2 | CY | 37.50 | 75 | — | — | — | — | 75 |
| 4. | 6" DIA. DRAIN TILE | 144 | LF | 1.87 | 269 | — | — | — | — | 269 |
| 5. | DRAIN GRAVEL | 21 | CY | 2.55 | 54 | — | — | 1.75 | 37 | 91 |
| 6. | BACKFILL BASEMENT | 157 | CY | 3.10 | 487 | — | — | 3.13 | 491 | 978 |
| 7. | GRAVEL UNDER S.O.G. | 18 | CY | 8.27 | 149 | — | — | 5.25 | 95 | 244 |
| | MATERIALS | | | | | | | | | |
| 8. | 6" DIA. DRAIN TILE | 158 | LF | — | — | 2.00 | 316 | — | — | 316 |
| 9. | DRAIN GRAVEL | 24 | CY | — | — | 20.00 | 480 | — | — | 480 |
| 10. | PIT RUN GRAVEL | 23 | CY | — | — | 24.28 | 561 | — | — | 561 |
| | TRANSPORTATION | | | | | | | | | |
| 11. | TRANSPORT EQUIPT. | ALLOW | | — | — | — | — | — | 150 | 150 |
| | TOTAL EXCAVATION TO SUMMARY | | | | 2133 | | 1357 | | 2361 | 5851 |

**Figure 10.6** House Example—Excavation and Backfill Recap

prices. Here a separate list of materials is included so that individual waste factors can be applied as detailed in the following.

5.  The quantity of drain tile has been increased by 10% to allow for wastage.

6.  The quantities of drain gravel and pit-run gravel materials are increased to allow for compaction and waste:

|  | *Net Quantity* | *Factor* | *Gross Quantity* |
|---|---|---|---|
| Drain Gravel | 21 cu. yd. | 15% | 24 cu. yd. |
| 3" Pit-Run Gravel | 18 cu. yd. | 25% | 23 cu. yd. |

The compaction factor on drain gravel is less because this material is not compacted as much as the gravel under slab-on-grade.

7.  Before excavation work can be priced, equipment prices and wage rates for the project have to be established. In this example the equipment prices shown in Figure 10.1 are used and the wage rates are as follows:

*Basic Hourly Wage*

| Equipment Operator | $30.00 |
|---|---|
| Truck Driver | $22.00 |
| Laborer | $21.00 |
| Labor Foreman | $24.00 |

8.  Productivity rates: In all the price calculations for this project, the productivity rates selected tend toward the low end of the scale because the relatively small quantities of each item prohibits higher rates from being attained.

9.

| Strip Topsoil Prices: | | *Labor* $ | *Equipment* $ |
|---|---|---|---|
| Operator | | 30.00 | — |
| Bobcat | $280.00/8 | — | 35.00 |
| | Per hr. | 30.00 | 35.00 |
| Per cu. yd. (Productivity: 20 cu. yd./hr.) | | 1.50 | 1.75 |

10. Before the basement excavation prices can be calculated, the number of trucks required for this operation has to be determined based on the following data:

| Truck capacity: | 9 cu. yd. bank measure |
|---|---|
| Distance to dump: | 3 miles one way |
| Time to off-load truck: | 5 minutes |
| Average truck speed: | 20 miles per hr. |
| Output of excavator: | 60 cu. yd./hr. |

$$\text{Number of trucks required} = \frac{\text{Unloading Time}}{\text{Loading Time}} + 1$$

where Unloading Time = Travel Time + 5 minutes to off-load

$$= \frac{(6 \text{ miles}) \times 60}{20 \text{ miles per hr.}} + 5$$

$$= 18 + 5$$

$$= 23 \text{ minutes}$$

and Loading Time $= \dfrac{\text{Truck Capacity}}{\text{Excavator Output}}$

$= \dfrac{9 \text{ cu. yd.} \times 60}{60 \text{ cu. yd./hr.}}$

$= 9$ minutes

Therefore, no. of trucks $= \dfrac{23}{9} + 1$

$= 3.6$

So allow for four trucks.

**11.**

| Excavate basement prices | | Labor $ | Equipment $ |
|---|---|---|---|
| Track Loader (1.5 cu. yd.) | $425.00/8 | — | 53.13 |
| Operator | 1.0 × $30.00 | 30.00 | — |
| Laborer | 0.5 × $21.00 | 10.50 | — |
| Dump trucks | 4.0 × $275.00/8 | — | 137.50 |
| Drivers | 4.0 x $22.00 | 88.00 | — |
| | Per hr. | 128.50 | 190.63 |
| Price per cu. yd. (Productivity: 60 cu. yd./hr.) | | 2.14 | 3.18 |

**12.**

| Hand Excavation prices | | Labor $ | Equipment $ |
|---|---|---|---|
| Foreman | 0.5 × $24.00 | 12.00 | — |
| Laborers | 3.0 × $21.00 | 63.00 | — |
| | Per hr. | 75.00 | — |
| Price per cu. yd. (Productivity: 2 cu. yd./hr.) | | 37.50 | — |

**13.**

| 6" Drain tile prices | | Labor $ | Equipment $ |
|---|---|---|---|
| Foreman | 0.5 × $24.00 | 12.00 | — |
| Skilled worker | 1.0 × $23.00 | 23.00 | — |
| Laborer | 1.0 × $21.00 | 21.00 | — |
| | Per hr. | 56.00 | — |
| Price per cu. yd. (Productivity: 30 cu. yd./hr.) | | 1.87 | — |

**14.**

| Drain gravel prices | | Labor $ | Equipment $ |
|---|---|---|---|
| Bobcat | 1.0 × $280.00/8 | — | 35.00 |
| Operator | 1.0 × $30.00 | 30.00 | — |
| Laborer | 1.0 × $21.00 | 21.00 | — |
| | Per hr. | 51.00 | 35.00 |
| Price per cu. yd. (Productivity: 20 cu. yd./hr.) | | 2.55 | 1.75 |

**15.**

| Backfill basement prices | | Labor $ | Equipment $ |
|---|---|---|---|
| 105-HP dozer | 1.0 × $650.00/8 | — | 81.25 |
| 3-Ton roller | 1.0 × $350.00/8 | — | 43.75 |
| Foreman/operator | 1.0 × $31.00 | 31.00 | — |
| Operator | 1.0 × $30.00 | 30.00 | — |
| Laborers | 3.0 × $21.00 | 63.00 | — |
| | Per hr. | 124.00 | 125.00 |
| Price per cu. yd. (Productivity: 40 cu. yd./hr.) | | 3.10 | 3.13 |

**16.**

| Gravel under S.O.G. prices | | Labor $ | Equipment $ |
|---|---|---|---|
| Bobcat | 1.0 × $280.00/8 | — | 35.00 |
| 3-Ton roller | 1.0 × $350.00/8 | — | 43.75 |
| Foreman/operator | 1.0 × $31.00 | 31.00 | — |
| Operator | 1.0 × $30.00 | 30.00 | — |
| Laborers | 3.0 × $21.00 | 63.00 | — |
| | Per hr. | 124.00 | 78.75 |
| Price per cu. yd. (Productivity: 15 cu. yd./hr.) | | 8.27 | 5.25 |

17. Drain gravel material price: $20.00 per cu. yd. (loose) is quoted by a supplier to deliver gravel to the site. This price can be used without adjustment as the quantity has been increased to allow for both compaction and waste.

18. 3" Pit-run gravel price: If the price of $15.00 per ton is offered to supply and deliver this material to the site, the price per cu. yd. (loose) is calculated using the weight per cubic yard of uncompacted gravel. Thus:

$$\text{Price per cu. yd.} = \text{Price per ton} \times \frac{\text{weight of loose cu. yd. (lbs)}}{2000 \text{ lbs/ton}}$$

$$= \$15.00 \times \frac{3250}{2000}$$

$$= \$24.38$$

19. Another item is added at the end of the recap for pricing the transportation cost for the major equipment used in the excavation work.

20. Note that all extensions (the amounts obtained when the unit prices are multiplied by the quantities) are rounded to the nearest whole number so there are no cents shown in any of the price totals.

## Excavation and Backfill Pricing Notes Example 2—Office/Warehouse Building

In practice, when using the Precision Estimating software, items will automatically be priced using the database prices at the time of the takeoff. Then all the estimator has to do at this stage is ensure that the prices used are appropriate for this particular project. Labor and equipment prices would be adjusted to reflect the difficulty of the work involved and firm price quotes should be obtained for the materials required for the job.

Since the database we are using here is incomplete, we will develop prices for each item on the spreadsheet using materials quotes, labor and equipment productivity rates, the equipment prices shown on Figure 10.1 and the following labor wage rates:

*Basic Hourly Wage*

| | |
|---|---|
| Equipment operator | $30.00 |
| Foreman/operator | $31.00 |
| Truck driver | $22.00 |
| Laborer | $21.00 |
| Labor foreman | $24.00 |

Figure 10.7 shows the estimate spreadsheet based on the prices calculated as follows. Unit prices for labor, materials and equipment are entered onto the spreadsheet by typing these prices into the unit price columns against each item of work. If there are existing prices in these fields, the required price can be typed over the previous price. Note that the material amounts on the spreadsheet for gravels includes the waste factors that were set up in the database.

| Group | Phase | Description | Takeoff Quantity | Labor Cost/Unit | Labor Amount | Material Price | Material Amount | Sub Amount | Equip Price | Equip Amount | Other |
|---|---|---|---|---|---|---|---|---|---|---|---|
| 2000.000 | | **SITEWORK** | | | | | | | | | |
| | 2144.000 | Gravel @ Slab | | | | | | | | | |
| | | Gravel under SOG | 402.90 CY | 6.20 /CY | 2,498 | 24.38 /CY | 11,787 | - | 3.94 /CY | 1,587 | - |
| | 2201.000 | Earthwk: Remove Topsoil | | | | | | | | | |
| | | Remove Topsoil | 1,087.04 CY | 0.60 /CY | 652 | - | - | - | 1.63 /CY | 1,772 | - |
| | 2205.000 | Earthwk: Excav Trench | | | | | | | | | |
| | | Excavate Trench | 248.82 CY | 0.81 /CY | 202 | - | - | - | 1.68 /CY | 418 | - |
| | 2210.000 | Earthwk: Excav Pits | | | | | | | | | |
| | | Excavate Pits | 13.22 CY | 1.62 /CY | 21 | - | - | - | 1.93 /CY | 26 | - |
| | 2220.000 | Earthwk: Bulk Cut | | | | | | | | | |
| | | Bulk Cut | 149.67 CY | 0.75 /CY | 112 | - | - | - | 2.03 /CY | 304 | - |
| | 2220.400 | Earthwk: Grade & Trim | | | | | | | | | |
| | | Grade & Trim Bottoms of Excav | 2,944.00 sf | 0.50 /sf | 1,472 | - | - | - | /sf | | - |
| | 2221.500 | Earthwk: Gravel Fill | | | | | | | | | |
| | | Gravel Fill | 504.26 CY | 3.85 /CY | 1,941 | 24.38 /CY | 13,523 | - | 5.58 /CY | 2,814 | - |
| | 2222.000 | Earthwk: Dispose Surplus | | | | | | | | | |
| | | Dispose Surplus | 211.00 CY | 2.61 /CY | 551 | - | - | - | 3.79 /CY | 800 | - |
| | 2225.000 | Backfill Trenches | | | | | | | | | |
| | | Bakfill Trenches | 190.82 CY | 6.20 /CY | 1,183 | - | - | - | 3.94 /CY | 752 | - |
| | 2230.000 | Backfill Pits | | | | | | | | | |
| | | Bakfill Pits | 9.89 CY | 8.27 /CY | 82 | - | - | - | 5.25 /CY | 52 | - |
| 3000.000 | | **CONCRETE** | | | | | | | | | |
| | 3110.100 | Forms: Footings | | | | | | | | | |
| | | Pad Footing Forms | 143.20 sf | 20.00 /sf | 2,864 | 0.78 /sf | 117 | - | - | - | - |
| | | Continuous Footing Forms | 329.41 sf | 20.00 /sf | 6,588 | 0.78 /sf | 270 | - | - | - | - |

**Figure 10.7** Estimate Spreadsheet for Pricing Notes Example 2

**1.**

| | | Labor $ | Equipment $ |
|---|---|---|---|
| Gravel under S.O.G. prices | | | |
| Bobcat | 1.0 × $280.00/8 | — | 35.00 |
| 3-Ton roller | 1.0 × $350.00/8 | — | 43.75 |
| Foreman/operator | 1.0 × $31.00 | 31.00 | — |
| Operator | 1.0 × $30.00 | 30.00 | — |
| Laborers | 3.0 × $21.00 | 63.00 | — |
| | Per hr. | 124.00 | 78.75 |
| Price per cu. yd. (Productivity: 20 cu. yd./hr.) | | 6.20 | 3.94 |

Pit-run gravel material will be used for this item and for granular fill. This material is quoted at a price of $15.00 per ton delivered to the site.

$$\text{Price per cu. yd.} = \frac{\text{Price per ton} \times \text{weight of loose cu. yd. (lbs)}}{2000 \text{ lbs/ton}}$$

$$= \frac{\$15.00 \times 3250}{2000}$$

$$= \$24.38 \text{ per cu. yd.}$$

**2.**

| | | Labor $ | Equipment $ |
|---|---|---|---|
| Remove topsoil prices | | | |
| 105-HP dozer | 1.0 × $650.00/8 | — | 81.25 |
| Operator | 1.0 × $30.00 | 30.00 | — |
| | Per hr. | 30.00 | 81.25 |
| Price per cu. yd. (Productivity: 50 cu. yd./hr.) | | 0.60 | 1.63 |

**3.**

| | | Labor $ | Equipment $ |
|---|---|---|---|
| Excavate trench prices | | | |
| 0.75 cu. yd. Backhoe | 1.0 × $670.00/8 | — | 83.75 |
| Operator | 1.0 × $30.00 | 30.00 | — |
| Laborer | 0.5 × $21.00 | 10.50 | — |
| | Per hr. | 40.50 | 83.75 |
| Price per cu. yd. (Productivity: 50 cu. yd./hr.) | | 0.81 | 1.68 |

**4.**

| | | Labor $ | Equipment $ |
|---|---|---|---|
| Excavate pits prices | | | |
| 0.5-cu. yd. backhoe | 1.0 × $385.00/8 | — | 48.13 |
| Operator | 1.0 × $30.00 | 30.00 | — |
| Laborer | 0.5 × $21.00 | 10.50 | — |
| | Per hr. | 40.50 | 48.13 |
| Price per cu. yd. (Productivity: 25 cu. yd./hr.) | | 1.62 | 1.93 |

| 5. | | | Labor $ | Equipment $ |
|---|---|---|---|---|
| | Bulk cut prices | | | |
| | 105-HP Dozer | 1.0 × $650.00/8 | — | 81.25 |
| | Operator | 1.0 × $30.00 | 30.00 | — |
| | | Per hr. | 30.00 | 81.25 |
| | Price per cu. yd. (Productivity: 40 cu. yd./hr.) | | .75 | 2.03 |

| 6. | | | Labor $ | Equipment $ |
|---|---|---|---|---|
| | Grade & trim bottom of excavation prices | | | |
| | Foreman | 0.5 × $24.00 | 12.00 | — |
| | Laborers | 3.0 × $21.00 | 63.00 | — |
| | | Per hr. | 75.00 | — |
| | Price per SF (Productivity: 150 SF/hr.) | | 0.50 | — |

| 7. | | | Labor $ | Equipment $ |
|---|---|---|---|---|
| | Granular fill prices | | | |
| | 105-HP dozer | 1.0 × $650.00/8 | — | 81.25 |
| | Motor grader | 1.0 × $785.00/8 | — | 98.13 |
| | 3-Ton roller | 1.0 × $350.00/8 | — | 43.75 |
| | Foreman/operator | 1.0 × $31.00 | 31.00 | — |
| | Operators | 2.0 × $30.00 | 60.00 | — |
| | Laborers | 3.0 × $21.00 | 63.00 | — |
| | | Per hr. | 154.00 | 223.13 |
| | Price per cu. yd. (Productivity: 40 cu. yd./hr.) | | 3.85 | 5.58 |

8. Because trucking is required to remove surplus excavated material, the number of trucks required for this operation is calculated from the following data:

| | |
|---|---|
| Truck capacity: | 9 cu. yd. (bank measure) |
| Distance to dump: | 5 miles one-way |
| Time to off-load truck: | 5 minutes |
| Average truck speed: | 15 miles per hr. |
| Output of loader: | 100 cu. yd. per hr. |

$$\text{Number of trucks} = \frac{\text{Unloading Time}}{\text{Loading Time}} + 1$$

$$\text{Unloading Time} = \text{Travel Time} + 5 \text{ minutes to off-load}$$

$$= \frac{10 \text{ miles} \times 60}{15 \text{ miles per hr.}} + 5$$

$$= 40 + 5$$

$$= 45 \text{ minutes}$$

$$\text{Loading Time} = \frac{\text{Truck Capacity}}{\text{Excavator Output}}$$

$$= \frac{9 \text{ cu. yd.}}{100 \text{ cu. yd./hr.}} \times 60$$

$$= 5.4 \text{ minutes}$$

$$\text{Number of trucks} = \frac{45}{5.4} + 1$$

$$= 9.33$$

So allow for 10 trucks.

**9.**

|  |  | Labor $ | Equipment $ |
|---|---|---|---|
| Dispose of surplus prices |  |  |  |
| Track loader (1.5 cu. yd.) | $280.00/8 | — | 35.00 |
| Operator | 1.0 × $30.00 | 30.00 | — |
| Laborer | 0.5 × $21.00/8 | 10.50 | — |
| Dump trucks | 10.0 × $275.00/8 | — | 343.75 |
| Drivers | 10.0 × $22.00 | 220.00 | — |
|  | Per hr. | 260.50 | 378.75 |
| Price per cu. yd. (Productivity: 100 cu. yd./hr.) |  | 2.61 | 3.79 |

**10.**

|  |  | Labor $ | Equipment $ |
|---|---|---|---|
| Backfill trenches prices |  |  |  |
| Bobcat | 1.0 × $280.00/8 | — | 35.00 |
| 3-Ton roller | 1.0 × $350.00/8 | — | 43.75 |
| Foreman/operator | 1.0 × $31.00 | 31.00 | — |
| Operator | 1.0 × $30.00 | 30.00 | — |
| Laborers | 3.0 × $21.00 | 63.00 | — |
|  | Per hr. | 124.00 | 78.75 |
| Price per cu. yd. (Productivity: 20 cu. yd./hr.) |  | 6.20 | 3.94 |

**11.**

|  |  | Labor $ | Equipment $ |
|---|---|---|---|
| Backfill pits prices |  |  |  |
| Bobcat | 1.0 × $280.00/8 | — | 35.00 |
| 3-Ton roller | 1.0 × $350.00/8 | — | 43.75 |
| Foreman/operator | 1.0 × $31.00 | 31.00 | — |
| Operator | 1.0 × $30.00 | 30.00 | — |
| Laborers | 3.0 × $21.00 | 63.00 | — |
|  | Per hr. | 124.00 | 78.75 |
| Price per cu. yd. (Productivity: 15 cu. yd./hr.) |  | 8.27 | 5.25 |

## SUMMARY

- There are many subcontractors performing excavation and backfill work. Most of these companies specialize in some aspect of this trade, often because they have equipment specifically for one particular type of work.
- The cost of transporting equipment should be considered when pricing excavation and backfill operations; this should include the cost of loading, unloading and trucking requirements.
- There are four common types of excavation methods:
  1. Site cut and fill operations—On large projects, scrapers, dozers, rollers, water trucks and graders may be used. Smaller jobs may use just a grader or a bob-cat with a small roller
  2. Basement-type excavations—A track backhoe or track loader and dump trucks are often used.
  3. Trench excavations—Trenching machines are available for shallow pipes; rubber-tired tractors and backhoes for small trenches. Larger trenches usually call for track-mounted backhoes.
  4. Pit and sump excavations—This is mostly performed using various sizes of backhoes.
- The productivity of excavation and backfilling operations depends on:
  a. Job factors
     —Type of material excavated or backfilled
     —Moisture content of materials
     —Weather conditions expected
     —Access to and around site
     —Project size and complexity
     —Distance to haul materials for disposal
     —Availability of gravels and fill materials
     —Wage and price levels at the job location
  b. Labor and management factors
     —Quality of supervision
     —Quality of job labor
     —Motivation and morale of workers
     —Type and quality of tools and equipment
     —Experience and records of similar past projects
- In order to determine the labor price per unit of excavation work, four elements have to be considered:
  1. The number of workers that comprise the work crew and their trades
  2. The wage rates of the trades involved
  3. The probable productivity rate of the crew or the equipment they are operating
  4. The productivity time factor (per hour, day or week)
- Productivity rates can be obtained from publications and from equipment manufacturers, but the estimator's best source of information is his or her own record of previous excavation jobs.
- Material prices for backfill have two components:
  1. The price of supplying the material to the site
  2. The price of spreading and compacting the material
- The estimator also has to account for swell factors and compaction factors when pricing excavation and backfill materials.
- Trucking requirements can be assessed using the following formula:

$$\text{Number of Trucks Required} = \frac{\text{Unloading Time}}{\text{Loading Time}} + 1$$

- Recap and pricing excavation and backfill can be done manually or via computer.

## REVIEW QUESTIONS

1. Describe the two methods of pricing equipment transportation expenses.
2. Give two examples of situations in which hand excavation may be required.
3. Describe how different materials excavated and different moisture content of these materials affect the price of excavation work.
4. Calculate the unit cost per cubic yard achieved in the following situation: A crew comprising 4 laborers and 0.5 of a labor foreman is observed to take 3 days (24 hours) to excavate by hand 40 cubic yards of heavy soil. (Note that 0.5 of a foreman is included because the foreman is attached to this crew for only 50% of the time.) The wage rate of laborers is $17.50 per hour and of foremen $19.00 per hour.
5. Based on the output shown in Figure 10.5, how long should it take a motor grader to grade a 10,000 square-foot area when the output is expected to be at the low end of the scale?
6. Calculate the cost per cubic yard (bank measure) of gravel delivered to the site for a price of $12.50 per ton when one cubic yard bank measure of this material weighs 3555 lbs.
7a. Calculate the price of pit-run gravel delivered to the site per cubic yard (bank measure) based on the following data:
    - The pit is located 10 miles from the site.
    - Trucks cost $35.00 per hour including fuel and maintenance; they have 12 cubic yards (loose material) capacity and travel at an average speed of 30 miles per hour.
    - The swell factor for this material is 15%.
    - Trucks take 5 minutes to unload at the site.
    - The loader costs $85.00 per hour and loads material at the pit at the rate of 40 cubic yards per hour.
    - The truck driver's wage is $22.00 per hour and the equipment operator's wage is $30.00 per hour.
7b. Calculate the price of crushed gravel delivered to the site per cubic meter (bank measure) based on the following data:
    - The gravel source is located 15 km from the site.
    - Trucks cost $32.00 per hour including fuel and maintenance; they have 9 cubic meters (loose material) capacity and travel an average speed of 50 km per hour.
    - The swell factor for this material is 12%.
    - Trucks take 4 minutes to unload at the site.
    - The loader costs $88.00 per hour and loads materials at the pit at the rate of 42 cubic meters per hour.
    - The truck driver's wage is $22.00 per hour and the equipment operator's wage is $30.00 per hour.
8. Based on the equipment prices given in Figure 10.1 and the crew described in Figure 10.5, calculate a price per cubic yard to strip topsoil from small areas when the expected productivity is 45 cubic yards per hour.

# 11

# PRICING CONCRETE WORK

## OBJECTIVES

*After reading this chapter and completing the review questions, you should be able to do the following:*

■ Describe the job factors and the labor and management factors that influence the cost of concrete work
■ Identify different methods of placing concrete.
■ Calculate and compare the costs of different concreting methods.
■ Explain how the slump of concrete influences the workability and cost of placing concrete.
■ Describe the factors that should be considered when pricing concrete materials.
■ Calculate the full price of concrete delivered to site from material supplier quotes.
■ Identify the factors to consider when pricing formwork systems.
■ Describe how formwork materials are priced.
■ Calculate the price per unit area of form systems.
■ Describe the factors that influence the cost of reinforcing steel.
■ Explain what influences the productivity of labor installing reinforcing steel.
■ Identify miscellaneous items associated with concrete work and explain how these are priced.
■ Complete the recap and pricing of concrete work using manual methods.
■ Use the Precision Estimating software to price concrete work.

## Cast-in-Place Concrete Work Generally

This chapter concerns the process of pricing the cast-in-place concrete work a general contractor would perform. Precast concrete operations are carried out primarily by specialist contractors and are considered in Chapter 13 together with the work of other subcontractors.

Concreting work encompasses a large array of activities that can be divided into four main categories:

1. Supply and placing *concrete*
2. Construction and removal of *formwork*
3. Supply and placing *reinforcing steel*
4. *Miscellaneous items* associated with concrete work

The nature and the quantity of the work included in the concreting operations has been described and measured in detail in the takeoff process, but the takeoff is unlikely to be neatly partitioned into these categories of concrete, formwork, reinforcing steel and miscellaneous items. So the first task in the pricing process is to recap the takeoff items into these divisions ready for pricing.

There are many components that must be examined in order to be able to price the work involved in each of the categories, so, in each case, we will briefly consider those aspects of the work that have an impact on cost. Then we will proceed to examine the process of pricing this work in terms of the productivity of the crews who perform the work and the cost of the materials required for the work.

## Supply and Placing Concrete

The cost of supplying concrete to a site includes the cost of cement, sand, aggregate, water and equipment to mix and transport the concrete. A concrete batching and mixing plant may be set up at the site of the work on projects where a great number of large-size concrete "pours" will create a demand for almost continuous concrete production, or where concrete is required in a remote area and there is no alternative. Otherwise most projects will use ready-mixed concrete obtained from a local concrete company, which quotes a price to mix and deliver to the site concrete that meets the project specifications.

If it is necessary for the contractor to operate a batch plant at the site, the estimator will have to calculate the quantity and cost of cement, sand, aggregate and water required for each class of concrete measured in the takeoff, together with the cost of equipment to mix this material. To determine the amounts of ingredients required to produce specific concrete strengths needed in the work, the estimator can consult tables that list the weights of materials required per cubic yard of concrete. Then the cost of procuring these materials can be determined from quotations obtained from suppliers who generally offer to provide and deliver the requirements to the site.

In the examples that follow, concrete is assumed to be obtained from ready-mixed concrete suppliers, which simplifies the pricing process but does introduce considerations that have to be addressed regarding ready-mix suppliers. These considerations are examined in the later section, "Concrete Materials."

### Productivity Placing Concrete

Once the concrete has been mixed and delivered to the site, the next cost to consider is the cost of placing the concrete in its required position. The productivity of this operation depends on a large number of factors that we can again classify under job factors and labor and management factors as we did with excavation operations. Many of the same job factors listed under excavation productivity in Chapter 10 also apply to concrete work, and certainly all five of the labor and management factors identified there will impact the productivity of any operation, but the following list of job factors introduces some items specific to the concrete placing operations:

*Job Factors*

1. Method of placing concrete
2. Rate of delivery of ready-mixed concrete
3. Properties of the concrete to be placed
4. Size and shape of concrete structures
5. Amount of rebar in the forms

*Method of Placing Concrete*

The factor that, perhaps, has the most influence on the cost of placing concrete is the method used to convey the concrete from the mixing trucks to its final location. When large continuous concrete "pours" are anticipated on a project, the estimator should first check on the maximum rate of delivery that can be sustained by the ready-mix supplier before deciding on a placing method. Using concrete pumps at an output of 120 cubic yards per hour may be possible, but if concrete can arrive at the site no faster than 50 cubic yards per hour, the decision on the method of placing concrete might be quite different.

There are a number of methods of placing concrete available including:

1. "Pouring" direct from mixing trucks by means of chutes
2. Using hand-operated or powered buggies
3. Using crane and bucket setup
4. Using concrete pumps
5. Using conveyors
6. Using a combination of two or more of these methods

Clearly, if the mixing trucks can get near to where the concrete is required, the most economical method of placing the concrete is to discharge it into position directly from the trucks. The concrete needs to be required at a lower level than the level at which the material is expelled from the truck if we are to use gravity as the motive force, but chutes may be used to increase the distance concrete can be conveyed away from a truck. This, however, has limitations because chutes over 8 feet long can be harmful to the concrete mix by causing segregation of concrete constituents.

The use of concrete buggies has declined significantly in recent years because using cranes or pumps to move concrete is usually less costly. The full price of buggies includes not only the cost of renting or owning the buggies, but also the cost of constructing runways for them to operate on. Buggies also require a large number of personnel to operate them if reasonable production rates are to be maintained in terms of cubic yards placed per hour. As wage rates increase, this labor intensiveness makes them even more expensive.

The use of a crane and bucket arrangement to place concrete is very popular especially on projects like highrise buildings where a central tower crane is employed for a variety of purposes in the construction process. The cost of placing concrete by means of a pump may be less in a straight comparison between "crane and bucket" and pumping, but, so the argument goes, if the crane is in place anyway, it is cheaper to make use of it for placing concrete and avoid the cost of operating a concrete pump. In other words, the expense of maintaining and operating a crane at the site is a "fixed cost" that will be incurred whether or not it is used for placing concrete.

Where a central tower crane is not in place on a site, using pumps to place concrete is usually the most economical alternative. Concrete pumps are readily available on hourly or daily rental terms in most areas of the country. Most contractors prefer to rent concrete pumps with an operator as they are costly items and require much maintenance to remain in good operating condition. Modern pumps are capable of high outputs, which usually means the limiting factor in determining output is not the pump but, rather, the rate at which concrete can be delivered or the amount that the labor crew placing the concrete can handle.

Conveying concrete in certain project situations may be best achieved using a combination of methods. For instance concrete may be raised to a central hopper on a high-level floor with a crane and bucket and then transported a short distance using hand buggies or even a small pump at that location. Sometimes an estimator may calculate the cost of placing concrete based on a certain placement method and then the project manager will use a quite different method on the job. There is nothing inherently wrong with this as the estimate provides a sum of money for an operation and if this work can be performed for less by using different methods than originally planned, then cost savings can be made. However, in order for the estimated price to be most competitive, the estimate has to be based on the most efficient methods, which, as we have previously stated, can only be determined, with any consistency, from frequent discourse between estimators and those that manage operations.

Conveying concrete using belt conveyor systems is only considered where large quantities of concrete have to be placed quickly on a project. This can occur in the construction of such structures as concrete dams that can call for several thousand cubic yards of concrete to be placed in a continuous single "pour." Occasionally other projects may include the placement of a large amount of concrete in one continuous operation, in which case the estimator may wish to consider a number of alternative methods of executing the work to determine, at least theoretically, which offers the lowest costs. An example of such a cost comparison follows.

## EXAMPLE

A highrise building is to be supported on a reinforced concrete "raft" foundation measuring 100 feet by 150 feet. This foundation contains 3,000 cubic yards of concrete to be placed in a single continuous "pour." Concrete is to be brought to the site in concrete mixers that can deliver up to 75 cubic yards of concrete per hour. Three alternative means of conveying the concrete from the trucks to the foundation are to be considered based on the following data:

*Wage Rates and Prices*

| | | |
|---|---|---|
| Laborer | $ 21.00 | per hour |
| Labor foreman | $ 24.00 | per hour |
| Equipment operator | $ 30.00 | per hour |
| 25-Ton mobile cranes | $1,015.00 | per day |
| 5" Concrete pump | $1,250.00 | per day |
| Conveyor system | $1,150.00 | per day |

1. *Crane and Bucket.* Using two mobile cranes with 1.25 cubic yard buckets and a crew consisting of 1 foreman, 12 laborers and 2 equipment operators, it is estimated that this concrete can be placed at the rate of 50 cubic yards per hour.

$$\text{Time required} = \frac{3,000 \text{ cu. yd.}}{50 \text{ cu. yd./hr}} + 3 \text{ hours for start-up and finishing}$$

$$= 63 \text{ hours}$$

| Costs | | Labor $ | Equipment $ |
|---|---|---|---|
| Mobile cranes | 2 × 3 days × $1,015.00 | — | 6,090.00 |
| Foreman | 1 × 63 hrs. × $ 24.00 | 1,512.00 | — |
| Laborers | 12 × 63 hrs. × $ 21.00 | 15,876.00 | — |
| Operators | 2 × 63 hrs. × $ 30.00 | 3,780.00 | — |

| | | Totals: | 21,168.00 | 6,090.00 |
|---|---|---|---|---|
| | | Price per cu. yd. (/3000): | 7.06 | 2.03 |

**2.** *Pumping*. Using two concrete pumps and a crew consisting of 1 foreman, 14 laborers and 2 equipment operators, it is estimated that this concrete can be placed at the rate of 60 cubic yards per hour.

$$\text{Time required} = \frac{3{,}000 \text{ cu. yd.}}{60 \text{ cu. yd./hr.}} + 3 \text{ hours for start-up and finishing}$$

$$= 53 \text{ hours}$$

| Costs | | | Labor $ | Equipment $ |
|---|---|---|---|---|
| Concrete pumps | 2 × 3  days × $1,250.00 | | — | 7,500.00 |
| Foreman | 1 × 53 hrs. × $ | 24.00 | 1,272.00 | — |
| Laborers | 14 × 53 hrs. × $ | 21.00 | 15,582.00 | — |
| Operators | 2 × 53 hrs. × $ | 30.00 | 3180.00 | — |
| | Totals: | | 20,034.00 | 7,500.00 |
| | Price per cu. yd. (/3000): | | 6.68 | 2.50 |

**3.** Using two belt conveyor systems and a crew consisting of 1 foreman, 18 laborers and 2 equipment operators, it is estimated that this concrete can be placed at the rate of 72 cubic yards per hour.

$$\text{Time required} = \frac{3{,}000 \text{ cu. yd.}}{72 \text{ cu. yd./hr.}} + 5 \text{ hours for start-up and finishing}$$

$$= 47 \text{ hours}$$

| Costs | | | Labor $ | Equipment $ |
|---|---|---|---|---|
| Conveyer system | 2 × 2  days × $1,150.00 | | — | 4,600.00 |
| Foreman | 1 × 47 hrs. × $ | 24.00 | 1,128.00 | — |
| Laborers | 18 × 47 hrs. × $ | 21.00 | 17,766.00 | — |
| Operators | 2 × 47 hrs. × $ | 30.00 | 2,820.00 | — |
| | Totals: | | 21,714.00 | 4,600.00 |
| | Price per cu. yd. (/3000): | | 7.24 | 1.53 |

So, on the basis of these calculations, the belt conveyor system of placing the concrete would appear to be the least-cost alternative in this situation. In practice these results would be carefully reviewed and a number of other factors would probably be taken into account before a final decision on which method to use is reached.

Whenever estimators and managers sit down to consider this type of decision, many "what if" questions can be raised. In order to better analyze the costing of these problems, calculations should be entered on a computer spreadsheet so that questions such as the following may be answered immediately at the discussion table.

This speeds up the decision-making process and can lead to better decisions being made.

1. What would be the effect of paying a 10% premium for shift work?
2. What productivity rate of the pumping alternative would take it below the cost of the conveyors?
3. If the number of workers on the crews were changed, how would this affect the overall costs?

### Properties of the Concrete to Be Placed

Some concrete mixes are easier to handle than others. Higher **slump** mixes of concrete that are sloppier and more liquid than low slump mixes will flow more easily and will take less time to consolidate. Consequently less labor should be required to place concrete with a higher slump. Unfortunately some concreting crews have been known to try to improve the workability of concrete that arrives on site by adding a little more water to the mix. This is an unacceptable practice as the addition of extra water to a mix can seriously impair the strength of the concrete. The task of the estimator is to recognize specifications that call for low slump mix concrete, appreciate the effect of this on the productivity of placing operations and make the necessary price adjustments for these.

Estimators should be aware of the use of **superplasticizer** additives to concrete. These ingredients are used to maintain a low water/cement ratio in a concrete mix for maximum strength but produce a high slump concrete mixture that has the good workability required for easy placing. Superplasticizers may be specified for a project, in which case the extra cost of these additives must be included in the concrete price. If they are not specified, analysis of a project might suggest possible net savings from introducing superplasticizers, in which case they would be used if the extra cost of the additives is more than offset by savings in the labor and equipment placing the concrete.

Another somewhat recent development in concrete technology is the use of fiber reinforced concrete (FRC). Randomly distributed fibers are incorporated into the concrete mix with the object of improving the tensile strength of the concrete and increasing the concrete's ability to resist cracking and deformation. The use of FRC reduces the need for some reinforcing steel bars and mesh that are in the concrete to attain these same objectives. Fiber reinforcing may be specified in the contract, which will require the estimator to account for the extra cost of this additive in the concrete prices, or the contractor may identify savings if allowed to substitute fiber reinforced concrete for steel reinforced concrete on the project.

### Size and Shape of Concrete Structures

Generally, large volume concrete "pours" will cost less on a unit price basis than small volume "pours"; thus, columns of greater cross-section will cost less per cubic yard than slender columns, and the cost of placing concrete in long lengths of wall or wider walls will be less than placing in short, thin walls, and suchlike.

Placing concrete continuously will be more economical than interrupting the placing operations to move from one location to another; therefore the unit cost of concrete in strip footings will normally be less than the cost of isolated footing per cubic yard placed.

The size and shape of the concrete structures to be built will also impact the decision previously discussed regarding which method to use in placing concrete. For instance, using a crane and bucket to place concrete where the work involves a large number of columns is usually more efficient than pumping the concrete into place because moving the pump hose from column to column hampers productivity.

### *Amount of Rebar in the Forms*

When placing concrete in columns, walls, pilasters and the like, the rate of placement can be seriously hampered if a great deal of reinforcing steel has been installed in the forms for these structures. A concrete mix that gives good workability (high slump) is called for so that concrete is able to flow into forms that are congested with large amounts of rebar. This condition also makes it difficult for workers to insert vibrating rods needed to consolidate and avoid air pockets in the concrete placed.

### Concreting Productivity Rates

Most contractors maintain a database of historic unit prices or productivity rates for concreting operations gathered on different projects in the past. The information contained in the database may be compiled from previous project cost reports (see Chapter 8) or from site observations of the work in progress as we discussed with regard to excavation work (see Chapter 10). Figure 11.1 shows an example of a concreting productivity database that we will use for pricing the following examples. As with the excavation productivity chart used in the Chapter 10, a range of historic production rates is given for each item. The estimator then decides where on this range the work under consideration lies by taking into account the factors affecting the work as previously listed.

## Concrete Materials

Prices for ready-mixed concrete are obtained from local suppliers who are interested in quoting the supply of all concrete required for a project. Most suppliers maintain a current price list of concrete, but it is always advisable to secure a price quotation specifically for the project being estimated for several reasons: discounts may be available, prices for concrete that meet the specifications are required and, perhaps most importantly, firm prices are required. The prices on a current price list are exactly that, that is, prices in effect at this time, but the concrete for the project being estimated may not be required until some time in the future. So, what the estimator needs is the price that will be charged for the concrete when it is used on the project; in other words, a price offer that, when it is accepted by the contractor, will bind the supplier to the prices quoted for the full duration of the project. This is what is referred to as a "firm price."

Price quotations obtained from ready-mix concrete suppliers usually have a large number of conditions and extra charges attached to them. Some of the issues that the estimator has to carefully consider when pricing concrete materials include:

1. Does the concrete described in the quote meet the specifications? The estimator should be alert to special mix requirements that are sometimes included in concrete specifications. These might be in the form of minimum cement requirements in addition to concrete strength provisions.
2. What are the extra charges for supplying special cements like sulfate-resisting or high-early cements? When using cements other than ordinary portland type I cement, concrete prices are invariably increased by a stipulated sum per cubic yard of material supplied.
3. What are the extra charges for air entrainment, calcium chloride or any other concrete additives required to meet specifications?

| CREWS | METHOD | CREW MEMBERS | |
|---|---|---|---|
| CREW A | Chute | 1.0 Foreman<br>4.0 Laborers<br>1.0 Cem. Finisher | |
| CREW B | Pumped | 1.0 Foreman<br>5.0 Laborers<br>1.0 Cem. Finisher<br>1.0 Operator | |
| CREW C | Crane and<br>Bucket | 1.0 Foreman<br>5.0 Laborers<br>1.0 Cem. Finisher | |
| CREW D | Hand Placed | 0.5 Foreman<br>1.0 Laborer<br>2.0 Cem. Finishers | |

| ITEM | OPERATION | METHOD | CREW | OUTPUT | |
|---|---|---|---|---|---|
| 1. | Continuous Strip Footings | Chute<br>Pumped | A<br>B | 10–20 cu. yd./hr.<br>8–15 cu. yd./hr. | 8–15 m³/hr.<br>6–12 m³/hr. |
| 2. | Isolated Footings and Pile Caps | Chute<br>Pumped | A<br>B | 6–12 cu. yd./hr.<br>5–10 cu. yd./hr. | 5–9 m³/hr.<br>4–8 m³/hr. |
| 3. | Grade Beams and Pilasters | Chute<br>Pumped | A<br>B | 16–25 cu. yd./hr.<br>14–18 cu. yd./hr. | 12–19 m³/hr.<br>11–14 m³/hr. |
| 4. | Foundation and Retaining Walls | Chute<br>Pumped | A<br>B | 10–13 cu. yd./hr.<br>9–12 cu. yd./hr. | 8–10 m³/hr.<br>7–9 m³/hr. |
| 5. | Above-grade Walls | Pumped<br>Crane | B<br>C | 8–13 cu. yd./hr.<br>7–12 cu. yd./hr. | 6–10 m³/hr.<br>5–9 m³/hr. |
| 6. | Columns | Pumped<br>Crane | B<br>C | 5–10 cu. yd./hr.<br>4–9 cu. yd./hr. | 4–8 m³/hr.<br>3–7 m³/hr. |
| 7. | Sumps and Manholes | Chute<br>Pumped | A<br>B | 6–10 cu. yd./hr.<br>5–9 cu. yd./hr. | 5–8 m³/hr.<br>4–7 m³/hr. |
| 8. | Slab-on-Grade | Chute<br>Pumped | A<br>B | 13–22 cu. yd./hr.<br>14–23 cu. yd./hr. | 10–17 m³/hr.<br>11–18 m³/hr. |
| 9. | Mud Slab | Chute<br>Pumped | A<br>B | 10–14 cu. yd./hr.<br>8–12 cu. yd./hr. | 8–11 m³/hr.<br>6–9 m³/hr. |
| 10. | Suspended Slab and Beams | Pumped<br>Crane | B<br>C | 13–19 cu. yd./hr.<br>11–17 cu. yd./hr. | 10–15 m³/hr.<br>8–13 m³/hr. |
| 11. | Stairs and Landings | Pumped<br>Crane | B<br>C | 5–9 cu. yd./hr.<br>4–8 cu. yd./hr. | 4–7 m³/hr.<br>3–6 m³/hr. |
| 12. | Stair Treads | Hand Placed | D | 25–60 sq. ft./hr. | 2–6 m²/hr. |
| 13. | Slab Topping—Separate<br>—Monolithic | Pumped<br>Pumped | B<br>B | 3–7 cu. yd./hr.<br>5–9 cu. yd./hr. | 2–5 m³/hr.<br>4–7 m³/hr. |
| 14. | Slab-on-Metal Deck | Pumped<br>Crane | B<br>C | 7–12 cu. yd./hr.<br>6–11 cu. yd./hr. | 5–9 m³/hr.<br>5–8 m³/hr. |
| 15. | Equipment Bases | Pumped | B | 6–10 cu. yd./hr. | 5–8 m³/hr. |
| 16. | Beams | Pumped<br>Crane | B<br>C | 5–8 cu. yd./hr.<br>6–9 cu. yd./hr. | 4–6 m³/hr.<br>5–7 m³/hr. |
| 17. | Sidewalks | Chute<br>Pumped | A<br>B | 15–25 cu. yd./hr.<br>15–25 cu. yd./hr. | 11–19 m³/hr.<br>11–19 m³/hr. |

**Figure 11.1**   Concrete Work Productivities

4.  Are there additional charges for cooling concrete in hot weather or heating concrete in cold weather to account for? In some areas the cost of heating or cooling will be charged automatically for all concrete delivered in certain months of the year.

5.  If small quantities of concrete are required on a project, what are the premiums charged on small loads of concrete?

6.  If delays are anticipated in unloading the concrete at the site, what will be the waiting time (demurrage) charges? Demurrage charges for long unloading times are seldom included in the estimate because estimators are usually optimistic about the crews' ability to unload concrete quickly.

7.  If concrete is required to be delivered to the job outside of normal working hours, what are the additional charges?

8.  If disposal of surplus concrete and cleaning of concrete trucks is restricted because of environmental restrictions, are there any additional fees payable to cover these activities? An "environmental fee" is sometimes charged by suppliers to cover the cost of complying with regulations.

### Waste Factors

Recall that the quantities of concrete taken off are the unadjusted net amounts shown on the drawings. Allowance for waste and spillage of this material can be made by increasing the takeoff quantities or by raising the price by the percentage factor considered necessary. In the examples that follow, the concrete material quantities have been increased as noted to account for wastage.

The value of waste factors usually lies between 1% and 5% for concrete placed in formwork and can be as much as 10% for concrete placed directly against soil. Even higher waste factors can be expected when concrete is to be placed in rock excavations where there is a possibility of "over break" present.

### EXAMPLE—CONCRETE SUPPLY PRICING

The example of a quotation for the supply of ready-mix concrete shown on Figure 11.2 is used to price the following concrete material requirements for work to be completed on a project between December and February:

**1.** 3000 psi Concrete, Type I Cement:

| | | |
|---|---|---:|
| Basic concrete price | | $ 87.00 |
| Environmental fee | | 1.00 |
| Extra for winter heat | | 6.00 |
| | | 94.00 |
| Sales tax | 7% | 6.58 |
| Total price per cu. yd. | | 100.58 |

**2.** 3500 psi Concrete, Type V Cement, Air Entrained:

| | |
|---|---:|
| Basic concrete price | $ 90.00 |
| Environmental fee | 1.00 |
| Extra for winter heat | 6.00 |
| Extra for type V | 4.50 |
| Extra for air entrainment | 3.50 |
| | 105.00 |

|                                 |     |        |
| ------------------------------- | --- | ------ |
| Sales tax                       | 7%  | 7.35   |
| Total price per cu. yd.         |     | 112.35 |

**3.** 4000 psi Concrete, Type V cement, Air Entrained, Fiber Reinforced:

|                                 |     |          |
| ------------------------------- | --- | -------- |
| Basic concrete price            |     | $  93.00 |
| Environmental fee               |     | 1.00     |
| Extra for winter heat           |     | 6.00     |
| Extra for type V                |     | 5.00     |
| Extra for air entrainment       |     | 4.50     |
| Extra for fiber reinforcement   |     | 17.00    |
|                                 |     | 126.50   |
| Sales tax                       | 7%  | 8.86     |
| Total price per cu. yd.         |     | 135.36   |

---

**ABC CONCRETE PRODUCTS INC.**

PROJECT: XYZ Office Building, Townville.

We are pleased to quote you as follows for the supply of ready-mix delivered to the above project:

| Mix | Strength (psi) | Aggregate Size | Cement Type | Delivered Price per CY | Additional for Type III or Type V | Additional for 4–6% Air Entrainment |
| --- | --- | --- | --- | --- | --- | --- |
| 1. | 3000 | ¾" | I | $ 87.00 | $4.00 | $2.00 |
| 2. | 3500 | ¾" | I | $ 90.00 | $4.50 | $3.50 |
| 3. | 4000 | ¾" | I | $ 93.00 | $5.00 | $4.50 |
| 4. | 4500 | ¾" | I | $ 97.50 | $5.50 | $5.50 |
| 5. | 5000 | ¾" | I | $104.00 | $6.00 | $6.00 |

The above prices are based on ABC Concrete's standard mix designs.

All products are subject to a municipal sales tax of 7%.

For ½" aggregates add $5.50 per CY to the above prices.
For polypropylene fibers add $17.00 for 2 lbs. per CY.
For pigments (red, black tan, or brown) add $4.50 per CY.

EXTRA CHARGES:        Calcium chloride (1%):  $2.00  Per CY

Calcium chloride (2%):  $4.00  Per CY

Winter heat between November 25 and March 15:  $6.00  Per CY

Environmental fee:  $1.00  Per CY

**Figure 11.2**   Concrete Materials Price Quote Example

**4.** 4500 psi Concrete, Type I Colored Cement, Air Entrained:

| | | |
|---|---|---|
| Basic concrete price | | $  97.50 |
| Environmental fee | | 1.00 |
| Extra for winter heat | | 6.00 |
| Extra for colored cement | | 4.50 |
| Extra for air entrainment | | 5.50 |
| | | 114.50 |
| Sales tax | 7% | 8.02 |
| Total price per cu. yd. | | 122.52 |

## Formwork

A wide range of methods of providing formwork for a project is available. At one end of the range, a formwork system may be custom built at the site for use on the one project only. At the other extreme, a system, which was prefabricated off-site at some time in the past, may be employed on the project for one or several uses and then be shipped off to other projects for further use. Whichever method of providing formwork is adopted, there are a number of costs that must be considered when pricing a formwork system:

1. The cost of building the form system (fabrication) and keeping it in good repair, which can be spread over the number of uses the form system is put to.
2. The cost of setting up the form system each time it is used.
3. The cost of removing the system after the concrete has been placed (stripping). The cost of cleaning and oiling the forms ready for reuse should be included with the stripping cost.
4. Other possible costs such as transportation, handling and storage of the form system between uses.

### Formwork Productivity

The factors that affect the productivity of crews fabricating and erecting formwork systems include the same general job factor and labor and management factors previously discussed, but there are a number of items that impact the productivity and cost of formwork, particularly:

1. Potential for reuse of a form system
2. Complexity of formwork design
3. Use of **"fly forms"** (**"gang forms"**)
4. Number of form ties required for a system

*Amount of Reuse*

A major factor in determining the cost of formwork is the number of times a set of forms can be reused. The cost of the materials used in a formwork system and the cost of constructing the system—the fabrication costs—are distributed over the total number of square feet (m²) formed by the form system over its life. Thus, increasing the number of times the system is used will reduce its unit cost per square foot (m²) of use. Also, reuse of a form system many times over can increase the productivity of the crew that sets up the forms each time they are used. As the crew becomes more familiar with the system after repeated uses, especially with the more complicated systems, the crew can maintain a learning rate that reflects increased efficiency every time the system is reused.

However, the estimator should note that as forms are used over and over again, the cost of their repair increases and offsets the advantage of reuse to some extent. In fact, a form system can be used so many times that it reaches such a state of disrepair that it is more economic to replace it.

Wherever possible, form systems are constructed of panels made in standard shapes and sizes so that they can be assembled in different configurations to produce a variety of designs. Using simple-to-build standard panels offers savings from both minimum material usage and maximum labor and equipment productivity. These panels often consist of a lumber or aluminum frame faced with plywood that will be in contact with the concrete. Panels made entirely of metal are available and the large number of reuses possible with metal panels can justify their high price. However, when metal panels have been used many times, the quality of concrete finish attainable from them deteriorates if they have been roughly handled in their use.

### Complexity of Formwork Design

Formwork can mold concrete to virtually any shape, but constructing complicated shapes significantly increases the cost of formwork. Where unique concrete shapes are called for on a project, forms will have to be custom made. This will not only result in a high initial construction cost but also, if the shape is not to be repeated again and the forms are not reusable, the cost per use will be extremely high because fabrication costs cannot be distributed over multiple uses.

The estimator has to always be alert to any concrete features that will demand special formwork on a project. The use of prices for "standard" forms in these situations can lead to considerable formwork cost overruns as the price of custom work can be many times the price of general formwork.

### Use of "Fly Forms" ("Gang Forms")

The cost of formwork operations can also be substantially reduced where formwork systems are constructed for large areas of floors or walls and reused many times over in the course of a project. These systems of "fly forms" or "gang forms" are fabricated in large modules, and "flown" from one setup to the next usually by means of a centrally located tower crane at the site. The construction of reinforced concrete highrise buildings is particularly suited to this method of construction.

### Number of Form Ties Required for a System

The most time-consuming activity involved in setting up a form system is usually the placing and adjusting of **form ties**. The function of form ties is to hold the form system together prior to concreting, then to resist the pressures exerted by concrete as it is placed in the forms. Form ties fall into two main categories: "snap ties," which are designed to break inside the concrete and are thus completely used up in the process, and she-bolt or coil ties systems that comprise bolts that are reused and connecting rods that remain in the concrete. The number of form ties required for a system depends on five variables:

1. The capacity of the form ties—Form ties have to resist the pressure of the concrete; higher capacity ties can resist more pressure so they can be spaced farther apart in the system.
2. The rate of filling the forms—Quickly filling forms for walls or columns will cause high pressures on the forms that will call for more ties especially at the lower levels.

3. The temperature of the concrete—Colder concrete takes longer to gain strength, so concreting in colder temperatures generally results in higher pressures on form ties.

4. The method of placing concrete—Methods that result in large loads of concrete being deposited quickly in the forms will cause high pressures.

5. The depth of drop and the distribution of reinforcing steel—Concrete that drops some distance into the forms will exert added pressure on forms, but if the concrete has to pass by rebar in its fall the effect will be reduced.

### Formwork Productivity Rates

Figure 11.3 shows an example of a formwork productivity database that might be maintained by a contractor who is engaged in pricing formwork operations based on the performance of crews on past projects. We will use the productivity rates shown in the figure for pricing the examples that follow.

| | CREWS | CREW MEMBERS | | |
|---|---|---|---|---|
| | CREW A | 1.0 Carp. Foreman | | |
| | | 6.0 Carpenters | | |
| | | 2.0 Laborers | | |
| | CREW B | 0.3 Carp. Foreman | | |
| | | 2.0 Carpenters | | |
| | CREW C | 1.0 Labor Foreman | | |
| | | 1.0 Carpenter | | |
| **ITEM** | **OPERATION** | **CREW** | **OUTPUT** | |
| 1. | Continuous Strip Footings | A | 115–150 sq. ft./hr. | 11–14 m²/hr. |
| 2. | 2 × 4 Keyways | B | 150–170 sq. ft./hr. | 46–52 m/hr. |
| 3. | Isolated Footings and Pile Caps | A | 95–130 sq. ft./hr. | 9–12 m²/hr. |
| 4. | Grade Beams | A | 110–130 sq. ft./hr. | 10–12 m²/hr. |
| 5. | 4" × 8" Void Forms | B | 150–170 ft./hr. | 46–52 m/hr. |
| 6. | Pilasters | A | 55–80 sq. ft./hr. | 5–7 m²/hr. |
| 7. | Foundation and Retaining Walls | A | 80–105 sq. ft./hr. | 7–10 m²/hr. |
| 8. | Bulkheads | A | 30–40 sq. ft./hr. | 3–4 m²/hr. |
| 9. | Blockouts up to 8 SF | B | 0.5–2.0 no./hr. | 0.5–2.0 no./hr. |
| 10. | Above-grade Walls | A | 60–90 sq. ft./hr. | 6–8 m²/hr. |
| 11. | Columns—Rectangular | A | 55–80 sq. ft./hr. | 5–7 m²/hr. |
| | —Circular | A | 25–50 sq. ft./hr. | 2–5 m²/hr. |
| 12. | Sumps and Manholes | A | 60–80 sq. ft./hr. | 6–7 m²/hr. |
| 13. | Edges of Slab-on-Grade | A | 80–100 sq. ft./hr. | 7–9 m²/hr. |
| 14. | Construction Joints—SOG | A | 45–60 sq. ft./hr. | 4–6 m²/hr. |
| 15. | Edges of Suspended Slab | A | 65–85 sq. ft./hr. | 6–8 m²/hr. |
| 16. | Soffit of Suspended Slabs | A | 90–120 sq. ft./hr. | 8–11 m²/hr. |
| 17. | Soffit of Stairs | A | 35–40 sq. ft./hr. | 3–4 m²/hr. |
| 18. | Edges and Risers of Stairs | A | 60–80 sq. ft./hr. | 6–7 m²/hr. |
| 19. | Edges of Slab-on-Metal Deck | A | 45–60 sq. ft./hr. | 4–6 m²/hr. |
| 20. | Edges of Equipment Bases and Curbs | A | 40–90 sq. ft./hr. | 4–8 m²/hr. |
| 21. | Sides and Soffits of Beams | A | 45–90 sq. ft./hr. | 4–8 m²/hr. |
| 22. | Edges of Sidewalks | A | 80–100 sq. ft./hr. | 7–9 m²/hr. |
| 23. | Stripping Forms | C | 120–350 sq. ft./hr. | 11–33 m²/hr. |
| 24. | Shoring Frames | B | 3–5 no./hr. | 3–5 no./hr. |

**Figure 11.3**  Formwork Productivities

### Pricing Formwork Materials

Formwork for cast-in-place concrete may be constructed of lumber, plywood, aluminum, steel and combinations of these materials. These materials are combined in a formwork system that can be reused a number of times on the project under construction and possibly on other projects. The cost of materials used in a formwork system would include the rental or purchase cost of all the components involved in its construction except for form ties and other hardware items that are priced separately. The cost of form hardware is often estimated as a percentage of the price of the form system it is required on. Any shoring required to support a form system is also usually priced separately from the form system.

After all the components of a form system have been priced, the total system price would be expressed as a *price per square foot* or square meter of formwork contact area. Recall that the "contact area" is the area of the form system that will be in contact with the concrete. This material price per square foot or square meter can then be applied to the total square foot or square meter area that is to be formed using this particular form system on the project as recorded in the quantity takeoff.

If a rented formwork system is to be considered for use on a project, pricing the materials is a relatively straightforward procedure. Prices are obtained from formwork rental companies quoting the cost of renting the required system for a specified time period. Any further charges necessary for such items as transportation, storing or handling the forms before and after their use are added to the rental prices. This total price is then divided by the total formwork contact area to determine the required price per square foot or square meter. Here is another situation in which the estimator should endeavor to secure from suppliers firm quotations whenever possible for the specific formwork requirements of a project. Using catalog prices for "off the shelf" form systems can lead to problems brought about by price increases that occur between the time a price is used in the estimate and the time a contract is concluded for the rental of the formwork.

Where form materials are to be purchased, calculating the price to use in the estimate can be a little more complicated than with rented systems. Quotations for formwork components such as lumber and plywood are readily available, but the price of the form system depends on its design and taking into account such factors as the size and type of components, the stud spacing, the amount of bracing required and so on. The price also depends on the amount of reuse the system is expected to get on this project *and on all possible future projects*. To obtain the most accurate prices, form materials are priced in detail by first considering all the components of the system needed to meet the required design, then adjusting the price for the cost of anticipated repairs and material wastage and, finally, accounting for the number of reuses it is estimated the system will have in its full lifetime.

When there are a large number of form systems required for a project, calculating each form material price in such detail can be a substantial chore. The "number crunching" can be reduced where a computer program is set up based on the form materials requirements of common form designs. This program can be designed to automatically calculate the price of an assortment of form systems from the input prices of the components used to construct the formwork systems considered. See Figure 11.4 for an example of form materials priced on a computer spreadsheet program.

Alternatively, the need for a multitude of calculations of form materials prices can be entirely avoided by using an "average price" for all form materials required on a project. Most contractors keep track of actual form materials costs incurred on their projects in much the same way they track productivities. From this information the average cost of form materials can be determined from past project records; then

**PRICING FORM MATERIALS ON A COMPUTER SPREADSHEET**

By setting up formulas that reflect the quantities of component parts required for a form system, the computer can be used to calculate the material price of that form system using the prices of the component parts as variables. Updating the prices of the component parts will automatically update the material price per SF for the system as shown below.

The formwork system prices generated by this program can be linked to other cells on the spreadsheet for pricing the quantities of formwork taken off.

**Continuous Footings:** From records of previous work: for 8 SF (contact area) of forms, 8 SF of formply is required; 15.33 BF of 2 × 4 is required; 6.0 BF of 2 × 6 is required.

Calculation for SET 1: Cost of formply = \$50.00/32　　= \$1.56/SF  
　　　　　　　　　　Cost of 2 × 4　= \$580.00/1000 = \$0.58/BF  
　　　　　　　　　　Cost of 2 × 6　= \$580.00/1000 = \$0.58/BF

Cost of Forms = 8 × \$1.56 + 15.33 × \$0.58 + 6 × \$0.58 = \$24.85 for 8 SF  
　　　　　　　　　　　　　　　Per SF = \$24.85/8 = \$3.11 /SF  
　　If these forms are used 6 times, cost per use = \$3.11/6 = \$0.52 /SF per use  
　　　　　　Add 20% for waste and repairs = \$0.52 × 1.20  
　　　　　　　　　　　　Total Cost = \$0.62 /SF per use

| SET 1: COMPONENTS | PRICES ($) |
|---|---|
| Form plywood: | 50.00　per sheet |
| 2 × 4 Lumber: | 580.00　per 1000 board feet |
| 2 × 6 Lumber: | 580.00　per 1000 board feet |

| SET 1: FORM SYSTEMS | PRICES ($) |
|---|---|
| Continuous Footings 1'0" high: | 0.622　per SF |
| Walls 8'0" high: | 0.559　per SF |
| Grade Beams 2'0" high: | 0.699　per SF |

| SET 2: COMPONENTS | PRICES ($) |
|---|---|
| Form plywood: | 65.00　per sheet |
| 2 × 4 Lumber: | 620.00　per 1000 board feet |
| 2 × 6 Lumber: | 720.00　per 1000 board feet |

| SET 2: FORM SYSTEMS | PRICES ($) |
|---|---|
| Continuous Footings 1'0" high: | 0.752　per SF |
| Walls 8'0" high: | 0.682　per SF |
| Grade Beams 2'0" high: | 0.835　per SF |

| SET 3: COMPONENTS | PRICES ($) |
|---|---|
| Form plywood: | 80.00　per sheet |
| 2 × 4 Lumber: | 680.00　per 1000 board feet |
| 2 × 6 Lumber: | 740.00　per 1000 board feet |

| SET 3: FORM SYSTEMS | PRICES ($) |
|---|---|
| Continuous Footings 1'0" high: | 0.872　per SF |
| Walls 8'0" high: | 0.797　per SF |
| Grade Beams 2'0" high: | 0.962　per SF |

**Figure 11.4**　Pricing Form Materials Example

this established "average price" can be used to estimate the average cost of form materials on a future project. This method of pricing form materials can be reasonably accurate if the same kind of formwork systems are used on past and future projects and necessary adjustments are made to the "average prices" for the price fluctuations of the lumber, plywood and suchlike components used in the formwork systems.

However, if a formwork system is to be used that is different from systems used on previous projects, an accurate price of the new system can be obtained only by pricing it in detail as previously described. Also, when a new formwork system is considered, the labor cost of the system has to be analyzed in detail since "historic" labor productivities apply only to reuse of established form systems. An example of pricing a new form system in detail follows.

### EXAMPLE—FORMWORK PRICING

Figure 11.5 shows the design of a formwork system for use in forming an elevated beam. It is assumed that a number of long sections are to be constructed and each section of formwork will be used 6 times. If the contractor has never used this form system before, a detailed analysis is required to obtain the most accurate assessment of the costs of the system. The labor price for this system is calculated using the labor productivity of a crew fabricating the forms, the productivity of a second crew setting up the forms each time the system is used and the productivity of a

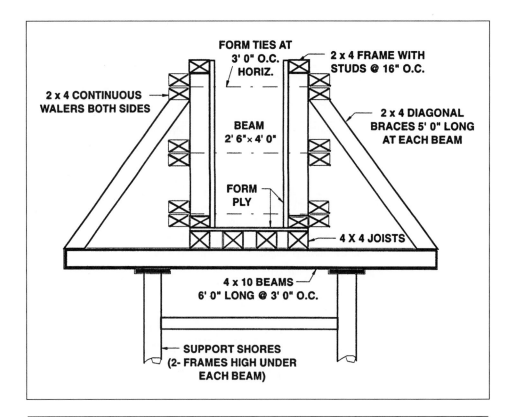

**Figure 11.5**   Formwork System to Elevated Beam

labor crew stripping, cleaning and oiling the forms ready for reuse. These productivities are based on the observation and cost analysis of similar operations involved in fabricating, erecting and stripping different form systems in the past.

The labor crews, their productivities and the prices of component materials used in this example are as follows:

| | | | |
|---|---|---|---|
| Labor Crew: | 1 Carpentry foreman | $31.00/hr. = | $ 31.00 |
| | 5 Carpenters | $28.00/hr. = | 140.00 |
| | 2 Laborers | $21.00/hr. = | 42.00 |
| Crew cost per hour: | | = | 213.00 |

Note: this crew will fabricate the form system, a crew of the same composition will erect the system each time it is used and a crew composed of laborers will strip the forms after each use.

| Component | Material Price | | Fabricating Productivity |
|---|---|---|---|
| ¾" Formply | $ 40.00/sheet | = $1.25/sq. ft. | 250 sq. ft./hr. |
| 2 × 4 Lumber | $480.00/1000 bd. ft. | = $0.48/bd. ft. | 200 bd. ft./hr. |
| 4 × 4 Lumber | $520.00/1000 bd. ft. | = $0.52/bd. ft. | 240 bd. ft./hr. |
| 4 × 10 Lumber | $580.00/1000 bd. ft. | = $0.58/bd. ft. | 300 bd. ft./hr. |
| Form oil | $ 4.00/gal. | = $0.04/sq. ft. | not applicable |

Productivity erecting the form system is 75 square foot (contact area)/hr. Productivity stripping the form system is 30 square foot/hr. per laborer.

The following materials are based on a section of forms 12'0" long. Note that this length of 12'0" is chosen merely for convenience; the actual length of the form system to be used on the job may not be known at the time of the estimate. Because the actual length of the system will probably be much longer than 12'0", the calculation of the number of beams will have to be modified. Usually, if beams are spaced 3'0" on center over a space of 12'0", we would divide 12 by 3 and add 1 for the end beam. In this case, however, we divide 12 by 3 and leave it as 4, which leaves spaces at each end of this 12'0" length for continuity with the remainder of the section.

|  |  |  |  |  | *Materials Price $* |
|---|---|---|---|---|---|
| *Components* | | | | | |
| 4 × 10 | Beams | 4 × 6'0" | = 24 ft. = 80 bd. ft. @ $0.58 | = | 46.40 |
| 4 × 4 | Joists | 4 × 12'0" | = 48 ft. = 64 bd. ft. @ $0.52 | = | 33.28 |
| 2 × 4 | Walers | 12 × 12'0" | = 144 ft. | | |
| | Frames | 4 × 12'0" | = 48 ft. | | |
| | | 2 × 10 × 4'0" = | 80 ft. | | |
| | Braces | 2 × 4 × 5'0" = | 40 ft. | | |
| | | | 312 ft. = 208 bd. ft. @ $0.48 | = | 99.84 |
| ¾" Formply—sides: | | 8'0" | | | |
| —bottom: | | +3'0" | | | |
| | | | = 11'0" × 12' = 132 sq. ft. @ $1.25 | = | 165.00 |
| | | | | | 344.52 |

|  |  |  |  |
|---|---|---|---|
| Waste and repairs add 20% |  | = | 68.90 |
|  |  |  | 413.42 |
| Price per use (/6 uses) |  | = | 68.90 |
| Material price per sq. ft. (/126 sq. ft. contact area) |  | = | 0.55 |
| Add the cost of form oil required for each use |  | = | 0.04 |
| Total material price per sq. ft. |  | = | 0.59 |

Again based on a section of forms 12'0" long the fabricating costs are as follows:

| Components |  |  | Labor Price $ |
|---|---|---|---|
| 4 × 10 Beams | 80 bd. ft.@ $213.00/300 bd. ft. | = | 56.80 |
| 4 × 4 Joists | 64 bd. ft.@ $213.00/240 bd. ft. | = | 56.80 |
| 2 × 4 Pieces | 208 bd. ft.@ $213.00/200 bd. ft. | = | 221.52 |
| ¾" Formply | 132 sq. ft.@ $213.00/250 sq. ft. | = | 112.46 |
|  |  |  | 447.58 |
|  | Waste and repairs add 20% | = | 89.52 |
|  |  |  | 537.10 |
|  | Price per use (/6 uses) | = | 89.52 |
|  | Fabrication price per sq. ft. (/126 sq. ft. contact area) | = | 0.71 |
|  | Erecting price per sq. ft. = $213.00/75 sq. ft. | = | 2.84 |
|  | Stripping price = $21.00/30 sq. ft. | = | 0.70 |
|  | Total labor price per sq. ft. | = | 4.25 |

The cost of form hardware and shoring frames required for this system would be estimated separately.

Form hardware is often computed at a rate of between 15% and 30% of the total value of formwork material included in the estimate. If a detailed analysis of hardware required for the above system is called for it may be calculated as follows.

The form design calls for 3 form ties to be placed at intervals of 3'0" horizontally along the length of the beam (see Figure 11.5), therefore, 12 ties are required for a 12'0" long section of beam. Thus 0.095 ties (12/126 sq. ft.) are required per sq. ft. of contact area.

| | | |
|---|---|---|
| If form ties cost $1.05 each, the cost per square foot for form ties = $1.05 × 0.095 ties/sq. ft. | = | $0.10 |
| Add for the cost of nails and other items of hardware | = | $0.05 |
| Then the total estimated cost of form hardware per sq. ft. | = | $0.15 |

The cost of shoring frames to support this beam form system includes the rental cost and the cost of labor erecting the frames. In order to estimate the cost of renting the shoring frames, the estimator needs to determine how long the shores will be used with each beam setup. Let us say, in this situation, that the shores will be required for an average of 9 workdays supporting one beam before they can be moved to the next beam location.

*Material Price:* If the shoring frames complete with accessories are rented at the price of $8.50/frame per month (22 workdays), the cost of shoring can be estimated as follows:

Considering a 12'0" length of beam forms once again,
the number of shoring frames required = 2 × (12'0"/3'0")      = 8 frames

Rental cost = 8 frames × 9 workdays × ($8.50/22)      = $27.82

Material cost per sq. ft. of contact area = $\dfrac{\$27.82}{126 \text{ sq. ft.}}$      = $ 0.22

*Labor Price:* With a crew comprising 0.3 carpenter foreman and 2 carpenters, the expected productivity to set up and take down shoring frames is 4 frames per hour.

Crew cost:    0.3 Carpentry foreman    $31.00/hr.  =  $  9.30

2.0 Carpenters    $28.00/hr.  =    56.00
                                    65.30

                                   130.00

Estimated cost for 8 frames = ($65.30/4) × 8    =   174.00

Labor cost per square foot of contact
area = $130.60/126 sq. ft.    =    1.04/sq. ft.

## Reinforcing Steel

Reinforcing steel for concrete includes welded wire mesh and reinforcing bars. Welded wire mesh is purchased in mats or rolls delivered to site for placing, so there are two costs associated with mesh: the supply cost, including delivery to site, and the labor cost for installation. The cost of supplying and placing reinforcing steel bars on a project has a number of additional constituents including:

1.  The cost of rebar shop drawings
2.  The cost of the raw steel bars
3.  The cost of handling, cutting, bending and identifying (tagging) rebar in the fabrication shop
4.  The cost of transporting the bars to the site
5.  The cost of spacers, chairs, saddles and ties used in the installation of rebar
6.  The labor cost of installation

Usually, costs 1 through 4 are included in the prices obtained from subcontractors who offer to supply rebar "cut, bent and tagged" to the site in accordance with the specifications. The general contractor completes the reinforcing steel pricing by adding an amount to the supply price for the installation materials (cost 5) and by estimating the labor cost for the installation of rebar (cost 6). In some cases, subcontractor prices for the rebar installation are available, which can simplify the task of the contractor's estimator. Alternatively, the estimator can make a detailed analysis of the installation cost as outlined in the following section.

### Reinforcing Steel—Installation Productivity

In addition to the general job factors and labor and management factors discussed in Chapter 8, the productivity of labor installing reinforcing steel bars is affected by the following factors:

1. The size and lengths of the reinforcing bars
2. The shapes of the bars
3. The complexity of the concrete design
4. The amount of tolerance allowed in the spacing of bars
5. The amount of tying required

Generally less time will be required to install a ton of large-size bars than a ton of small bars, so productivity should be higher when installing the large-size bars.

Where bars are bent into complicated shapes or bars have to be laid in elaborate configurations, extra time spent sorting out the bars and interpreting requirements will add to the cost of the installation process.

Having to work to small tolerances will hinder productivity and where the amount of tying is increased because the number of bar intersections is more than usual, the rate of labor output will again be reduced.

Figure 11.6 indicates expected labor productivity rates for the installation of reinforcing steel.

## Miscellaneous Concrete Work Items

There are a number of items of work that are associated with concreting operations but not covered in the preceding concrete, formwork and reinforcing steel sections. These items include such work as concrete finishes, grouting and installing accessories like anchor bolts and **waterstops** that are required to be embedded in concrete. See Figure 11.7 for the labor productivity rates of a sample of miscellaneous concrete work items.

### Pricing Miscellaneous Concrete Materials

The first thing that needs to be considered before an item can be priced is the question of what materials are required for that item. Although this question appears to be quite trivial, in some cases it is not clear exactly what materials are required and a decision has to be made before the estimator can proceed. The item of setting slab screed is an example. Screeds are set in the process of preparing for placing concrete in slabs; their function is to provide a means of obtaining the required elevations in the top surface of the slab. Slabs often have to provide a slope so that any spills on the surface of the slab flow to drains, for instance. A number of materials can be used for screeds including lumber, steel "T-bar" or steel pipes, together with a variety of adjustable support bases.

Most contractors tend to use the same screed system on all their jobs so the estimator may be able to obtain cost information from a previous job and use it to price the next job. However, if a new system is to be implemented, the estimator may have to get information about the price and method of use of the proposed system so that a detailed estimate of its cost can be prepared.

After the estimator has established what materials are required for an item, he or she usually proceeds to obtain prices for the supply of these materials. The next problem that often arises is that the prices received are for units of measurement that are different from the takeoff units. Curing compound to be sprayed on concrete slabs, for example, might be quoted as a price per gallon (per 5-gallon or per 40-gallon

**1. STEEL BAR REINFORCING**

CREW: 1 Foreman
5 Rodmen

| ITEM | OPERATION | BAR SIZE | |
| | | #3 TO #6 | #7 AND OVER |
|---|---|---|---|
| 1. | Footings | 0.39—0.40 tons/hr. | 0.42–0.68 tons/hr. |
| 2. | Walls | 0.56—0.57 tons/hr. | 0.58–0.75 tons/hr. |
| 3. | Columns | 0.28—0.29 tons/hr. | 0.30–0.43 tons/hr. |
| 4. | Beams | 0.30—0.31 tons/hr. | 0.32–0.51 tons/hr. |
| 5. | Suspended Slabs | 0.54—0.55 tons/hr. | 0.56–0.75 tons/hr. |
| 6. | Slabs-On-Grade | 0.43—0.44 tons/hr. | 0.43–0.72 tons/hr. |

| ITEM | OPERATION | METRIC BAR SIZE | |
| | | 10 M TO 20 M | 25 M AND OVER |
|---|---|---|---|
| 1. | Footings | 0.35—0.36 tonnes/hr. | 0.38–0.62 tonnes/hr. |
| 2. | Walls | 0.51—0.52 tonnes/hr. | 0.53–0.68 tonnes/hr. |
| 3. | Columns | 0.25—0.26 tonnes/hr. | 0.27–0.39 tonnes/hr. |
| 4. | Beams | 0.27—0.28 tonnes/hr. | 0.29–0.46 tonnes/hr. |
| 5. | Suspended Slabs | 0.49—0.50 tonnes/hr. | 0.51–0.68 tonnes/hr. |
| 6. | Slabs-On-Grade | 0.39—0.40 tonnes/hr. | 0.39–0.65 tonnes/hr. |

**2. WIRE MESH REINFORCING**

CREW: 1 Foreman
3 Laborers

| ITEM | SIZE OF MESH | OUTPUTS | |
| | | SMALL AREAS | LARGE AREAS |
|---|---|---|---|
| 1. | 6 × 6—10/10 | 800 sq. ft./hr. | 1480 sq. ft./hr. |
| 2. | 6 × 6—8/8 | 750 sq. ft./hr. | 1370 sq. ft./hr. |
| 3. | 6 × 6—6/6 | 680 sq. ft./hr. | 1260 sq. ft./hr. |
| 4. | 6 × 6—4/4 | 630 sq. ft./hr. | 1160 sq. ft./hr. |
| 5. | 4 × 4—10/10 | 720 sq. ft./hr. | 1330 sq. ft./hr. |
| 6. | 4 × 4—8/8 | 660 sq. ft./hr. | 1200 sq. ft./hr. |
| 7. | 4 × 4—6/6 | 600 sq. ft./hr. | 1100 sq. ft./hr. |
| 8. | 4 × 4—4/4 | 540 sq. ft./hr. | 1000 sq. ft./hr. |

| ITEM | SIZE OF MESH | OUTPUTS (METRIC UNITS) | |
| | | SMALL AREAS | LARGE AREAS |
|---|---|---|---|
| 1. | 150 × 150—W1.4 × W1.4 | 74 m²/hr. | 137 m²/hr. |
| 2. | 150 × 150—W2.1 × W2.1 | 70 m²/hr. | 127 m²/hr. |
| 3. | 150 × 150—W2.9 × W2.9 | 63 m²/hr. | 117 m²/hr. |
| 4. | 150 × 150—W4.0 × W4.0 | 59 m²/hr. | 108 m²/hr. |
| 5. | 100 × 100—W1.4 × W1.4 | 67 m²/hr. | 124 m²/hr. |
| 6. | 100 × 100—W2.1 × W2.1 | 61 m²/hr. | 111 m²/hr. |
| 7. | 100 × 100—W2.9 × W2.9 | 56 m²/hr. | 102 m²/hr. |
| 8. | 100 × 100—W4.0 × W4.0 | 50 m²/hr. | 93 m²/hr. |

**Figure 11.6**   Reinforcing Steel Productivities

| | CREWS | CREW MEMBERS | | |
|---|---|---|---|---|
| | CREW A | 0.3 Carp. Foreman<br>2.0 Carpenters | | |
| | CREW B | 0.3 Foreman<br>2.0 Laborers | | |
| | CREW C | 0.3 Foreman<br>2.0 Cement Finishers | | |
| **ITEM** | **OPERATION** | **CREW** | **OUTPUT** | |
| 1. | Set Screeds | A | 1350–2000 sq. ft./hr. | 125–186 m²/hr. |
| 2. | Curing Slabs | B | 1000–1200 sq. ft./hr. | 93–111 m²/hr. |
| 3. | Finish Slabs—Screed | C | 200–250 sq. ft./hr. | 19–23 m²/hr. |
| | —Float | C | 160–200 sq. ft./hr. | 15–19 m²/hr. |
| | —Broom | C | 155–185 sq. ft./hr. | 14–17 m²/hr. |
| | —Steel Trowel | C | 140–180 sq. ft./hr. | 13–17 m²/hr. |
| 4. | Floor Hardener | C | 140–210 sq. ft./hr. | 13–20 m²/hr. |
| 5. | Rub Finish Walls | C | 70–110 sq. ft./hr. | 7–10 m²/hr. |
| 6. | Grout Base Plates | C | 0.90–1.67 hr./cu. ft. | 35–59 hr./m³ |
| 7. | Install Anchor Bolts—½" to ¾" | A | 100–180 no./hr. | 100–180 no./hr. |
| | —over ¾" | A | 24–120 no./hr. | 24–120 no./hr. |
| 8. | Waterstops—6" Wide | A | 36–38 ft./hr. | 11–12 m/hr. |
| | —9" Wide | A | 33–35 ft./hr | 10–11 m/hr. |
| 9. | Install Anchor Slot | A | 88–114 ft./hr. | 27–35 m/hr. |
| 10. | Rigid Insulation to Walls—1" | B | 200–250 sq. ft./hr. | 19–23 m²/hr. |
| | —2" | B | 180–225 sq. ft./hr. | 17–21 m²/hr. |
| 11. | Sprayed Damp-Proofing—1 Coat | B | 200–220 sq. ft./hr. | 19–20 m²/hr. |
| | —2 Coat | B | 120–140 sq. ft./hr. | 11–13 m²/hr. |
| 12. | 6 Mil Polyethylene Vapor Barrier | B | 900–1000 sq. ft./hr. | 84–93 m²/hr. |
| 13. | ½" × 6" Expansion Joint Filler | A | 80–100 ft./hr. | 24–30 m/hr. |

**Figure 11.7** Miscellaneous Concrete Work Productivities

container), but the takeoff unit is square feet of area to be cured. See Figure 11.8 for prices and conversion factors of miscellaneous concrete materials. Note that these conversion factors can vary according to packaging sizes and different rates of application.

## Wage Rates

The wage rates of workers in the following examples are:

| *Basic Hourly Wage* | $ |
|---|---|
| Equipment operator | 30.00 |
| Labor foreman | 24.00 |
| Laborer | 21.00 |
| Cement finisher | 22.00 |
| Carpenter foreman | 31.00 |
| Carpenter | 28.00 |
| Rebar rodman | 27.00 |

| ITEM | MATERIAL | SAMPLE PRICE | CONVERSION FACTOR | UNIT PRICE |
|------|----------|--------------|-------------------|------------|
| 1. | Curing Compound | $50.00/5 gal. | 200–400 sq. ft./gal. | $0.03–$0.05/sq. ft. |
|    | in metric units | $50.00/5 gal. | 18.5–37 m²/gal. | $0.27–$0.54/m² |
| 2. | Nonshrink Grout | $19.00/bag | 5 bags/cu. ft. | $95.00/cu. ft. |
|    | in metric units | $19.00/bag | 176.5 bags/m³ | $3354/m³ |
| 3. | Waterstop—⅜" × 6" | $240.00/roll | 48 ft./roll | $5.00/ft. |
|    | in metric units | $240.00/roll | 14.6 m/roll | $16.44/m |
|    | —⅜" × 9" | $375.00/roll | 48 ft./roll | $7.81/ft. |
|    | in metric units | $375.00/roll | 14.6 m/roll | $25.68/m |
| 4. | Rigid Insulation—1" | $10.50/sheet | 16 sq. ft./sheet | $0.66/sq. ft. |
|    | in metric units | $10.50/sheet | 1.5 m²/sheet | $7.00/m² |
|    | (SM Styrofoam)—2" | $20.50/sheet | 16 sq. ft./sheet | $1.28/sq. ft. |
|    | in metric units | $20.50/sheet | 1.5 m²/sheet | $13.67/m² |
| 5. | Sprayed Damp-Proofing | $5.00/gal. | 20–40 sq. ft./gal. | $0.12–$0.25/sq. ft. |
|    | in metric units | $5.00/gal. | 1.85–3.70 m²/gal. | $1.35–$2.70/m² |
| 6. | 6-Mil Polyethylene | $72.00/roll | 2000 sq. ft./roll | $0.04/sq. ft. |
|    | in metric units | $72.00/roll | 186 m²/roll | $0.39/m² |
| 7. | Floor Hardener | $18.00/bag | 50 lb/bag | $0.36/lb. |
|    | in metric units | $18.00/bag | 22.7 kg/bag | $0.79/kg |
| 8. | ½" × 6" Expansion Joint Filler | $12.00/piece | 16 ft./piece | $0.75/ft. |
|    | in metric units | $12.00/piece | 4.9 m/piece | $2.45/m |

**Figure 11.8**    Miscellaneous Concrete Material Prices

## Concrete Work Recap and Pricing Notes Example 1—House

### Concrete Work Figure 11.9

1. It is recommended that each trade be recapped onto a separate page as in the example because this allows extra items to be more easily inserted and provides space to make notes on the pricing for that trade.
2. On the Concrete Work Recap, the three items of concrete from the takeoff are listed and against these the labor and equipment prices are entered. Below, the concrete materials required for these items are listed for pricing separately.
3. The footings and the slab-on-grade concrete are to be placed by chute directly from the concrete trucks, so Crew A is selected from Figure 11.1 at the following price per hour:

|  |  |  | *Labor* |
|--|--|--|---------|
| *Concrete Crew A* |  |  | $ |
| 1 Labor foreman | $24.00/hr. | = | 24.00 |
| 4 Laborers | $21.00/hr. | = | 84.00 |
| 1 Cement finisher | $22.00/hr. | = | 22.00 |
|  |  |  | $130.00/hr. |

**PRICING SHEET**

JOB  *HOUSE EXAMPLE*   DATE ...................

ESTIMATED  *A B F*                                          Page No. $\boxed{2 \ of \ 14}$

| No. | DESCRIPTION | QUANTITY | UNIT | UNIT PRICE | LABOR | UNIT PRICE | MATERIALS | UNIT PRICE | EQUIPMENT | TOTAL |
|---|---|---|---|---|---|---|---|---|---|---|
|  | *CONCRETE WORK* |  |  |  |  |  |  |  |  |  |
| 1. | FOOTINGS | 7 | CY | 18⁵¹ | 130 | — | — | — | — | 130 |
| 2. | WALLS | 26 | CY | 18¹⁰ | 471 | — | — | 17³⁰ | 450 | 921 |
| 3. | SLAB-ON-GRADE | 12 | CY | 10⁸⁵ | 130 | — | — | — | — | 130 |
|  |  |  |  |  |  |  |  |  |  |  |
|  | *MATERIALS* |  |  |  |  |  |  |  |  |  |
| 4. | 3000# CONC. TYPE V WITH AIR. | 35 | CY | — | — | 94⁰⁰ | 3290 | — | — | 3290 |
| 5. | 3000# CONC. TYPE I WITH AIR. | 13 | CY | — | — | 90⁰⁰ | 1170 | — | — | 1170 |
| 6. | EXTRA FOR WINTER HEAT | 48 | CY | — | — | 6⁰⁰ | 288 | — | — | 288 |
|  |  |  |  |  |  |  |  |  |  |  |
|  | TOTAL CONCRETE WORK TO SUMMARY |  |  |  | 731 |  | 4748 |  | 450 | 5929 |

**Figure 11.9**   House Example—Pricing Concrete Work

4.  A full hour from this crew will probably be needed to place the 7 cu. yd. in the footings and to place the 12 cu. yd. in the slab-on-grade so the unit prices will be:

*Labor*
$

| | | | |
|---|---|---|---|
| Footings: | Crew A | $ 30.00/7 | = 18.57/cu. yd. |
| Slab-on-grade: | Crew A | $130.00/12 | = 10.83/cu. yd. |

5.  The walls are to be placed using a concrete pump, so Crew B will be used at the following price per hour:

*Labor*

| *Concrete Crew B* | | | $ |
|---|---|---|---|
| 1 Labor foreman | $24.00/hr. | = | 24.00 |
| 5 Laborers | $21.00/hr. | = | 105.00 |
| 1 Cement finisher | $22.00/hr. | = | 22.00 |
| 1 Equipment operator | $30.00/hr. | = | 30.00 |
| | | | $181.00/hr. |

6.  The rental price of the concrete pump is $150.00 per hour. If the 26 cu. yds. is to be placed in the walls at the rate of 10 cu. yd./hr., this pump will be needed for about 3 hours. Using this rate of output the unit prices will be:

| | *Labor* | *Equipment* |
|---|---|---|
| | $ | $ |
| Pump 3 hrs.<br>    @ $150.00 = $450.00/26 cu. yd.    = | — | 17.31 |
| Crew B        = $181.00/10 cu yd. per hr.  = | 18.10 | — |
| Unit price for concrete walls:        = | 18.10/cu. yd. | 17.31/cu. yd. |

7.  The concrete in contact with soil, footing and wall concrete is required to be sulfate resisting and also to be air entrained, so this is priced separately from the slab-on-grade concrete.
8.  Concrete material quantities were increased by 5% to allow for wastage and prices are taken from the sample quote shown on Figure 11.2.

## Formwork Figure 11.10

9.  On the formwork recap, a materials price and a labor price will be entered against each item and 15% will be added to the total materials price to allow for rough hardware in the last item. A number of hand tools will also be used for this work; the cost of these tools is included on the general expenses under the item: "Small Tools."
10.  Figure 11.3 indicates that with a formwork crew A, the productivity forming footings should be between 115 and 150 sq. ft./hr. In this case the

**PRICING SHEET**

JOB ......*House Example*......

ESTIMATED ......*A.B.F.*......  DATE ......................

Page No. | 3 of 14 |

| No. | DESCRIPTION | QUANTITY | UNIT | UNIT PRICE | LABOR | UNIT PRICE | MATERIALS | UNIT PRICE | EQUIPMENT | TOTAL |
|---|---|---|---|---|---|---|---|---|---|---|
| | FORMWORK | | | | | | | | | |
| 1. | Strip Footings | 179 | SF | 2.01 | 360 | .55 | 98 | — | \| | 458 |
| 2. | Pad Footings | 27 | SF | 2.54 | 69 | .55 | 15 | — | \| | 84 |
| 3. | 2 x 4 Keyway | 133 | LF | 0.44 | 59 | .10 | 13 | — | \| | 72 |
| 4. | Walls | 2133 | SF | 2.41 | 5141 | .42 | 896 | — | \| | 6037 |
| 5. | Bulkheads | 22 | SF | 10.88 | 239 | 1.20 | 26 | — | \| | 265 |
| 6. | Strip, clean & oil | 2361 | SF | 0.51 | 1204 | .05 | 118 | — | \| | 1322 |
| 7. | Rough Hardware | 15% MATLS | | | — | | 175 | — | \| | 175 |
| | | | | | | | | | | |
| | Total Formwork To Summary | | | | 7072 | | 1341 | | — | 8413 |

**Figure 11.10**   House Example—Pricing Formwork

quantity involved is small so the productivity will not be high, so a rate of 120 sq. ft./hr. is adopted. This translates into the following unit price:

|  |  |  | Labor $ |
|---|---|---|---|
| *Framework Crew A* |  |  |  |
| 1 Carpentry foreman | $31.00/hr. | = | 31.00 |
| 6 Carpenters | $28.00/hr. | = | 168.00 |
| 2 Laborers | $21.00/hr. | = | 42.00 |
|  |  |  | $241.00/hr. |

| Unit price for continuous footings | = | $241.00/21 |
|---|---|---|
| Labor price: | = | $ 2.01/sq. ft. |

11.   Forming keyways uses crew B and should be at a rate of about 150 ft./hr.:

|  |  |  | Labor $ |
|---|---|---|---|
| *Framework Crew B* |  |  |  |
| 0.3 Carpentry foreman | $31.00/hr. | = | 9.30 |
| 2 Carpenters | $28.00/hr. | = | 56.00 |
|  |  |  | $65.30/hr. |

| Unit price for keyway to footings | = | $65.30/150 |
|---|---|---|
| Labor price: | = | $ 0.44/sq. ft. |

12.   Forming a small area of pad footings would be at a low rate of productivity; but straightforward wall forms would be at an average productivity rate:

| Unit price for pad footings: | = | $241.00/95 |
|---|---|---|
| Labor price: | = | $ 2.54/sq. ft. |
| Unit price for walls: | = | $241.00/100 |
| Labor price: | = | $2.41/sq. ft. |

13.   Generally the form materials prices used are based on past records of the contractor using his or her own forms except for walls. Wall forms are assumed to be rented in 2'0" × 8'0" panels for the price of $6.40 per panel. This price includes the cost of walers and accessories for securing snap ties. This gives the following unit price:

| Rental of panels: | = | $\dfrac{\$6.40}{16 \text{ sq. ft.}}$ |
|---|---|---|
|  | = | $0.40 |
| Add for waste and repairs: 5% | = | $0.02 |
| Materials price: | = | $0.42/sq. ft. |

14.   Because there is such a small area of bulkheads in this example, a crew B is thought to be more appropriate at a rate of about 6 sq. ft./hr.:

| Unit price for forming bulkheads: | = | $65.30/6 |
|---|---|---|
| Labor price: | = | $10.88/ft. |

15.   Crew C is used for stripping forms, a high productivity rate should be attained on a simple job such as this:

|  |  | | Labor |
| --- | --- | --- | --- |
| *Framework Crew C* | | | $ |
| 1 Labor foreman | $24.00/hr. | = | 24.00 |
| 1 Carpenter | $28.00/hr. | = | 28.00 |
| 6 Laborers | $21.00/hr. | = | 126.00 |
|  |  | | $178.00/hr. |

| | | |
| --- | --- | --- |
| Unit price for stripping forms: | | $178.00/350 |
| Labor price: | = | $  .51/sq. ft. |

## Miscellaneous Work Figure 11.11

16. With a rebar crew consisting of 1 foreman and 5 rodmen, the output shown in Figure 11.6 is between 780 and 800 lbs/hr. installing #5 bars in footings. But, because the quantity is so small, an even lower output is expected, so a rate of 680 lbs/hr. is used:

|  |  | | Labor |
| --- | --- | --- | --- |
| *Rebar Crew* | | | $ |
| 1 Foreman | $30.00/hr. | = | 30.00 |
| 5 Rodmen | $27.00/hr. | = | 135.00 |
|  |  | | $165.00/hr. |

| | | |
| --- | --- | --- |
| Unit price for rebar to footings: | = | $65.00/680 lbs |
| Labor price: | = | $ 0.24/lb |

17. The unit rate for installing rebar in walls is expected to be the same as in footings.

18. If the quote to supply reinforcing steel is $500.00 per ton, "cut bent and tagged," then the

| | |
| --- | --- |
| Price per lb | $= \dfrac{\$500.00}{2000 \text{ lbs}}$ |
| | = $  0.25 |
| Add for waste 10% | = $  0.02 |
| Rebar material price: | $  0.27/lb |

19. The trowel finish requires the rental of a power trowel at $60.00 per day. From Figure 11.7, a crew C should be able to finish the slab-on-grade at the rate of at least 140 sq. ft./hour:

| | | |
| --- | --- | --- |
| Time to finish slab: | = | 991 sq. ft./140 |
| | = | 7.08 hrs. |
| So, the trowel is required for one day: | = | $ 60.00/991 sq. ft. |
| Equipment price: | = | $  0.06/sq. ft. |

PRICING SHEET

JOB _HOUSE EXAMPLE_                 DATE_____

ESTIMATED __A.B.F.__

Page No. [4 of 14]

| No. | DESCRIPTION | QUANTITY | UNIT | UNIT PRICE | LABOR | UNIT PRICE | MATERIALS | UNIT PRICE | EQUIPMENT | TOTAL |
|-----|-------------|----------|------|------------|-------|------------|-----------|------------|-----------|-------|
| | MISCELLANEOUS WORK | | | | | | | | | |
| 1. | #5 REBAR IN FTGS. | 329 | lbs | 0.24 | 79 | 0.27 | 89 | — | — | 168 |
| 2. | #4 REBAR IN WALLS | 420 | lbs | 0.24 | 101 | 0.27 | 113 | — | — | 214 |
| 3. | TROWEL FINISH S.O.G. | 991 | SF | 0.37 | 367 | — | — | 0.06 | 59 | 426 |
| 4. | SCREEDS TO S.O.G. | 991 | SF | 0.05 | 50 | 0.05 | 50 | — | — | 100 |
| 5. | CURING S.O.G. | 991 | SF | 0.05 | 50 | 0.05 | 50 | — | — | 100 |
| 6. | 6 MIL POLY. VAPOR BARRIER | 1090 | SF | 0.05 | 55 | 0.04 | 44 | — | — | 99 |
| 7. | 8×8×8/8 WELDED WIRE MESH | 1090 | SF | 0.21 | 229 | 0.07 | 76 | — | — | 305 |
| 8. | ½"×9" ANCHOR BOLTS | 23 | No. | 0.65 | 15 | 1.00 | 23 | — | — | 38 |
| 9. | WASTE ON MATERIALS | 10% | | | — | | 45 | — | | 45 |
| | | | | | 946 | | 490 | | 59 | 1495 |

**Figure 11.11**  House Example—Pricing Miscellaneous Work

|                          |              |   | Labor<br>$ |
|--------------------------|--------------|---|------------|
| *Miscellaneous Crew C*   |              |   |            |
| 0.3 Labor foreman        | $24.00/hr.   | = | 24.00      |
| 2 Cement Finishers       | $22.00/hr.   | = | 44.00      |
|                          |              |   | $ 68.00/hr. |

| Unit price for trowel finish: | = | $ 68.00/140 sq. ft. |
|-------------------------------|---|---------------------|
| Labor price:                  | = | $ 0.48/sq. ft.      |

20. The miscellaneous materials prices are based on quotes and conversion to appropriate unit rates as per Figure 11.8.

21. Setting screeds uses a crew A with an output of 1350 sq. ft. per hour:

|                          |              |   | Labor<br>$ |
|--------------------------|--------------|---|------------|
| *Miscellaneous Crew A*   |              |   |            |
| 0.3 Carpenter foreman    | $31.00/hr.   | = | 9.30       |
| 2 Carpenters             | $28.00/hr.   | = | 56.00      |
|                          |              |   | $ 65.30/hr. |

| Unit price for setting screeds | = | $ 65.00/1350 sq. ft. |
|--------------------------------|---|----------------------|
| Labor price:                   | = | $ 0.05/sq. ft.       |

22. Curing slabs uses a crew B with an output of 1000 sq. ft. per hour:

|                          |              |   | Labor<br>$ |
|--------------------------|--------------|---|------------|
| *Miscellaneous Crew B*   |              |   |            |
| 0.3 Labor foreman        | $24.00/hr.   | = | 7.20       |
| 2 Laborers               | $21.00       | = | 42.00      |
|                          |              |   | $ 49.20/hr. |

| Unit price for curing slabs: | = | $ 49.20/1000 sq. ft. |
|------------------------------|---|----------------------|
| Labor price:                 | = | $ 0.05/sq. ft.       |

23. Crew B is used to install vapor barrier at the rate of 1,000 sq. ft. per hour:

| Unit price for vapor barrier: | = | $ 49.20/1000 sq. ft. |
|-------------------------------|---|----------------------|
| Labor price:                  | = | $ 0.50/sq. ft.       |

24. The rebar crew is used for installing welded wire mesh and its productivity for this small area should be about 800 sq. ft. per hour:

| Unit price for 8 × 8 × 8/WWM | = | $165.00/800 sq. ft. |
|------------------------------|---|---------------------|
| Labor price:                 | = | $ 0.21/sq. ft.      |

25. Crew A can set anchor bolts at a rate of 100 bolts per hour:

| Unit price installing anchor bolts: | = | $ 65.30/100 no. |
|-------------------------------------|---|-----------------|
| Labor price:                        | = | $ 0.65/no.      |

26. Crew A can install a termite shield on the foundation wall at the rate of 85 ft. per hour:

| Unit price installing termite shield: | = | $65.30/85 ft. |
|---------------------------------------|---|---------------|
| Labor price:                          | = | $ 0.77/ft.    |

### Concrete Work Pricing Notes Example 2—Office/Warehouse Building

1. Figure 11.12 shows the Precision Estimating spreadsheet with the unit prices for labor, materials and equipment based on the following calculations.
2. Labor crews used for concrete work, formwork and miscellaneous work:

|  |  |  | *Labor* |
|---|---|---|---|
| *Formwork Crew A (FA)* |  |  | $ |
| 1 Carpenter foreman | $31.00/hr. | = | 31.00 |
| 6 Carpenters | $28.00/hr. | = | 168.00 |
| 2 Laborers | $21.00/hr. | = | 42.00 |
|  |  |  | $241.00/hr. |

|  |  |  |  |
|---|---|---|---|
| *Formwork Crew B (FB)* |  |  |  |
| 0.3 Carpenter foreman | $31.00/hr. | = | 9.30 |
| 2 Carpenters | $28.00/hr. | = | 56.00 |
|  |  |  | $ 65.30/hr. |

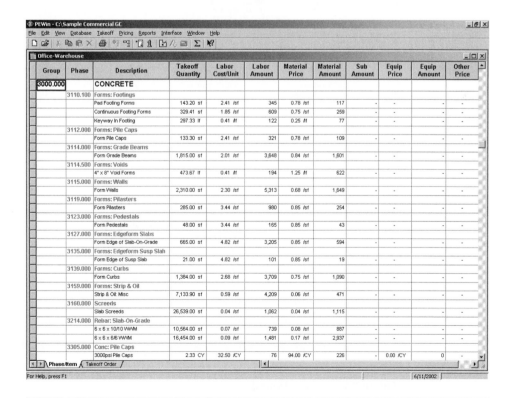

**Figure 11.12**    Spreadsheet—Concrete Prices

(Continued on next page)

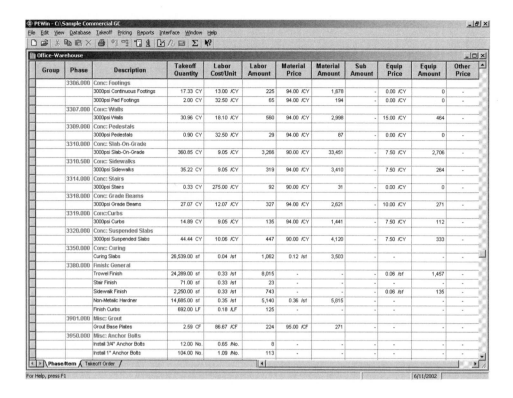

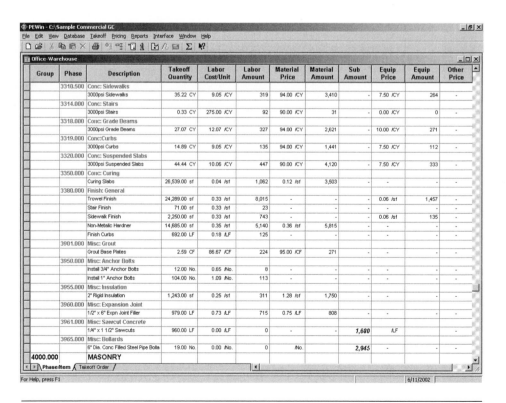

**Figure 11.12    continued**

|  |  |  | *Labor* $ |
|---|---|---|---|
| *Formwork Crew C (FC)* | | | |
| 1 Labor foreman | $24.00/hr. | = | 24.00 |
| 1 Carpenter | $28.00/hr. | = | 28.00 |
| 6 Laborers | $21.00/hr. | = | 126.00 |
|  |  |  | $178.00/hr. |
| | | | |
| *Reinforcing Crew A (RA)* | | | |
| 1 Foreman | 30.00/hr. | = | 30.00 |
| 5 Rodmen | 27.00/hr. | = | 135.00 |
|  |  |  | $165.00/hr. |
| | | | |
| *Reinforcing Crew B (RB)* | | | |
| 1 Labor foreman | $24.00/hr. | = | 24.00 |
| 3 Laborers | $21.00/hr. | = | 63.00 |
|  |  |  | $ 87.00/hr. |
| | | | |
| *Concrete Crew A (CA)* | | | |
| 1 Labor foreman | $24.00/hr. | = | 24.00 |
| 4 Laborers | $21.00/hr. | = | 84.00 |
| 1 Cement finisher | $22.00/hr. | = | 22.00 |
|  |  |  | $130.00/hr. |
| | | | |
| *Concrete Crew B (CB)* | | | |
| 1 Labor foreman | $24.00/hr. | = | 24.00 |
| 5 Laborers | $21.00/hr. | = | 105.00 |
| 1 Cement finisher | $22.00/hr. | = | 22.00 |
| 1 Equipment operator | $30.00/hr. | = | 30.00 |
|  |  |  | $181.00/hr. |
| | | | |
| *Concrete Crew D (CD)* | | | |
| 0.5 Labor foreman | $24.00/hr. | = | 12.00 |
| 1 Laborer | $21.00/hr. | = | 21.00 |
| 2 Cement finishers | $22.00/hr. | = | 44.00 |
|  |  |  | $ 77.00/hr. |
| | | | |
| *Miscellaneous Crew A (MA)* | | | |
| 0.3 Carpenter foreman | $31.00/hr. | = | 9.30 |
| 2 Carpenters | $28.00/hr. | = | 56.00 |
|  |  |  | $ 65.30/hr. |

|  | | Labor |
|---|---|---|
| *Miscellaneous Crew B (MB)* | | $ |
| 0.3 Labor foreman | $24.00/hr.  = | 7.20 |
| 2 Laborers | $21.00/hr.  = | 42.00 |
| | | $ 49.20/hr. |

| *Miscellaneous Crew C (MC)* | | |
|---|---|---|
| 0.3 Labor foreman | $24.00/hr.  = | 24.00 |
| 2 Cement finishers | $22.00/hr.  = | 28.00 |
| | | $ 52.00/hr. |

3.  Labor unit prices:

| Item | Crew | Productivity | | Labor Price |
|---|---|---|---|---|
| Form Pad Ftgs. & Pile Caps | FA | $241.00/100 sq. ft. | = | $2.41/sq. ft. |
| Form Strip Footings | FA | $241.00/130 sq. ft. | = | $1.85/sq. ft. |
| 2 × 4 Keyway | FB | $65.30/160 sq. ft. | = | $0.41/ft. |
| Form Grade Beams | FA | $241.00/120 sq. ft. | = | $2.01/sq. ft. |
| 4 × 8 Void Forms | FB | $65.30/160 ft. | = | $0.41/ft. |
| Form Walls | FA | $241.00/105 sq. ft. | = | $2.30/sq. ft. |
| Form Pilasters/Pedestals | FA | $241.00/70 sq. ft. | = | $3.44/sq. ft. |
| Form Edges of Slab-on-Grade | FA | $241.00/50 sq. ft. | = | $4.82/sq. ft. |
| Form Edges of Susp. Slab | FA | $241.00/50 sq. ft. | = | $4.82/sq. ft. |
| Form Curbs and Sidewalks | FA | $241.00/90 sq. ft. | = | $2.68/sq. ft. |
| Stripping Forms | FC | $178.00/300 sq. ft. | = | $0.59/sq. ft. |
| Slab Screeds | MA | $65.30/1500 sq. ft. | = | $0.04/sq. ft. |
| 6 × 6 × 10/10 WWM | RB | $87.00/1200 sq. ft. | = | $0.07/sq. ft. |
| 6 × 6 × 6/6 WWM | RB | $87.00/970 sq. ft. | = | $0.09/sq. ft. |
| Conc. Pile Caps/Pad Footings | CA | $130.00/4 cu. yd. | = | $32.50/cu. yd. |
| Conc. Continuous Footings | CA | $130.00/10 cu. yd. | = | $13.00/cu. yd. |
| Conc. Walls | CB | $181.00/10 cu. yd. | = | $18.10/cu. yd. |
| Conc. Pedestals | CA | $130.00/4 cu. yd. | = | $32.50/cu. yd. |
| Conc. Slab-on-Grade | CB | $181.00/20 cu. yd. | = | $9.05/cu. yd. |
| Conc. Sidewalks/Curbs | CB | $181.00/20 cu. yd. | = | $9.05/cu. yd. |
| Conc. Stair Treads | CD | $77.00/0.28 cu. yd. | = | $275.00/cu. yd. |
| Conc. Grade Beams | CB | $181.00/15 cu. yd. | = | $12.07/cu. yd. |
| Conc. Suspended Slabs | CB | $181.00/18 cu. yd. | = | $10.06/cu. yd. |
| Curing Slabs | MB | $49.20/1200 sq. ft. | = | $0.04/sq. ft. |

| | | | | |
|---|---|---|---|---|
| Trowel Finish | MC | $52.00/160 sq. ft. | = | $0.33/sq. ft. |
| Applying Floor Hardener | MC | $52.00/150 sq. ft. | = | $0.35/sq. ft. |
| Grout Base Plates | MC | $52.00/0.60 cu. ft. | = | $86.67/cu. ft. |
| Install ¾" Anchor Bolts | MA | $65.30/100 no. | = | $0.65/no. |
| Install 1" Anchor Bolts | MA | $65.30/60 no. | = | $1.09/no. |
| 2" Rigid Insulation | MB | $49.20/200 sq. ft. | = | $0.25/sq. ft. |
| ½" × 6" Expansion Joint | MA | $65.30/90 ft. | = | $0.73/ft. |

4. Concrete material prices are based on the quote shown on Figure 11.2.
5. The equipment price for the concrete pump is based on renting a pump for $150.00 per hour that, together with the placing productivities, gives the following prices:

| | | | |
|---|---|---|---|
| Conc. Pumping—Walls | $150.00/10 cu. yd. | = | $15.00/cu. yd. |
| Conc. Pumping—Grade Beams | $150.00/15 cu. yd. | = | $10.00/cu. yd. |
| Conc. Pumping—Slab-on-Grade | $150.00/20 cu. yd. | = | $ 7.50/cu. yd. |
| Conc. Pumping—Side-walks and Curbs | $150.00/20 cu. yd. | = | $ 7.50/cu. yd. |

6. Form material prices are obtained from previous records and can be calculated in the manner described on Figure 11.4.
7. Sales tax on materials will be included as an add-on. See Chapter 14.
8. Waste factors have been built into the database items so prices do not have to be increased to allow for waste.
9. Miscellaneous material prices:

| | | | |
|---|---|---|---|
| 6 × 6 × 10/10 WWM | $7.50/100 sq. ft. | = | $0.08/sq. ft. |
| 6 × 6 × 6/6 WWM | $16.50/100 sq. ft. | = | $0.17/sq. ft. |
| Nonmetallic hardener (1 lb per sq. ft.) | $0.36/lb | = | $0.36/sq. ft. |

## SUMMARY

■ This chapter concerns the pricing of concrete work that includes four main categories:
1. Supply and placing *concrete*
2. Construction and removal of *formwork*
3. Supply and placing *reinforcing steel*
4. *Miscellaneous items* associated with concrete work

■ The productivity of crews placing concrete is influenced by the following factors:

*Job Factors*

1. Methods of placing concrete
2. Rate of delivery of ready-mixed concrete
3. Properties of the concrete placed
4. Size and shape of concrete structures
5. Amount of rebar in the forms

*Labor and Management Factors*

 **1.** Quality of supervision
 **2.** Quality of job labor
 **3.** Motivation and morale of workers
 **4.** Type and quality of tools and equipment
 **5.** Experience and records of similar past projects

- Concrete can be placed using a number of different methods:
  **1.** "Pouring" direct from mixing trucks by means of chutes
  **2.** Using hand-operated or powered buggies
  **3.** Using crane and bucket setup
  **4.** Using concrete pumps
  **5.** Using conveyors
  **6.** Using a combination of two or more of these methods

- The cost of supplying concrete includes the cost of the ingredients plus the cost of mixing and transporting the mixture to the site. The basic price offered by ready-mix concrete suppliers generally covers these components and goes on to quote prices for a number of possible other requirements such as air entrainment and sulphate-resisting concrete.

- An allowance for waste and spillage of concrete is added as waste factor to concrete takeoff quantities.

- There are a number of different costs to consider when pricing a formwork system:
  **1.** Cost of building the system (fabrication)
  **2.** Cost of setting up the system each time it is used
  **3.** Cost of removing the system after concrete is placed (stripping)
  **4.** Other possible costs including transportation and storage expenses

- The factors that affect the productivity of crews fabricating and erecting formwork systems include the same general job factor and labor and management factors previously discussed plus the following:
  **1.** Potential for reuse of a system
  **2.** Complexity of form design
  **3.** Use of fly forms (gang forms)
  **4.** Number of form ties required for a system

- The cost of materials used in a formwork system would include the rental or purchase cost of all the components involved in its construction except for form ties and other hardware items, which are priced separately.

- Reinforcing steel for concrete includes welded wire mesh and reinforcing bars. Welded wire mesh is purchased in mats or rolls delivered to site for placing so there are two costs associated with mesh: the supply cost, including delivery to site, and the labor cost for installation. The cost of supplying and placing reinforcing steels bars on a project has a number of additional constituents including:
  **1.** The cost of rebar shop drawings
  **2.** The cost of the raw steel bars
  **3.** The cost of handling, cutting, bending and identifying (tagging) rebar in the fabrication shop
  **4.** The cost of transporting the bars to the site
  **5.** The cost of spacers, chairs, saddles and ties used in the installation of rebar
  **6.** The labor cost of installation

- In addition to the general job factors and labor and management factors discussed in Chapter 8, the productivity of labor installing reinforcing steel bars is affected by the following factors:
  **1.** The size and lengths of the reinforcing bars
  **2.** The shapes of the bars

3. The complexity of the concrete design
4. The amount of tolerance allowed in the spacing of bars
5. The amount of tying required

■ There are a number of items of work that are associated with concreting operations but not covered in the specific concrete, formwork and reinforcing steel sections in this chapter. These items include such work as concrete finishes, grouting and installing accessories like anchor bolts and waterstops that are required to be embedded in concrete.

■ Pricing and recapping concrete work can be done manually or via computer.

## REVIEW QUESTIONS

1. In what circumstances would a contractor consider setting up a concrete batch plant at the site of the work?

2. List five job factors that are specific to concrete work and should be considered whenever this type of work is priced.

3. What are demurrage charges sometimes imposed by concrete suppliers?

4. Discuss the impact of superplasticizers on the cost of placing concrete.

5. What cost advantages may be obtained from the use of fiber-reinforced concrete?

6. What are the dangers of using suppliers' current price lists to price concrete materials in an estimate?

7. Using the concrete price quotation shown on Figure 11.2, calculate the full price of 4500 psi concrete with type III cement, ½" aggregate and air entrainment to be delivered to the site in December.

8. List four job factors that are specific to formwork and should be considered whenever formwork is priced.

   *Note:* For questions 9 through 14, use the wage rates listed on page 256 under the section headed Wage Rates.

9. Calculate the cost per cubic yard to place concrete using each of the following two alternatives described. The work in both cases consists of placing 140 cubic yards of concrete in footings.
   a. Using a crane and bucket at a price of $120 per hour including an operator, with a crew consisting of 6 laborers, a cement finisher and a foreman. The concrete will be placed in 7 hours using this setup, and transportation of the crane costs $195.
   b. Using a concrete pump at a price of $193 per hour including an operator and a crew consisting of 5 laborers, a cement finisher and a foreman. With this setup, concrete is placed in 6 hours and transportation of the pump costs $145.

10. Calculate a labor price for placing concrete in columns where the columns are slender and widely spaced and, so, the productivity expected is low.
    a. Using a concrete pump
    b. Using a crane and bucket

11. Calculate a labor price per cubic yard for placing concrete in sidewalks where the material can be deposited directly from the mixer truck and high productivity is expected.

12. Calculate a labor price per square foot for forming the sides of large-size sumps where a medium productivity is expected.

13. Calculate a labor price per square foot for placing 4 × 4 × 10/10 mesh reinforcement in medium-size areas.

14. Calculate a labor price for installing 1" diameter anchor bolts when a low productivity is expected.

# 12

# Pricing Masonry, Carpentry and Finishes Work

## OBJECTIVES

*After reading this chapter and completing the review questions, you should be able to do the following:*

- Identify what is included in the price of masonry work.
- Describe the specific job factors that affect the price of masonry work.
- Explain what influences the material price of masonry bricks and blocks.
- Describe the specific job factors that affect the price of rough carpentry work.
- Describe how lumber sizes are specified.
- Explain what influences the material price of lumber.
- Explain how lumber is graded.
- Describe how hardware for rough carpentry is priced.
- Describe how finish carpentry is priced.
- Describe how interior and exterior finishes are priced.
- Complete the recap and pricing of masonry work, rough carpentry, finish carpentry and interior and exterior finishes using manual methods.
- Use the Precision Estimating software to price masonry work.

## Introduction

In this chapter we will consider the process of pricing masonry work, rough carpentry work, finish carpentry work and interior and exterior finishes.

Although much of this work is subcontracted to specialist subtrades these days, some contractors still like to use their own forces to perform part, if not all, of the work of these trades. Also, some of the specialized companies engaged in this work have a somewhat questionable reliability record. It is not uncommon to receive subtrade prices that are clearly in error—either too high or too low—and on more than one occasion an estimator has been left having to prepare a hasty estimate because subtrades have failed to deliver the prices they promised. So, the general contractor's estimator should remain acquainted with estimating these trades so that he or she can at least be able to put together a check price when required.

This section may also be of interest to subtrade estimators who normally do not estimate the labor cost of work in as much detail as shown herein. For instance, in the residential industry especially, a subcontractor will offer to perform carpentry framing work at a price per square foot of building area. This price is normally no more than the "going rate" for framing and often fails to reflect the complexity or any other job factors associated with the specific work involved. The practice of quoting "the going rate" like this can be defended on the grounds that residential work does not vary as much as commercial construction does, and that if a higher price were quoted, it would not be competitive. While these observations may be quite valid, the estimator who would like to forecast the cost of work more accurately should consider using the more detailed approach described in this chapter.

## Masonry

The cost of masonry work includes the cost of labor crews laying the masonry units and installing accessories such as horizontal ladder reinforcing and **wall ties**; masonry materials including accessories and mortar; equipment used in the mixing and hoisting of masonry materials; and erecting and dismantling temporary scaffolding and work platforms required by the work crews. Modularization of masonry work has been experimented with on some projects. This involves factory construction of masonry panels that are erected at the site employing techniques that are similar to those used in precast concrete work. But this practice is not common. Instead, most masonry work is executed on-site in the time honored way by masonry crews.

### Masonry Productivity

The job factors that affect the productivity of crews engaged in masonry work includes the general factors discussed in Chapter 8 and also a number of factors that are specific to masonry work including:

1. What is being built with the masonry units
2. The standards of workmanship required
3. The size and weight of the masonry units being used
4. The kind of mortar being used
5. The kind of bond being used
6. The complexity of the design of the work
7. The presence or lack of repetition in the work
8. The availability of workers skilled in the masonry trade

Higher outputs can be expected when constructing long lengths of plain walls rather than columns, pilasters and curved work.

When we discussed excavation work and concrete work in previous chapters, little was mentioned about the standard of workmanship and its influence on productivity. Producing good quality work does generally take longer, but there is a certain accepted level of workmanship required in excavation and concreting operations. To come short of this standard may save time in the short term, but in the long run, the cost of necessary rework will often negate the initial savings. Producing a higher than required level of quality is also usually uneconomic. While these statements can also be made of masonry work, some masonry construction is finishing work exposed to public view so more exacting standards may be demanded of this than would be required if the work is to be covered up. In consequence, where high standards are insisted on, productivity will be lower than where less rigid standards would be satisfactory.

The time taken per unit is usually longer with larger masonry units than with smaller, but the time taken to construct 100 square feet of a certain wall should be less using the larger units. Where the units are the same size, such as a standard brick or block size, productivity will be lower when building with heavier units.

Using some types of mortar can reduce productivity and production rates will be lower where complicated bonds or mixtures of various bonds are required in brickwork.

See Figure 12.1 for the productivity rates of a selection of items of masonry work.

### Masonry Materials

The prices of masonry units and accessories can vary greatly depending on requirements. Facing bricks and blocks for use in exposed areas are available from many sources in a wide variety of types and colors, which provides an extensive choice to the owner over a wide range of prices. Masonry specifications will normally spell out the precise types of bricks, blocks and accessories required to be used in the work, and the estimator, as usual, will endeavor to obtain firm prices for the supply of materials to meet these specifications.

Where specifications are vague or not provided for a project, the estimator can try to determine what is acceptable to the owner and base the price on the materials agreed on. Perhaps a better way to proceed in this situation is to prepare a price based on the use of certain materials defined in the bid and offer to adjust the price to meet the owner's specific requirements when they are known.

| CREW: | 1.0 Foreman | | |
| | 3.0 Bricklayers | | |
| | 2.0 Laborers | | |

| ITEM | OPERATION | OUTPUT | |
|------|-----------|--------|--|
| 1. | Brickwork Backup Walls | 0.210—0.275 M units/hr. | 0.210—0.275 M units/hr. |
| 2. | 4" Facing Bricks | 0.190–0.225 M units/hr. | 0.190–0.225 M units/hr. |
| 3. | 8" Block Walls | 45–103 Blocks/hr. | 45–103 Blocks/hr. |
| 4. | 10" Block Walls | 42–98 Blocks/hr. | 42–98 Blocks/hr. |
| 5. | 12" Block Walls | 40–92 Blocks/hr. | 40–92 Blocks/hr. |
| 6. | Extra Over for 8" Bond Beams | 100–150 ft/hr. | 30–46 m/hr. |
| 7. | Extra Over for 10" Bond Beams | 70–110 ft/hr. | 21–34 m/hr. |
| 8. | Extra Over for 12" Bond Beams | 45–80 ft/hr. | 14–24 m/hr. |
| 9. | Masonry Ties | 700–800 Ties/hr. | 700–800 Ties/hr. |
| 10. | Control Joint in 8" Wall | 220–300 ft./hr. | 67–91 m/hr. |
| 11. | Horizontal Ladder Reinforcing | 2100–2200 ft./hr. | 640–670 m/hr. |
| 12. | Loose Insulation to Blockwork | 7.0–9.0 cu. yd./hr. | 5.4–6.9 m³/hr. |

**Figure 12.1**   Masonry Work Productivities

Because supply lots do not always provide the exact number of bricks or blocks required, more units will usually be delivered to the site than can be used. This is accounted for by rounding up the takeoff quantities to the nearest supply lots or by applying a waste factor to the quantities or price of the materials. Wastage should also be allowed for on the quantity of mortar, accessories and all other materials used in masonry work. The average waste factor is generally between 5% and 15% of the total cost of masonry materials.

## Rough Carpentry

Increasingly these days, rough carpentry components are fabricated off-site and merely assembled at the job site in the same way as steel construction work is undertaken. Nevertheless, on many projects carpenters working on site in a temporary setup still perform all the operations necessary to transform raw lumber into the carpentry items required for the project.

There are generally only labor and material costs to consider when pricing rough carpentry work. Lumber or fabricated components are usually delivered to the site by the suppliers and, on low-rise construction, most pieces are then manhandled into position. The use of small mobile cranes may be necessary when heavier items such as glue-lam beams or large trusses are being used. On highrise projects the project crane, usually a centrally located tower crane, will normally hoist carpentry materials and components to where they are to be installed. Sometimes the use of mobile cranes may be required in addition to the tower crane for assembly of the larger size beams and trusses in situations in which the central crane cannot be used.

Labor and material prices are applied directly to the takeoff quantities listed on the rough carpentry recap, while crane and hoisting requirements are considered with the site overhead costs on the General Expenses pricing sheet.

### Rough Carpentry Productivity

The job factors that specifically affect the rate of productivity of labor fabricating and installing rough carpentry include:

1. The nature of the carpentry components being used
2. The size of cross-section and length of the pieces of lumber involved
3. The extent to which power tools are used in the work
4. The complexity of the design of the work
5. The presence or lack of repetition in the work
6. The extent to which components and assemblies are fabricated off-site

Fabricating and installing roof components such as hips, valleys, ridges and rafters requires far more time than fabricating and installing plain joists.

The time taken to fabricate and install 1000 board feet of lumber of the smaller cross-sections 2 × 4 and 2 × 6 will be greater than the time taken for the larger sizes because of the extra cutting and fitting required per board foot. More fabrication time will be required if lumber has to be cut into short lengths; this is required with some types of blocking.

Also, when short lengths of lumber are being used, the installation productivity per board foot will be less than when handling pieces of longer lengths. This last observation can be difficult for student estimators to accept since a longer piece of material will usually take longer to install than a shorter piece. The following example shows how the productivity in terms of board foot installed per hour can be lower while, at the same time, the number of pieces installed per hour is higher.

**EXAMPLE**

100 pieces of 20-foot-long 2 × 10 joists (3333 BF) are installed in 20 hours:

$$\text{Productivity rate} = \frac{100 \text{ pieces}}{20} \quad \text{or:} \quad = \frac{3333 \text{ BF}}{20}$$

$$= 5 \text{ pieces/hr.} \quad\quad\quad\quad = 167 \text{ BF/hr.}$$

140 pieces of 14-foot-long 2 × 10 joists (3267 BF) are installed in 20 hours:

$$\text{Productivity rate} = \frac{140}{20} \quad \text{or:} \quad = \frac{3267 \text{ BF}}{20}$$

$$= 7 \text{ pieces/hr.} \quad\quad\quad\quad = 163 \text{ BF/hr.}$$

Thus, when using smaller pieces of lumber, the number of pieces installed per hour increased while the amount of board feet installed per hour decreased.

---

The use of power drills and saws is now standard practice in construction, but where pneumatic nailers, staple guns and the like can also be used, productivity will be better than where these tools are not used.

Having to frame floor systems with complicated steps or unusual shapes will reduce productivity as will dealing with irregular wall systems and roof systems that have multitudes of hips and valleys. Simple designs will generally always facilitate the highest rate of production.

Where a design is repeated or merely where the operations involved in the work are repetitive, higher productivities can be expected. This can occur in situations in which the crew is installing the same kind of window or door a number of times in a project. Fabricating a large number of identical components like rafters can also lead to high productivity. Clearly, if each component of carpentry has to be custom made and a machine such as a radial arm saw has to be set differently for each piece cut, the production rate will be very low.

Many of the efficiencies discussed and the reduced costs associated with them are more likely to be achieved if rough carpentry components are fabricated off the site in factory-like conditions. While this is the way most trusses are now produced, off-site fabrication is also becoming more common with items such as beams, joists, roof components and even wall systems in some situations. This simplifies the estimating process as the only labor to consider is that required to assemble the components on-site.

See Figure 12.2 for the labor productivity rates of a selection of rough carpentry work items.

**Rough Carpentry Materials**

Specifications for lumber used in rough carpentry operations often define the species of lumber required and its use classification within that species. The grade of lumber required can also be specified; this is important where the lumber is to be used structurally.

Softwood lumber rather than hardwood is more commonly used for rough carpentry in most of North America. Species used in the construction industry include Douglas fir, balsam fir, Pacific Coast hemlock, eastern hemlock, Sitka spruce, white spruce, southern pine, white pine, lodgepole pine, western larch and various cedars and redwoods. Softwood lumber is classified by use into three groups:

1.  Yard lumber—Lumber used for general building purposes.
2.  Structural Lumber—Lumber of at least 2" nominal width that will be exposed to structural stresses.
3.  Factory and shop lumber—Lumber selected for remanufacturing use.

Softwood lumber is also classified according to the extent of processing in its manufacture:

1.  Rough lumber—Sawn lumber that has not been surfaced.
2.  Surfaced (dressed) lumber—Lumber that has been surfaced by a planer to provide a smooth finish on one or more sides. This lumber is referred to as S1S if one side is dressed, S2S if two sides are dressed and so on.
3.  Worked lumber—In addition to being surfaced, worked lumber has been further shaped to produce such products as tongue and groove boards.

| CREW: | 0.3 Carpenter Foreman | | | | | | |
| | 2.0 Carpenters | | | | | | |

| ITEM | OPERATION | OUTPUT | | ITEM | OPERATION | OUTPUT | |
|---|---|---|---|---|---|---|---|
| 1. | 3" Diameter Telescopic Post | 1.50–2.50 no./hr. | 1.50–2.50 no./hr. | 17. | Trusses 24'–40' Span | 2–3 no./hr. | 2–3 no./hr. |
| 2. | 2 × 3 Plates | 50–54 BF/hr. | 30–33 m/hr. | 18. | 2 × 2 Cross-Bridging | 30–34 Sets/hr. | 30–34 Sets/hr. |
| 3. | 2 × 4 Plates | 63–68 BF/hr. | 29–31 m/hr. | 19. | 1 × 3 Rafter Ties (Ribbons) | 25–35 BF/hr. | 30–34 m/hr. |
| 4. | 2 × 6 Plates | 90–95 BF/hr. | 27–29 m/hr. | 20. | 1 × 6 Rafter Ties (Ribbons) | 30–40 BF/hr. | 18–24 m/hr. |
| 5. | 2 × 3 Studs | 79–92 BF/hr. | 48–56 m/hr. | 21. | 2 × 4 Rafters | 60–110 BF/hr. | 27–50 m/hr. |
| 6. | 2 × 4 Studs | 100–115 BF/hr. | 46–53 m/hr. | 22. | 2 × 6 Rafters | 65–120 BF/hr. | 20–37 m/hr. |
| 7. | 2 × 6 Studs | 110–125 BF/hr. | 34–38 m/hr. | 23. | 2 × 4 Lookouts | 30–50 BF/hr. | 14–23 m/hr. |
| 8. | 2 × 8 in Built-Up Beam | 150–260 BF/hr. | 34–59 m/hr. | 24. | 2 × 4 Rough Fascia | 30–50 BF/hr. | 14–23 m/hr. |
| 9. | 2 × 10 in Built-Up Beam | 200–310 BF/hr. | 37–57 m/hr. | 25. | 2 × 6 Rough Fascia | 35–60 BF/hr. | 11–18 m/hr. |
| 10. | 2 × 8 Joists | 150–180 BF/hr. | 34–41 m/hr. | 26. | T & G Floor Sheathing | 150–160 sq. ft./hr. | 14–15 m²/hr. |
| 11. | 2 × 10 Joists | 160–185 BF/hr. | 29–34 m/hr. | 27. | Wall Sheathing | 130–150 sq. ft./hr. | 12–14 m²/hr. |
| 12. | 2 × 2 Blocking | 12–25 BF/hr. | 11–23 m/hr. | 28. | Roof Sheathing | 160–180 sq. ft./hr. | 15–17 m²/hr. |
| 13. | 2 × 4 Blocking | 20–35 BF/hr. | 9–16 m/hr. | 29. | ½" Ply Soffit | 40–60 sq. ft./hr. | 4–6 m²/hr. |
| 14. | 2 × 6 Blocking | 25–45 BF/hr. | 8–14 m/hr. | 30. | Aluminum Soffit | 40–60 sq. ft./hr. | 4–6 m²/hr. |
| 15. | 2 × 8 Blocking | 30–50 BF/hr. | 7–12 m/hr. | 31. | 1 × 6 Fascia Board | 15–20 BF/hr. | 9–12 m/hr. |
| 16. | 2 × 10 Blocking | 35–55 BF/hr. | 6–10 m/hr. | 32. | 6" Aluminum Fascia | 30–40 sq. ft./hr. | 9–12 m/hr. |

**Figure 12.2**    Rough Carpentry Productivities Work

The size of lumber is stated in accordance with its nominal size, which is the dimension of the cross-section of the piece before it has been surfaced. Thus $2 \times 4$ size pieces were approximately $2" \times 4"$ in cross-section before they were surfaced but now have a cross-section of $1\frac{1}{2}" \times 3\frac{1}{2}"$ after the lumber is dressed. Softwood lumber is further classified according to its nominal size:

1. Boards—Lumber less than 2" in thickness and 2" or more in width.
2. Dimension—Lumber from 2" up to, but not including, 5" in thickness and 2" or more in width.
3. Timber—Lumber 5" or over in its least dimension.

*Lumber Lengths and Waste Factors*

Lumber is generally available in lengths that are multiples of 2 feet. Studs, however, can be obtained precut to the required length. When taking off joist quantities, the estimator will usually allow for length of the joists used even thought the span may not be a multiple of 2-feet. For example, if the span of the joists is 12'6", the estimator will take off 14-foot joists. If this is not done, a waste factor of 12% (14 – 12.5/12.5) should be allowed on the material for these joists in the pricing process.

Even where the 2-foot multiple lengths are allowed for in the takeoff, a waste factor is still required as components such as plates, beams, headers and suchlike usually require some cutting and there has to be allowance for some poor quality lumber and error. This waste factor is usually between 5% and 15%.

*Lumber Grades*

Within lumber's use classification, lumber is graded in order to provide information to users so that they can assess its ability to meet requirements.

Yard lumber is graded as being either Selects or Commons in which Selects are pieces of good appearance and finishing qualities suitable for natural or painted finishes, and Commons are pieces suitable for general construction and utility purposes.

Structural lumber is stress graded in order to be able to specify the safe working stresses that can be applied to it. Tables are published that list the allowable stresses for structural lumber in each of the following use categories including:

1. Light framing
2. Structural light framing
3. Structural joists and planks
4. Appearance framing
5. Decking
6. Beams and stringers
7. Posts and columns

Within each of these categories there are one or more grades. Allowable stresses are specified according to species of lumber and grade within each of the categories. Light framing, for example, is divided into Construction Grade, Standard Grade and Utility Grade for each species of lumber used in this capacity. Structural light framing generally has five grades for each species:

1. Select Structural
2. No. 1
3. No. 2
4. No. 3
5. Stud

A full specification for structural lumber would call for a certain category, grade and species of lumber to be used. For instance, a housing specification may state: "All loadbearing stud walls shall be framed using Structural Light Framing graded lumber of No. 2 or better Hem-Fir." More often, abbreviated versions of such a specification are found calling merely for "No. 2 or better." The task of the estimator, as usual, is to ensure that the lumber prices used in the estimate reflect the specification requirements for lumber.

An example of a lumber company price list is shown as Figure 12.3. This price list will be used for pricing lumber in the examples that follow.

*Rough Hardware*

Rough hardware for lumber mostly consists of nails, but other fasteners and such items as joist hangers may also be required. Many estimators allow for rough carpentry hardware in the same way we allowed for formwork hardware. Recall that this was done by keeping track of hardware costs of previous jobs in terms of a percentage of total lumber costs, then applying this historical rate to the estimated cost of lumber for future jobs. An alternative method of estimating hardware costs, which

**ABC LUMBER INC.**

**DOUGLAS FIR AND LARCH S4S**
Dimension and Timbers: 2x4 & 4x4 (Std. & Btr.)
2x6, 4x6 and wider (#2 & Btr.)

| SIZES | LENGTH | "GREEN" | "DRY" |
|---|---|---|---|
| 2x4 | 8 - 14' | | $600.00 |
| | 16 - 20' | $650.00 | $650.00 |
| 2x6 | 8 - 14' | | $600.00 |
| | 16 - 20' | | $650.00 |
| 2x8 | 8 - 16' | $540.00 | |
| | 18' | $580.00 | |
| | 20' | $580.00 | |
| 2x10 | 8 - 16' | $680.00 | |
| | 18 - 20' | $700.00 | |

**PLANKS AND TIMBERS**

| | | |
|---|---|---|
| S4S | 4x4 | $480.00 |
| S4S | 4x6 | $500.00 |
| ROUGH | 6x6 | $420.00 |
| ROUGH | 8x8 | $560.00 |
| ROUGH | 3x12 | $700.00 |

**"S-DRY SPRUCE"**

| | | |
|---|---|---|
| 2x3 - 92 5/8" Studs | | $460.00 |
| 2x4 - 92 5/8" Studs | | $460.00 |
| 2x6 - 92 5/8" Studs | | $460.00 |
| 2x3 | 8' - 12' | $460.00 |
| | 14 -16' | $480.00 |
| 2x4 | 8 -12' | $480.00 |
| | 14 -16' | $500.00 |
| | 18 - 20' | $530.00 |
| 2x6 | 8 -12' | $480.00 |
| | 14 -16' | $500.00 |
| | 18 - 20' | $520.00 |
| 2x8 | 8 -12' | $500.00 |
| | 14 -16' | $530.00 |
| | 18 - 20' | $600.00 |
| 2x10 | 8 - 20' | $650.00 |

**P.W.F. LUMBER (.50 LBS. CU.FT.)**

| | | |
|---|---|---|
| 2x4 | 8 - 14' | $620.00 |
| | 16' | $650.00 |
| 2x6 | 10 - 14' | $620.00 |
| | 16' | $650.00 |
| 2x8 | 10 - 14' | $620.00 |
| | 16' | $650.00 |

**STRAPPING - BRIDGING** — **Per LF**

| | | |
|---|---|---|
| 1x1 | Spruce | $0.08 |
| 1x2 | Spruce | $0.09 |
| 1x3 | Spruce | $0.15 |
| 1x6 | Spruce | $0.29 |
| 2x2 | D. Fir | $0.24 |

Bridging 2x2 Precut for
2x8, 2x10 at 16" on Center per Length          0.35

| SHEATHING | STD. FIR PER SHEET | SELECT FIR PER SHEET |
|---|---|---|
| 3/8" | $15.75 | $18.00 |
| 1/2" | $21.00 | $23.20 |
| 5/8" | $25.25 | $28.50 |
| 3/4" | $30.25 | $33.70 |
| 5/8" T & G | $26.95 | $29.30 |
| 3/4" T & G | $31.95 | $34.50 |

**Figure 12.3** Quote for Supply of Lumber

is perhaps more precise, is to take off the quantity of specific items of hardware required or estimate the quantity of nails required; then price these quantities.

The amount of nails required for rough carpentry can be roughly determined by allowing 7 lbs of nails for every 1000 board feet of lumber and 10 lbs of nails for every 1000 square feet of sheathing.

## Finish Carpentry and Millwork

The process of pricing finish carpentry and millwork consists of applying materials and labor prices to the items measured in the takeoff process. Many finish carpentry components such as stairs, doors, windows and cabinets are manufactured goods that are constructed off-site, delivered to the site by the supplier and merely installed by the site labor crews. Hand tools and some crane time for unloading and hoisting is often required with the installation of finish carpentry. Both of these items are priced with the site overheads on the General Expenses sheet.

The productivity of labor crews installing finish carpentry and millwork is influenced by job factors similar to those discussed for rough carpentry. Expected productivity rates for crews installing finish carpentry and millwork are shown on Figure 12.4.

## Exterior and Interior Finishes

The most dependable way of estimating the cost of exterior and interior finishing work follows the same procedure we have adopted with the preceding trades in this text. That is to measure the quantity of each individual item of the trade in the takeoff process, list the takeoff items on a recap and apply separate materials and labor prices to the items listed. Material prices are determined from suppliers who are required to quote prices to supply and deliver items that meet specifications. Some manipulation of material prices is again often necessary to convert prices for supply units of measurement to prices for takeoff units of measurement as we discussed when considering concrete items.

Labor productivities are once again affected by generally the same job factors and labor and management factors we considered with the other trades examined previously. A list of expected productivity rates for labor crews on exterior finish and interior finish work is shown on Figure 12.5.

## Wage Rates

The wage rates of workers in the following examples are:

| *Basic Hourly Wage* | *$* |
|---|---|
| Masonry foreman | 31.00 |
| Mason | 28.00 |
| Equipment operator | 30.00 |
| Labor foreman | 24.00 |
| Laborer | 21.00 |
| Cement finisher | 22.00 |
| Carpentry foreman | 31.00 |
| Carpenter | 28.00 |
| Painting foreman | 28.00 |
| Painter | 25.00 |

The outputs given represent productivity rates for the INSTALLATION of stock items of Millwork and Finish Carpentry using the following crew:

CREW  0.3 Carpenter Foreman
2.0 Carpenters

| ITEM | OPERATION | OUTPUT | |
|------|-----------|--------|--|
| 1. | 3'0" Wide Stair with 13 risers | 0.4–0.6 units/hr. | |
| 2. | 2 × 2 Handrail | 12–20 ft./hr. | 3.6–6.1 m/hr. |
| 3. | Wood Railings | 8–12 ft./hr. | 2.4–3.7 m/hr. |
| 4. | Exterior Doors with Frame and Trim | | |
| | —2'8" × 6'8" × 1¾" | 0.75–1.25 units/hr. | |
| | —3'0" × 6'8" × 1¾" | 0.65–1.00 units/hr. | |
| 5. | Interior Doors with Frame and Trim | | |
| | —1'6" × 6'8" × 1⅜" | 0.90–1.35 units/hr. | |
| | —2'0" × 6'8" × 1⅜" | 1.85–1.30 units/hr. | |
| | —2'6" × 6'8" × 1⅜" | 0.80–1.25 units/hr. | |
| | —2'8" × 6'8" × 1⅜" | 0.75 × 1.20 units/hr. | |
| 6. | Bifold Closet Doors with Jambs and Trim | | |
| | —2'0" × 6'8" × 1⅜" | 0.85–1.00 units/hr. | |
| | —3'0" × 6'8" × 1⅜" | 0.80–0.85 units/hr. | |
| | —4'0" × 6'8" × 1⅜" | 0.70–0.90 units/hr. | |
| | —5'0" × 6'8" × 1⅜" | 0.65–0.85 units/hr | |
| 7. | Attic Access Hatch | 1.00–2.00 units/hr. | |
| 8. | Finish Hardware to Doors | | |
| | —Passage Sets | 2.00–3.00 no./hr. | |
| | —Privacy Sets | 2.00–3.00 no./hr. | |
| | —Dead Bolts | 1.50–2.50 no./hr. | |
| | —Key-in-knob Lock/Latch | 1.75–2.75 no./hr. | |
| 9. | Window Trim | 40–60 ft./hr. | 12.2–18.3 m/hr |
| 10. | Windows: Installing Manufactured Units | | |
| | —3'0" × 1'0" | 2.20–3.20 units/hr. | |
| | —2'0" × 3'6" | 1.00–3.00 units/hr. | |
| | —4'0" × 3'6" | 1.75–2.75 units/hr. | |
| | —6'0"–3'6" | 1.50–2.50 units/hr. | |
| | —4'0" × 4'0" | 1.50–2.50 units/hr. | |
| | —4'0" × 6'0" | 1.30–2.30 units/hr. | |
| | —4'0" × 8'0" | 1.20–2.20 units/hr. | |
| 11. | Base Board | 50–60 ft./hr. | 15.2–18.3 m/hr. |
| 12. | Floor Cabinets Including Countertop | 2.0–4.0 ft./hr. | 0.6–1.2 m/hr. |
| 13. | 1'0" × 26" Wall Cabinets | 3.0–8.0 ft./hr. | 0.9–2.4 m/hr. |
| 14. | 1'0" × 1'10" Wall Cabinets | 3.5–8.5 ft./hr. | 1.0–2.6 m/hr. |
| 15. | 1'0" × 26" Ceiling Hung Cabinets | 2.5–6.0 ft./hr. | 0.8–1.8 m/hr. |
| 16. | 2'0" × 26" Bathroom Vanities | 4.0–5.5 ft./hr. | 1.2–1.7 m/hr. |
| 17. | 12" Wide Closet Shelves | 24.0–28.0 ft. hr | 7.3–8.5 m/hr. |
| 18. | 16" Wide Closet Shelves | 22.0–26.0 ft./hr. | 6.7–7.9 m/hr. |
| 19. | Adjustable Closet Rods | 4.00–7.00 units/hr. | 4.00–7.00 no./hr. |
| 20. | Bathroom Accessories | | |
| | —Toilet Roll Holders | 7.00–8.00 no./hr. | |
| | —2'0" × 3'0" Mirrors | 3.00–4.00 no./hr. | |
| | —Medicine Cabinets | 2.00–3.00 no./hr. | |
| | —Shower Curtain Rod | 4.00–6.00 no./hr. | |

**Figure 12.4** Finish Carpentry and Millwork Productivities

CREW A: 0.3 Carpenter Foreman
2.0 Carpenters

CREW B: 0.3 Labor Foreman
2.0 Laborers

CREW C: 0.3 Labor Foreman
2.0 Cement Finishers

CREW D: 0.3 Painter Foreman
2.0 Painters

## EXTERIOR FINISHES

| ITEM | OPERATION | CREW | OUTPUT | |
|---|---|---|---|---|
| 1. | Sprayed Damp-Proofing | | | |
| | —1 Coat | B | 200–220 sq. ft./hr. | 18.5–20.4 m²/hr. |
| | —2 Coats | B | 120–140 sq. ft./hr. | 11.2–13.0 m²/hr. |
| 2. | ½" Parging | C | 28–34 sq. ft./hr. | 2.6–3.2 m²/hr. |
| 3. | Aluminum Siding | A | 60–65 sq. ft./hr. | 5.6–6.0 m²/hr. |
| 4. | Building Paper | A | 850–900 sq. ft./hr. | 79–84 m²/hr. |
| 5. | 4" Wide Flashing | A | 90–130 ft./hr. | 27.4–39.6 m/hr. |
| 6. | 5" Eaves Gutter | A | 20–30 ft./hr. | 6.1–9.1 m/hr. |
| 7. | 3" Downspouts | A | 30–48 ft./hr. | 9.1–14.6 m/hr. |
| 8. | 210# Asphalt Shingles | A | 125–138 sq. ft./hr. | 11.6–12.8 m²/hr. |
| 9. | Ridge Cap | A | 90–100 ft./hr. | 27.4–30.5 m/hr. |
| 10. | Drip-Edge Flashing | A | 90–100 ft./hr. | 27.4–30.5 m/hr. |
| 11. | 6-Mil Poly. Eaves Protection | A | 800–900 sq. ft./hr. | 74–84 m²/hr. |
| 12. | Precast Concrete Steps | | | |
| | —2 Risers | B | 0.4–0.6 no./hr. | 0.4–0.6 no./hr. |
| | —4 Risers | B | 0.3–0.5 no./hr. | 0.3–0.5 no./hr. |
| 13. | Wrought Iron Railing | A | 2.0–4.0 ft./hr. | 0.6–1.2 m/hr. |
| 14. | Paint Iron Railing | D | 10–14 ft./hr. | 3.0–4.3 m/hr. |

## INTERIOR FINISHES

| ITEM | OPERATION | CREW | OUTPUT | |
|---|---|---|---|---|
| 15. | 32 oz. Carpet | A | 120–130 sq. ft./hr. | 11–12 m²/hr. |
| 16. | Sheet Vinyl Flooring | A | 48–62 sq. ft./hr. | 4.5–5.8 m²/hr. |
| 17. | ½" Drywall Ceilings | A | 90–95 sq. ft./hr. | 8.4–8.8 m²/hr. |
| 18. | Textured Ceiling Finish | D | 180–200 sq. ft./hr. | 17–19 m²/hr. |
| 19. | R35 Loose Cellulose Insln. | B | 150–160 sq. ft./hr. | 14–15 m²/hr. |
| 20. | 6-Mil Poly. Vapor Barrier | A | 850–950 sq. ft./hr. | 79–88 m²/hr. |
| 21. | Insulation Stops | A | 20–30 no./hr. | 20–30 no./hr. |
| 22. | R20 Batt Insulation | A | 260–340 sq. ft./hr. | 24–32 m²/hr. |
| 23. | R12 Batt Insulation | A | 300–400 sq. ft./hr. | 28–37 m²/hr. |
| 24. | ½" Drywall Walls | A | 110–120 sq. ft./hr. | 10–11 m²/hr. |
| 25. | Spray Paint Walls | D | 375–425 sq. ft./hr. | 35–39 m²/hr. |
| 26. | Paint Exterior Doors | D | 40–45 sq. ft./hr. | 3.7–4.2 m²/hr. |
| 27. | Stain Doors | D | 50–55 sq. ft./hr. | 4.6–5.1 m²/hr. |
| 28. | Paint Attic Access Hatch | D | 3.0–4.0 no./hr. | 3.0–4.0 no./hr. |
| 29. | Paint Handrail | D | 90–95 ft./hr. | 27–29 m/hr. |
| 30. | Paint Wood Railing | D | 12–14 ft./hr. | 3.7–4.3 m/hr. |
| 31. | Paint Shelves | D | 60–64 sq. ft./hr. | 5.6–5.9 m²/hr. |

**Figure 12.5** Exterior and Interior Finishes Productivities

## Masonry, Rough Carpentry and Finish Carpentry Recap and Pricing Notes Example 1—House

### Masonry Work Figure 12.6

1. Masonry items are listed from the takeoff and two further items are added: scaffold and waste.

2. All the materials prices used on this recap are based on quotes to supply and deliver to site materials that meet specifications.

3. It is assumed that this house is one of a number of jobs in progress so that materials can be shared among projects. If this were not the case, the material prices and waste factor would be greater to reflect the minimum charges made by suppliers when very small quantities are ordered.

4. The labor prices are based on the following masonry crew:

| Masonry Crew | | | Labor $ |
|---|---|---|---|
| 1.0 Masonry foreman | $31.00/hr. | = | 31.00 |
| 3.0 Masons | $28.00/hr. | = | 84.00 |
| 2.0 Laborers | $21.00/hr. | = | 42.00 |
| | | | $157.00/hr. |

5. The productivity of this crew is taken to be on the low end of the range because this is a small project. For example the productivity for laying facing bricks is taken to be 220 bricks per hour:

   Thus the labor unit price for brickwork is:     $157.00/220 units

   = $ 0.71/unit

6. Because this is such a small-scale project, it is not considered necessary to provide a full-size scaffold system; so only a nominal amount has been allowed for scaffold.

**PRICING SHEET**

JOB....HOUSE EXAMPLE........................ DATE.....................

ESTIMATED.......................

Page No. | 5 of 14 |

| No. | DESCRIPTION | QUANTITY | UNIT | UNIT PRICE | LABOR | UNIT PRICE | MATERIALS | UNIT PRICE | EQUIPMENT | TOTAL |
|---|---|---|---|---|---|---|---|---|---|---|
| | MASONRY WORK | | | | | | | | | |
| 1. | FACING BRICKS | 1267 | No. | 0.60 | 805 | 0.45 | 604 | | — | 1409 |
| 2. | MORTAR | 7 | CF | | 1 | 4.00 | 28 | | — | 28 |
| 3. | BRICK TIES | 91 | No. | 0.22 | 20 | 0.14 | 13 | | — | 33 |
| 4. | 3"x3"x¼" ANGLE | 36 | LF | 1.25 | 45 | 3.00 | 108 | | — | 153 |
| 5. | ½" DRILLED ANCHORS | 12 | No. | 3.00 | 36 | 2.50 | 30 | | — | 66 |
| 6. | 12" WIDE FLASHING | 36 | LF | 0.90 | 32 | 1.00 | 36 | | — | 68 |
| 7. | SCAFFOLD | ALLOW | | | 75 | | 40 | | — | 115 |
| 8. | WEEP HOLES | 7 | No. | | 1 | 1.25 | 9 | | — | 9 |
| 9. | WASTE | 10% MATL. | | 868 | 1 | | 87 | | — | 87 |
| | TOTAL MASONRY TO SUMMARY | | | | 1013 | | 955 | | — | 1968 |

**Figure 12.6**   House Example—Masonry Pricing

### Rough Carpentry Work Figure 12.7

7. Rough carpentry items are listed in the order in which they were taken off, but where the same type of work appears more than once in the takeoff, quantities are added together so that there is just one quantity of this item to price. Note that the takeoff quantities are summed in the margin beside the item on the recap.

8. All lumber prices used on the recap are based on the quote shown on Figure 12.3. S-dry spruce is used for all $2 \times 4$ and $2 \times 6$ lumber, and Douglas fir and larch is used for larger sizes.

   Other prices used for teleposts, trusses and the like are assumed to be based on quotes obtained from suppliers.

9. P.W.F. (preserved wood foundation) lumber is used for the sill plate at the price of $620.00 per 1000 board feet as quoted.

10. The labor unit prices for rough carpentry work are based on using the following crew:

| Rough Carpentry Crew | | | *Labor* $ |
|---|---|---|---|
| 0.3 Carpentry foreman | $31.00/hr. | = | 9.30 |
| 2.0 Carpenters | $28.00/hr. | = | 56.00 |
| | | | $63.50/hr. |

11. Labor productivities close to the middle of the range shown on Figure 12.2 for each item are adopted except where noted. So, for installing teleposts, a productivity of 2.00 posts/hr. is used.

    Thus the labor unit price for teleposts is:     $65.30/2 no.
          =   $32.65/no

12. Because $2 \times 4$ cleats are a form of blocking, a productivity of 30 BF/hr. is used:

    Labor unit price for $2 \times 4$ Cleats     =   $65.30/30
          =   $2.18/BF

13. The price of precut $2 \times 2$ bridging is quoted as $0.35 per length on Figure 12.3. Two lengths are required for a set of bridging, so the unit price per set is $0.70.

14. ¾" T & G (tongue and groove) sheathing in standard fir is quoted at $31.95 per 4'0" × 8'0" sheet.

    Thus the material unit price for ¾" sheathing   =   $31.95/32
          =   $1.00/SF

15. A productivity of 55 BF/hr. is used for $2 \times 10$ lintels since this work is similar to installing long pieces of blocking; 55 BF/hr. is on the high end of the range of productivities for $2 \times 10$ blocking.

16. The $2 \times 4$ lintels are not as long as the $2 \times 10$ lintels, so 30 BF/hr. is used which is *toward* the high end of the $2 \times 4$ blocking productivity range.

17. $1 \times 3$ spruce is quoted as $0.15 per LF. As a price per BF is required for the $1 \times 3$ ribbons, this price has to be converted in this way:

$$\$0.15 \times \frac{12}{3} = \$0.60/BF$$

PRICING SHEET
JOB...... HOUSE EXAMPLE ...... DATE......
ESTIMATED......

| No. | DESCRIPTION | QUANTITY | UNIT | UNIT PRICE | LABOR | UNIT PRICE | MATERIALS | UNIT PRICE | EQUIPMENT | TOTAL |
|---|---|---|---|---|---|---|---|---|---|---|
| | ROUGH CARPENTRY | | | | | | | | | |
| 1. | 2x6 SILL PLATE | 136 | BM | 0 72½ | 98 | 0 62½ | 84 | | — | 182 |
| 2. | 3" DIA. STEEL TELEPOSTS | 3 | No. | 32 65 | 98 | 27 50 | 83 | | — | 181 |
| 3. | 2x10 IN BUILTUP BEAM | 267 | BM | 0 25 | 67 | 0 68 | 182 | | — | 249 |
| 4. | 2x10 JOISTS | 1600 | BM | 0 36 | 576 | 0 68 | 1088 | | — | 1664 |
| 5. | 2x4 CLEATS | 41 | BM | 2 18 | 89 | 0 48½ | 20 | | — | 109 |
| 6. | 2x2 X-BRACING | 60 | SETS | 1 98 | 119 | 0 70 | 42 | | — | 161 |
| 7. | 3/4" T&G PLY FLOOR SHEATHING | 1080 | SF | 0 42 | 454 | 1 00 | 1080 | | — | 1534 |
| 8. | 2x6 PLATES | 424 | BM | 0 70 | 297 | 0 48½ | 204 | | — | 501 |
| 9. | 2x6 STUDS | 1176 | BM | 0 55 | 647 | 0 46½ | 541 | | — | 1188 |
| 10. | 3/8" WALL SHEATHING | 1264 | SF | 0 47 | 594 | 0 49 | 619 | | — | 1213 |
| 11. | 2x10 LINTELS | 214 | BM | 2 18 | 467 | 0 68½ | 146 | | — | 613 |
| 12. | 2x4 PLATES | 652 | BM | 1 00 | 652 | 0 48½ | 313 | | — | 965 |
| 13. | 2x4 STUDS | 1531 | BM | 0 60 | 919 | 0 46½ | 704 | | — | 1623 |
| | CARRIED FWD: | | | | 5077 | | 5106 | | | 10183 |

**Figure 12.7**    House Example—Rough Carpentry Pricing

**PRICING SHEET**

JOB: HOUSE EXAMPLE   DATE: ..........

ESTIMATED: ..........

Page No. | 7 of 14 |

| No. | Description | Quantity | Unit | Unit Price | Labor | Unit Price | Materials | Unit Price | Equipment | Total |
|---|---|---|---|---|---|---|---|---|---|---|
| | BROUGHT FWD: | | | | 5077 | | 5106 | | | 10183 |
| 1. | 2x4 LINTELS | 35 | BM | 2/18 | 76 | .048 | 17 | | | 93 |
| 2. | "W" TRUSSES 28' SPAN | 19 | No. | 26/10 | 496 | 55.00 | 1045 | | | 1541 |
| 3. | GABLE ENDS | 2 | No. | 27/00 | 54 | 60.00 | 120 | | | 174 |
| 4. | 1x3 RIBBONS | 50 | BM | 2/18 | 109 | .60 | 30 | | | 139 |
| 5. | 2x4 RIDGE BLOCKING | 30 | BM | 2/25 | 68 | .048 | 14 | | | 82 |
| 6. | 2x4 BARGE RAFTER | 46 | BM | 0/76 | 35 | .048 | 22 | | | 57 |
| 7. | 2x4 LOOKOUTS | 85 | BM | 2/18 | 185 | .048 | 41 | | | 226 |
| 8. | 2x6 ROUGH FASCIA | 88 | BM | 1/30 | 114 | .048 | 42 | | | 156 |
| 9. | 2x4 CEILING BLOCKING | 91 | BM | 2/18 | 198 | .048 | 44 | | | 242 |
| 10. | 3/8" 12y ROOF SHEATHING | 1533 | SF | 0/38 | 583 | .49 | 751 | | | 1334 |
| 11. | VENTED ALUM. SOFFIT | 342 | SF | 1/32 | 451 | 1.80 | 616 | | | 1067 |
| 12. | ALUM. "J" MOLD | 152 | LF | 0/85 | 129 | .50 | 76 | | | 205 |
| 13. | ALUM. FASCIA 6" WIDE | 158 | LF | 1/90 | 300 | .90 | 142 | | | 442 |
| 14. | ROUGH HARDWARE | ALLOW | | | | | 120 | | | 120 |
| 15. | WASTE 10% MATL. | | | | 1 | | | | | |
| | TOTAL ROUGH CARP. TO SUMMARY: | | | | 7875 | | 8186 | | | 16061 |

**Figure 12.7 continued**

18. A productivity of about 80 LF/hr. is expected from this carpentry crew installing the "J" mold for the aluminum soffit.

19. Rough hardware is mostly for nails, and the price is assessed using the following analysis:

| | | |
|---|---|---|
| Total BM of lumber on recap | = | 6446 |
| Total SF of sheathing | = | 4219 |
| | | 10,685 |

Allowing 10 lbs per 1000 BF or SF gives: 106.85 lbs

At the price of $110.00 per 100 lbs this gives:

$$106.85 \text{ lbs} \times \frac{\$110.00}{100} = \$117.54$$

So, $120.00 is allowed for rough hardware.

20. An allowance of 10% for wastage of rough carpentry materials is considered to be appropriate for this project.

### Finish Carpentry Work Figure 12.8

21. Unit prices for finish carpentry materials are assumed to be obtained from suppliers who have been asked to quote prices for items that meet the required specifications.

22. The labor unit prices for finish carpentry work are based on the cost of using a crew again comprised of 0.3 of a carpenter foreman and 2 carpenters, and productivity rates are taken to be in the middle of the ranges shown in Figure 12.4 except when noted otherwise.

23. The prices of the interior doors are established from the following component prices:

| Width of Door: | 1'6" | 2'0" | 2'6" | 2'8" |
|---|---|---|---|---|
| Slab door | $30.00 | $31.00 | $35.00 | $37.00 |
| 1 × 5 Jambs | 23.50 | 23.50 | 23.50 | 23.50 |
| Doorstop | 3.50 | 3.50 | 3.50 | 3.50 |
| Casing both sides | 12.20 | 12.20 | 12.20 | 12.20 |
| | $69.20 | $70.20 | $74.20 | $76.20 |

24. The prices of bifold closet doors are established from the following component prices:

| Width of Bifold Door: | 2'0" | 3'0" | 4'0" | 5'0" |
|---|---|---|---|---|
| Door and hardware | $35.00 | $47.00 | $70.00 | $85.00 |
| Half Jambs | 12.50 | 12.50 | 14.00 | 16.00 |
| Casing one side | 6.10 | 6.10 | 6.50 | 6.90 |
| | $53.60 | $65.60 | $90.50 | $107.90 |

**PRICING SHEET**

JOB .......... HOUSE EXAMPLE .......... DATE ..........

ESTIMATED ..........

Page No. | 8 of 14 |

| No. | DESCRIPTION | QUANTITY | UNIT | UNIT PRICE | LABOR | UNIT PRICE | MATERIALS | UNIT PRICE | EQUIPMENT | TOTAL |
|---|---|---|---|---|---|---|---|---|---|---|
|  | FINISH CARPENTRY |  |  |  |  |  |  |  |  |  |
| 1. | 3'-0" WIDE STAIR | 1 | No. |  | 130 |  | 600 |  | — | 730 |
| 2. | 2" x 2" HANDRAIL | 12 | LF | 4.35 | 52 | 1.25 | 15 |  | — | 67 |
| 3. | RAILING c/w 2" x 6" HANDRAIL | 12 | LF | 6.55 | 79 | 8.00 | 96 |  | — | 175 |
|  | DOORS |  |  |  |  |  |  |  |  |  |
| 4. | 2'-8" x 6'-8" x 1¾" EXTERIOR | 1 | No. |  | 90 | 475.00 | 475 |  | — | 565 |
| 5. | 3'-0" x 6'-8" x 1¾" DITTO | 1 | No. |  | 95 | 500.00 | 500 |  | — | 595 |
| 6. | 2'-0" x 6'-8" x 1⅜" INTERIOR | 1 | No. |  | 65 | 70.00 | 70 |  | — | 135 |
| 7. | 2'-6" x 6'-8" x 1⅜" DITTO | 4 | No. | 70.00 | 280 | 74.25 | 297 |  | — | 577 |
| 8. | 1'-6" x 6'-8" x 1⅜" DITTO | 1 | No. |  | 60 | 69.00 | 69 |  | — | 129 |
| 9. | 2'-8" x 6'-8" x 1⅜" DITTO | 1 | No. |  | 75 | 76.00 | 76 |  | — | 151 |
| 10. | 2'-0" x 6'-8" BIFOLD | 1 | No. |  | 65 | 54.25 | 54 |  | — | 119 |
| 11. | 5'-0" x 6'-8" DITTO | 2 | No. | 80.00 | 160 | 108.00 | 216 |  | — | 376 |
| 12. | 4'-0" x 6'-8" DITTO | 2 | No. | 75.00 | 150 | 90.50 | 181 |  | — | 331 |
| 13. | 3'-0" x 6'-8" DITTO | 2 | No. | 70.00 | 140 | 66.00 | 132 |  | — | 272 |
| 14. | ATTIC ACCESS HATCH | 1 | No. |  | 60 | 80.00 | 80 |  | — | 140 |
|  | DOOR HARDWARE |  |  |  |  |  |  |  |  |  |
| 15. | 4" BUTT HINGES | 3 | Pr. |  | — | 13.50 | 41 |  |  | 41 |
| 16. | 3½" DITTO | 7 | Pr. |  | — | 12.50 | 88 |  | — | 88 |
|  | CARRIED FWD: |  |  |  | 1501 |  | 2990 |  |  | 4491 |

**Figure 12.8**   House Example—Finish Carpentry Pricing

PRICING SHEET

JOB ... House Example ... DATE ...

ESTIMATED ... A.B.F.

| No. | DESCRIPTION | QUANTITY | UNIT | UNIT PRICE | LABOR | UNIT PRICE | MATERIALS | UNIT PRICE | EQUIPMENT | TOTAL |
|---|---|---|---|---|---|---|---|---|---|---|
| | BROUGHT FWD. | | | | 1501 | | 2990 | | | 4491 |
| 1. | KEY-IN-KNOB LOCKSET | 2 | No. | 30⁰⁰ | 60 | 70⁰⁰ | 140 | | ‖ | 200 |
| 2. | DEAD BOLTS | 2 | No. | 35⁰⁰ | 70 | 30⁰⁰ | 60 | | | 130 |
| 3. | PASSAGE SETS | 3 | No. | 30⁰⁰ | 90 | 40⁰⁰ | 120 | | | 210 |
| 4. | PRIVACY SETS | 3 | No. | 30⁰⁰ | 90 | 50⁰⁰ | 150 | | | 240 |
| | WINDOWS | | | | | | | | | |
| 5. | 36"x 12' IN CONC. WALL | 4 | No. | 25⁰⁰ | 50 | 180⁰⁰ | 360 | | | 410 |
| 6. | COMPRISING: 2 x 18" CSMTS + 54" FXD SASH x 48" HIGH | 1 | No. | 45⁰⁰ | 45 | | 780 | | | 825 |
| 7. | DITTO: 24" CSMT + 24" FXD SASH x 42" HIGH | 1 | No. | 30⁰⁰ | 30 | | 385 | | | 415 |
| 8. | DITTO: 24" CSMT + 2x 24" FXD SASH x 48" HIGH | 1 | No. | 40⁰⁰ | 40 | | 505 | | | 545 |
| 9. | DITTO: 24" CSMT x 42" HIGH | 1 | No. | 30⁰⁰ | 30 | | 260 | | | 290 |
| 10. | DITTO: 24" CSMT + 2x 24" FXD SASH x 42" HIGH | 1 | No. | 35⁰⁰ | 35 | | 480 | | | 515 |
| 11. | DITTO: 24" CSMT + 24" FXD SASH x 42" HIGH | 2 | No. | 30⁰⁰ | 60 | 360⁰⁰ | 720 | | | 780 |
| 12. | 3/4" x 1 5/8" WINDOW TRIM | 118 | LF | 1³⁵ | 159 | 0⁴⁰ | 47 | | | 206 |
| | CABINETS | | | | | | | | | |
| 13. | 2'0½"x 3'-0" FLOOR MOUNTED c/w COUNTERTOP | 23 | LF | 22⁰⁰ | 506 | 150⁰⁰ | 3450 | | | 3956 |
| | CARRIED FWD: | | | | 2766 | | 10447 | | | 13213 |

Figure 12.8 contunued

PRICING SHEET
JOB: House Example
ESTIMATED: A.B.F
DATE:
Page No. 10 of 14

| No. | DESCRIPTION | QUANTITY | UNIT | UNIT PRICE | LABOR | UNIT PRICE | MATERIALS | UNIT PRICE | EQUIPMENT | TOTAL |
|---|---|---|---|---|---|---|---|---|---|---|
| | Brought Fwd: | | | | 2766 | | 10447 | | | 13213 |
| 1. | 1'-0" x 2'-8" Wall Mounted | 17 | LF | 12.00 | 112 | 75.00 | 701 | | | 813 |
| 2. | 1'-0" x 1'-10" Ditto | 5 | LF | 11.00 | 55 | 55.00 | 275 | | | 330 |
| 3. | 1'-0" x 2'-8" Ceiling Hung | 5 | LF | 16.00 | 80 | 80.00 | 400 | | | 480 |
| 4. | 2'-0" x 2'-6" Vanity | 9 | LF | 14.00 | 126 | 85.00 | 765 | | | 891 |
| 5. | 12" wide closet shelving | 18 | LF | 2.50 | 45 | 3.50 | 63 | | | 108 |
| 6. | 16" wide Ditto | 35 | LF | 4.00 | 140 | 4.00 | 140 | | | 280 |
| 7. | Closet Rods | 5 | No. | 11.00 | 55 | 15.00 | 75 | | | 130 |
| 8. | Bathroom Accessories / Toilet Roll Holder | 2 | No. | 9.00 | 18 | 15.00 | 30 | | | 48 |
| 9. | 2'-0" x 3'-0" Mirror | 2 | No. | 20.00 | 40 | 110.00 | 220 | | | 260 |
| 10. | Medicine Cabinets | 2 | No. | 25.00 | 50 | 80.00 | 160 | | | 210 |
| 11. | Shower Curtain Rail | 1 | No. | 10.00 | 10 | 25.00 | 25 | | | 35 |
| 12. | 3/4" x 2 1/2" Baseboard | 432 | LF | 1.20 | 518 | 1.15 | 497 | | | 1015 |
| 13. | Rough Hardware | Allow | | | | | 75 | | | 75 |
| 14. | Waste | 2% Matl. 13,875 | | | | | 278 | | | 278 |
| | Total Finish Carp. To Summary | | | | 4016 | | 14151 | | | 18166 |

**Figure 12.8 continued**

25. There is no labor price against hinges because the cost of installation is included in the door installation price.
26. Windows are taken to be prefinished vinyl-clad units.
27. Prices of cabinets and countertops vary a great deal. The prices used here are for medium-quality cabinets with a plastic laminate countertop.
28. Some nails and screws are required for installing finish carpentry items, so there is an allowance made based on costs of previous jobs.
29. A 20% allowance for waste is considered appropriate on finish carpentry materials.

### Exterior and Interior Finishes Figure 12.9 and Figure 12.10

30. The price of bituminous paint for damp-proofing is $4.25 per gallon and a coverage of 24 SF/gal. is expected for spraying two coats on concrete walls. This gives a unit price of:

$$\frac{\$4.25}{24} = \$0.18/SF$$

31. The following crews are used for pricing exterior and interior finishes:

|  |  |  | Labor $ |
|---|---|---|---|
| **Crew A** | | | |
| 0.3 Carpentry foreman | $31.00/hr. | = | 9.30 |
| 2.0 Carpenters | $28.00 | = | 56.00 |
| | | | $65.30/hr. |
| | | | |
| **Crew B** | | | |
| 0.3 Labor foreman | $24.00/hr. | = | 7.20 |
| 2.0 Laborers | $21.00 | = | 42.00 |
| | | | $49.20/hr. |
| | | | |
| **Crew C** | | | |
| 0.3 Labor foreman | $24.00 | = | 7.20 |
| 2.0 Cement finishers | $22.00 | = | 44.00 |
| | | | $51.20/hr. |
| | | | |
| **Crew D** | | | |
| 0.3 Painter foreman | $28.00 | = | 8.40 |
| 2.0 Painters | $25.00 | = | 50.00 |
| | | | $58.40/hr. |

32. With Crew B at a productivity of 130 SF per hour, the unit price for damp-proofing is:

$$\frac{\$49.20}{130\ SF} = \$0.38/SF$$

PRICING SHEET

JOB: House Example ............ DATE ............

ESTIMATED: A.B.F.

Page No. | 11 of 14 |

| No. | DESCRIPTION | QUANTITY | UNIT | UNIT PRICE | LABOR | UNIT PRICE | MATERIALS | UNIT PRICE | EQUIPMENT | TOTAL |
|---|---|---|---|---|---|---|---|---|---|---|
| | EXTERIOR FINISHES | | | | | | | | | |
| 1. | Asphalt Dampproofing | 918 | SF | 0.38 | 349 | 0.18 | 165 | | — | 514 |
| 2. | ½" Parging | 193 | SF | 1.71 | 330 | 0.25 | 48 | | — | 378 |
| 3. | Horiz. Aluminum Siding | 1175 | SF | 1.04 | 1222 | 1.00 | 1175 | | — | 2397 |
| 4. | Building Paper | 1484 | SF | 0.08 | 119 | 0.10 | 148 | | — | 267 |
| 5. | 4" Wide Galv. Flashing | 45 | LF | 0.65 | 29 | 0.33 | 15 | | — | 44 |
| 6. | 4" Eaves Gutter | 90 | LF | 2.60 | 234 | 1.10 | 99 | | — | 333 |
| 7. | 3" Down Spouts | 29 | LF | 1.75 | 51 | 0.75 | 22 | | — | 73 |
| 8. | 210 lb Asphalt Shingles | 1656 | SF | 0.50 | 828 | 0.41 | 679 | | — | 1507 |
| 9. | Ridge Cap | 45 | LF | 0.60 | 27 | 0.65 | 29 | | — | 56 |
| 10. | 4" Galv. Drip-Edge Flashing | 90 | LF | 0.75 | 68 | 0.30 | 27 | | — | 95 |
| 11. | 6 Mil Poly. Eaves Protection | 360 | SF | 0.08 | 29 | 0.04 | 14 | | — | 43 |
| 12. | 48" x 42" PC Conc. Steps w. 4 Risers | 1 | No | 150.00 | 150 | 420.00 | 420 | | — | 570 |
| | CARRIED FWD: | | | | 3436 | | 2841 | | 1 | 6277 |

**Figure 12.9** House Example—Exterior Finishes Pricing

**PRICING SHEET**

JOB...... *House Example* .......... DATE......

ESTIMATED...... *ABF*

Page No. **12 of 14**

| No. | DESCRIPTION | QUANTITY | UNIT | UNIT PRICE | LABOR | UNIT PRICE | MATERIALS | UNIT PRICE | EQUIPMENT | TOTAL |
|---|---|---|---|---|---|---|---|---|---|---|
| | BROUGHT FWD: | | | | 3436 | | 2841 | | | 6277 |
| 1. | 48"x24" PC Conc. Steps w. 2-risers | 1 | No. | 110.00 | 110 | 375.00 | 375 | | | 485 |
| 2. | Wrought Iron Rail. | 7 | LF | 25.00 | 175 | 45.00 | 315 | | | 490 |
| 3. | Paint W.I. railing | 7 | LF | 6.52 | 46 | 0.12 | 1 | | | 47 |
| 4. | Rough Hardware | ALLOW | | | | | 75 | | | 75 |
| 5. | Waste | 10% matls. 3607 | | | | | 361 | | | 361 |
| | | | | | | | | | | |
| | Total Extr. Finishes to Summary | | | | 3767 | | 3968 | | | 7735 |

**Figure 12.9 continued**

PRICING SHEET

JOB...... _HOUSE EXAMPLE_ ...... DATE......

ESTIMATED...... _A.B.F._ ......

Page No. | 13 of 14 |

| No. | DESCRIPTION | QUANTITY | UNIT | UNIT PRICE | LABOR | UNIT PRICE | MATERIALS | UNIT PRICE | EQUIPMENT | TOTAL |
|---|---|---|---|---|---|---|---|---|---|---|
| | _INTERIOR FINISHES_ | | | | | | | | | |
| 1. | 32 oz Carpet | 676 | SF | 0.50 | 338 | 2.50 | 1690 | | — | 2028 |
| 2. | Sheet Vinyl Flooring | 404 | SF | 1.15 | 465 | 1.50 | 606 | | — | 1071 |
| 3. | 1/2" Drywall Ceiling | 1080 | SF | 0.70 | 756 | 0.32 | 324 | | — | 1080 |
| 4. | Textured Finish | 1080 | SF | 0.35 | 378 | 0.35 | 376 | | — | 756 |
| 5. | R35 Loose Insulation | 1080 | SF | 0.40 | 432 | 0.45 | 486 | | — | 918 |
| 6. | 6 Mil. Poly. Vapor Barrier | 3510 | SF | 0.08 | 281 | 0.04 | 140 | | — | 421 |
| 7. | Insulation Stops | 46 | No. | 2.50 | 115 | 1.50 | 69 | | — | 184 |
| 8. | R20 Batt Insulation | 1088 | SF | 0.25 | 272 | 0.40 | 435 | | — | 707 |
| 9. | R12 Batt Insulation | 1023 | SF | 0.20 | 205 | 0.25 | 256 | | — | 461 |
| 10. | 1/2" Drywall Walls | 5007 | SF | 0.60 | 3004 | 0.30 | 1502 | | — | 4506 |
| 11. | Paint Walls | 3456 | SF | 0.17 | 588 | 0.14 | 484 | | — | 1072 |
| 12. | Paint Extr. Doors | 80 | SF | 1.50 | 120 | 0.10 | 8 | | — | 128 |
| 13. | Stain Intr. Doors | 621 | SF | 1.25 | 776 | 0.08 | 50 | | — | 826 |
| | CARRIED FWD: | | | | 7730 | | 6428 | | | 14158 |

**Figure 12.10** House Example—Interior Finishes Pricing

**PRICING SHEET**

JOB... *House Example*    DATE.................

ESTIMATED... *ABF*

Page No. | 14 of 14 |

| No. | DESCRIPTION | QUANTITY | UNIT | UNIT PRICE | LABOR | UNIT PRICE | MATERIALS | UNIT PRICE | EQUIPMENT | TOTAL |
|---|---|---|---|---|---|---|---|---|---|---|
| | Brought Fwd: | | | | 7730 | | 6428 | | | 14158 |
| 1. | Paint Attick Access Hatch | 1 | No. | $20^{00}$ | 20 | $2^{22}$ | 2 | | $\mid$ | 22 |
| 2. | Paint Handrail | 12 | LF | $0^{75}$ | 9 | $0^{.02}$ | 1 | | $\mid$ | 10 |
| 3. | Paint Wood Railing | 12 | LF | $5^{00}$ | 60 | $0^{.25}$ | 3 | | $\mid$ | 63 |
| 4. | Paint Shelves | 139 | SF | $1^{05}$ | 146 | $0^{.08}$ | 11 | | $\mid$ | 157 |
| 5. | Rough Hardware | Allow | | | — | | 100 | | $\mid$ | 100 |
| 6. | Waste | 15% Matl. 6545 | | | — | | 982 | | | 982 |
| | Total Intr. Finishes To Summary | | | | 7965 | | 7527 | | $\mid$ | 15492 |

**Figure 12.10** continued

33. With Crew C at a productivity of 30 SF per hour, the unit price for parging is:

$$\frac{\$51.20}{30 \text{ SF}} = \$1.71/\text{SF}$$

34. With Crew A at a productivity of 62.5 SF per hour, the unit price for aluminum siding is:

$$\frac{\$65.30}{62.5 \text{SF}} = \$1.04/\text{SF}$$

35. The remaining labor unit prices are calculated in the same fashion using average outputs from Figure 12.5.

36. The price quoted for asphalt shingles is $13.50 per bundle. three bundles are required to cover 1 square foot of roof, so the price per square foot is:

$$\frac{\$13.50 \times 3}{100} = \$0.41/\text{SF}$$

37. In addition to the labor crew, a crane is also required for the installation of the precast steps. The price of the crane will be included on the General Expenses sheet under the Site Equipment item.

38. An allowance for rough hardware is added to Exterior Finishes, and 10% is considered to be necessary for wastage of these materials.

39. The rough hardware allowance on Interior Finishes is mainly for drywall screws. This price is calculated by allowing 1.2 screws per SF of drywall at a price of $9.00 per 1000 screws:

| | | | |
|---|---|---|---|
| Total drywall area from recaps | | = | 6087 SF |
| Number of screws required | $6087 \times 1.2$ | = | 7304 |
| Cost of screws | $7304 \times \dfrac{\$9.00}{1000}$ | = | $66.00 |
| For other miscellaneous hardware add | | | $34.00 |
| Total hardware allowance | | | $100.00 |

## Masonry Work Pricing Notes Example 2—Office/Warehouse Building

### Masonry Work Figure 12.11

1. The labor prices are based on the following masonry crew:

| Masonry Crew | | | Labor $ |
|---|---|---|---|
| 1 Masonry foreman | $31.00/hr. | = | 31.00 |
| 3 Masons | $28.00/hr. | = | 84.00 |
| 2 Laborers | $21.00/hr. | = | 42.00 |
| | | | $157.50/hr. |

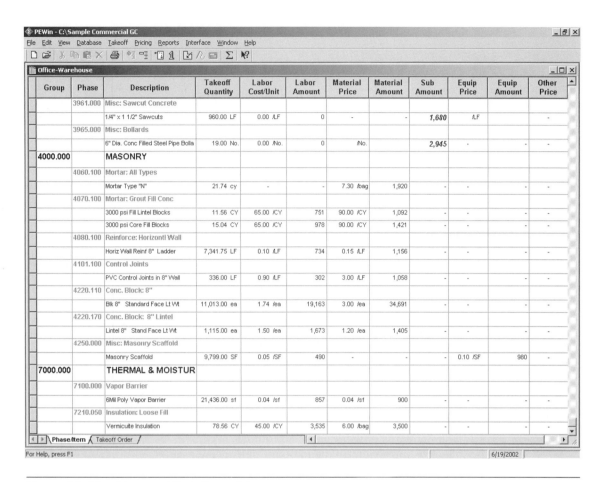

**Figure 12.11**    Masonry Work Spreadsheet

2. The productivity of this crew on this larger project is expected to be at the average of the range, so for laying 8" concrete blocks, an output of 90 blocks per hour is used:

   Thus the labor unit price for blockwork is        $157.00/90

                                                    =  $1.74/unit

3. There is no labor price against mortar since the cost of laying the mortar is included in the labor price for laying the blocks.

4. When the item for mortar was set up in the database, the order unit for material was defined as "bags" of mortar and a conversion factor of 11 bags per CY was used to define the relationship between the takeoff units (CY) and the order units. So, here we enter the price of a bag of premixed mortar that, in this case, is $7.30 and the program then uses the conversion factor to compute the total cost of the mortar based on this price per bag.

5. A new phase and item was set up so that we could add the price of masonry scaffold.

New phase:     4250.00     Misc: Masonry Scaffold

New item:         10     Masonry Scaffold

6. The takeoff units for this new item are square feet of wall area, and labor and equipment cost items are needed to price the cost of setting up and dismantling the scaffold and the cost of renting the equipment.

7. An order unit of "bags" was set up for the loose fill insulation with a conversion factor of 6.75 bags per CY and here the price of a bag of insulation is $6.00.

## SUMMARY

- This chapter concerns the pricing of masonry work, rough carpentry work, finish carpentry work, and interior and exterior finishes.
- This work is usually performed by subcontractors who estimate the cost of their work and quote prices to general contractors.
- The cost of masonry work includes the cost of labor crews laying the masonry units and installing accessories such as horizontal ladder reinforcing and wall ties; masonry materials including accessories and mortar; equipment used in the mixing and hoisting of masonry materials; and erecting and dismantling temporary scaffolding and work platforms required by the work crews.
- The job factors that affect the price of masonry include:
  1. What is being built with the masonry units
  2. The standards of workmanship required
  3. The size and weight of the masonry units being used
  4. The kind of mortar being used
  5. The kind of bond being used
  6. The complexity of the design of the work
  7. The presence or lack of repetition in the work
  8. The availability of workers skilled in the masonry trade
- The price of masonry bricks and blocks varies over a wide range depending on the size, type and color of units required.
- Many rough carpentry components are fabricated off-site and merely assembled at the job site.
- The job factors that influence the price of rough carpentry include:
  1. The nature of the carpentry components being used
  2. The size of cross-section and length of the pieces of lumber involved
  3. The extent to which power tools are used in the work
  4. The complexity of the design of the work
  5. The presence or lack of repetition in the work
  6. The extent to which components and assemblies are fabricated off-site
- The size of lumber is stated in accordance with its nominal size. This is the dimension of the cross-section of the piece before it has been surfaced. Thus $2 \times 4$ size pieces were approximately $2" \times 4"$ in cross-section before they were surfaced but now have a cross-section of $1\frac{1}{2}" \times 3\frac{1}{2}"$ after the lumber is dressed.
- The material price of lumber for rough carpentry depends on the species of work and the grade of lumber required for the particular use.

- Lumber is graded according to these categories:
  1. Light framing
  2. Structural light framing
  3. Structural joists and planks
  4. Appearance framing
  5. Decking
  6. Beams and stringers
  7. Posts and columns
- Hardware for rough carpentry, including nails and other fastening devices, is usually priced as a percentage of the price for lumber.
- Most finish carpentry and millwork items are manufactured goods that are produced off-site. A price is obtained to supply these items to the site; to this is added the price of installation by the on-site crews.
- Materials for interior and exterior finishes are priced in the same way as the materials of other trades.
- Labor productivities for rough carpentry, finish carpentry and millwork are affected by generally the same job factors and labor and management factors we considered with the other trades examined previously.
- Recapping and pricing masonry, carpentry and finishes work can be done manually or via computer.

## REVIEW QUESTIONS

1. List seven factors, specific to masonry work, which influence the level of productivity of crews performing masonry work.
2. Explain why the productivity rate per board foot is lower when working with short pieces of lumber than with long pieces.
3. List and describe the three use classifications of softwood lumber.
   NOTE: For questions 4 through 9, use the wage rates listed on page 281 in the Wage Rates section.
4. With a crew comprising 1 masonry foreman, 3 masonry workers and 2 laborers, what would be the labor price to lay 1000 facing bricks if the output was expected to be at the low end of the scale?
5. Use the crew listed on Figure 12.2 to calculate a labor price per SF to install a ½" plywood soffit when the output is expected to be about average.
6. From the information provided on Figure 12.3, calculate the material price of 2 × 3 "S-Dry Spruce":
   a. per board foot
   b. per linear foot
7. From the information provided on Figure 12.4, calculate the installation price of a 1 × 6 base board when the output is expected to be low:
   a. per linear foot
   b. per board foot
8. If the output is expected to be on the high end of the scale, what would be the labor price to install 210 lb asphalt shingles based on Figure 12.5 data?
9. Why is it considered necessary to add a waste factor to the estimated material's price of masonry work?

CHAPTER

# 13

# PRICING SUBCONTRACTORS' WORK

## OBJECTIVES

*After reading this chapter and completing the review questions, you should be able to do the following:*

- Describe how subtrade work is usually estimated and explain the role of the contractor's estimator in this process.
- Identify potential problems with subcontractors on a project.
- Describe what can be done to minimize problems with subcontractors on a project.
- Explain how the competency of subtrades can be evaluated and list what should be considered in this evaluation.
- Explain how surety bonds can be used to help manage subtrades.
- Describe subtrade pre-bid proposals and explain their use with subcontractor bids.
- Describe how subtrade bids are evaluated.
- Describe the main features of bid depositories and explain their use with subcontractor bids.

## Introduction

Subcontractors often perform 80% or more of the work on building construction projects. In some situations all of the project work is put in place by subtrades, leaving the general contractor to merely coordinate the activities of the subtrades as construction manager. Consequently, the contractor has to rely on subtrades and to work closely with them, first to assemble an accurate estimate that meets project requirements at a competitive price, and later to successfully complete the work of the project. Therefore, there has to be a good relationship between the contractor and each of the subtrades. This relationship must be built on mutual trust, a willingness to cooperate and, perhaps above all, effective two-way communications. A dialogue between project participants needs to be established in the estimating process and maintained throughout the course of the work. The contractor and subtrades have to identify common goals and together strive to meet those goals; they have to be willing to share ideas and unite forces in order to overcome problems and find creative ways of dealing with project issues.

### Potential Problems

In the estimating process, subtrades bidding a job will normally be responsible for the takeoff and pricing of their own portions of the work but, to do this effectively, they need to keep in close contract with the general contractor's estimator who is coordinating the estimate. On the face of it, the work of the general contractor's estimator would appear to be quite straightforward: First she or he has to ensure that subtrades are aware of the project and the requirement for prices, then simply find the subcontractor with the best price for each trade and, finally, gather together these prices to determine the total price of the work. However, in reality it never seems to be quite that simple. Problems can emanate from the practice of subcontracting for a number of reasons including:

1. Unreliability of some subcontractors
2. Errors in subtrade bids
3. Overlap or underlap in the scope of work subcontracted
4. Conditional bids from subcontractors
5. Bid closing congestion
6. Compliance with government positive action requirements regarding minorities

*Unreliable Subcontractors*

While most subtrades can be relied on to deliver the bids they have promised to submit and, at least, to try to meet their contractual obligations, there are still occasions when some subcontractors break their agreements leaving the contractor in much difficulty.

Subcontractor defaults can occur in numerous ways and at various stages in the construction process: During the bid period a subtrade may be late or completely fail to prepare the expected bid; just after the bid period a subtrade may try to withdraw a bid that has been used by the contractor in the bid to the owner; when the project is underway a subtrade may not perform in accordance with contractual obligations or may not even show up when the contracted work is required to begin. Certainly there can be legal recourse against a defaulting subtrade, but this often provides little compensation to the contractor who has suffered the consequences of the defaults.

Part of the estimator's role is to try to assess the risk involved in accepting the price of and then employing a particular subcontractor. This assessment is discussed later in the "Evaluating Subcontractors" section.

*Errors in Subtrade Bids*

One of the recurring reasons for the deficient performance of subtrades is the presence of errors in the estimates prepared by the subtrades. Accepting a subtrade bid that contains a mistake can lead to endless trouble for a contractor, so the estimator has to make every effort to ferret out subtrade bids that contain errors if the consequential problems are to be avoided. When a bid is received from a subtrade that is obviously out of line with the other bids, it is advisable for the estimator to contact the bidder before the price is used in the contractor's bid. There is no need to discuss details of prices with this subtrade; merely indicating that its estimate should be examined for errors should be sufficient warning.

*Overlap and Underlap of the Scope of Work Subcontracted*

Ensuring that all the work of the project is properly covered, either by subtrades or by the contractor's own forces, can be difficult since it is easy for a section of work to be excluded from the subtrade's scope and be overlooked in the contractor's takeoff. Similarly, discovering that some work has been covered more than once in the estimate is not uncommon. These cases of underlap and overlap occur when work is not clearly defined as being part of the scope of a single trade or as being the general contractor's work. Avoiding these problems calls for a lot of investigative work by the estimator, which is discussed later under "Scope of Work."

### Conditional Subtrade Bids

The estimator's task becomes far more complicated when subtrades attach conditions to their bids. Where all of the subtrades quoting a certain trade base their bids on the same scope of work and comply with the same set of specifications, selecting a subtrade price should not be difficult. But when each subtrade has a different set of bid inclusions and exclusions and each expects the general contractor to supply a variety of different items, or the subtrades bid according to various alternative specifications, selecting a subtrade can become a horrendous task for the contractor's estimator.

### Bid Closing Congestion

Compounding all of these problems is the congestion that often occurs in the prime contractor's office when the subtrade bids are received on the bid closing day. If 30 or more trades are involved in a project and numerous bids are submitted for each trade in the space of a few hours, a chaotic situation can easily develop at the place where these bids are being received and processed. A well-organized bid closing procedure in the prime contractor's office is essential to preserve any control over this stage of the estimating process. Bid closings are discussed in Chapter 15.

### Positive Action Programs

On some publicly funded projects, the general contractor may be required by government regulations to award a certain percentage of the work to women-owned and minority-owned businesses. There may also be an official local preference policy stipulating that businesses established in a defined geographical area be given preference in contract awards in which their bids are within a certain percentage of the price of "outside" subcontractors. Complying with these regulations further complicates the process of selecting subtrades and, if not handled properly, can put the prime contractor into awkward situations vis-à-vis government agencies.

## Managing Subcontract Pricing Procedures

The problems just described can be controlled only where the general contractor applies effective management practices throughout the estimating process. The underlying objective of the subcontract pricing procedures detailed in the following is to provide a well-organized system of dealing with subtrades in the estimating process. Following this system allows the prime contractor first to clearly identify the subtrade needs for the project being estimated and, second, to ensure that the subtrades selected to perform the work are capable of meeting their contractual obligations.

## List of Subtrades

As soon as bid drawings and specifications are obtained for a project and the estimator has completed the general review of bid documents described in Chapter 2, a list of the subtrades required for the project should be prepared. This list is compiled and subtrades are notified early in the estimating process to give the subtrade bidders as much time as possible to put together their estimates. Many of the design questions raised in the estimating process relate to subtrade work, so subtrades have to be encouraged to make an early start on their takeoffs so that the problems they encounter can be properly dealt with in the time period when estimates are being prepared.

The list of subtrades required for the project is compiled by carefully examining the project specifications and drawings. Some estimators like to make a new list of subtrades each time they go through the bid documents of a project. Others prefer to check off the subtrades required from a document that lists all of the trades a contractor may need for its projects. Figure 13.1 is an example of the kind of checklist of trades used by an estimator to indicate the particular trades needed on a project. The list shown is derived from the CSI MasterFormat of construction trade categories, which is also the basic source of the code numbers indicated against each item. While this list may represent the usual subtrades required by a contractor engaged in a certain type of construction work, the estimator should be aware that such a list can never be entirely comprehensive; other trades might have to be added to the list to meet specific requirements of a project.

The subtrade list is first used to determine which trades have to be contacted for bid requests on the project; later the list can be incorporated into the estimate summary and used to ensure that all subtrade work necessary for the project is duly priced as required.

### Contacting Subcontractors

Once the subtrade needs of a project have been defined, the next task is to inform subcontractors that the general contractor is going to prepare a bid for the project and that bids for their trade will be required. Most contractors maintain a directory of the subtrades they prefer to deal with. These trades may be notified of a new project by postcard or, more commonly these days, by e-mail or fax.

One way to organize this process is to have a subtrade checklist like the one shown as Figure 13.1 set up on a computer spreadsheet and then proceed as follows:

1. The estimator checks off the trades required for the project and adds trades that are not on the basic list.
2. The completed checklist is then compiled.
3. A copy of this checklist is e-mailed to all the subtrades on the contractor's mailing list.
4. The subcontractors are asked to return the checklist with their name against all the trades on the checklist that they are prepared to submit prices for.

By this means subtrades are quickly notified of a project and the estimator promptly determines if all the subtrade prices necessary for a project are going to be obtained from the subcontractors who normally work for that general contractor.

When the contractor's subtrade directory does not cover all the trades necessary for the job or where prices from additional subtrades are needed to improve competition, soliciting prices by advertising in trade publications or newspapers may have to be considered. This, however, introduces the problem of receiving bids from unknown subcontractors.

### Unknown Subcontractors

Having to award contracts to companies from outside their select group of preferred subcontractors can be a stressful experience for some general contractors; however, it may be necessary for a number of reasons. For instance, simply complying with bid conditions can necessitate that a contractor deal with an unknown subcontractor.

| | | Req'd | Comments |
|---|---|---|---|
| **SUBTRADE LIST** | | | |
| Project: | Estimator: | | |
| Bid Closing Date: | B.D.* Closing Date: | | |
| 2100 | Demolition | | |
| 2200 | Earthwork | | |
| 2300 | Piling | | |
| 2350 | Shoring | | |
| 2480 | Landscaping | | |
| 2700 | Paving and Surfacing | | |
| 3010 | Concrete Materials | | |
| 3050 | Concrete Placement | | |
| 3055 | Concrete Finishing | | |
| 3100 | Formwork | | |
| 3150 | Falsework | | |
| 3200 | Reinforcing Steel—Supply and Place | | |
| 3210 | —Supply only | | |
| 3220 | —Placing only | | |
| 3410 | Precast Concrete—Structural | | |
| 3420 | —Architectural | | |
| 3430 | —Paving | | |
| 4000 | Masonry | | |
| 5100 | Structural Steel | | |
| 5200 | Steel Joists | | |
| 5300 | Metal Decking | | |
| 5500 | Misc. Metals | | |
| 5800 | Expansion Joints | | |
| 6100 | Rough Carpentry | | |
| 6180 | Glue-Lam Construction | | |
| 6400 | Architectural Woodwork | | |
| 7100 | Waterproofing | | |
| 7160 | Damp-proofing | | |
| 7200 | Insulation | | |
| 7500 | Membrane Roofing | | |
| 7900 | Sealants | | |
| 8100 | Metal Doors and Frames | | |
| 8200 | Wood and Plastic Doors | | |
| 8300 | Overhead Doors | | |
| 8500 | Metal Windows | | |
| 8800 | Glazing | | |
| 9200 | Plaster and Stucco | | |
| 9250 | Gypsum Wallboard | | |
| 9300 | Ceramic and Quarry Tile | | |
| 9500 | Acoustical Treatment | | |
| 9650 | Resilient Flooring | | |
| 9680 | Carpeting | | |
| 9900 | Painting | | |
| 10200 | Toilet Partitions | | |
| 10400 | Washroom Accessories | | |
| 10500 | Specialties - Miscellaneous | | |
| 11000 | Equipment | | |
| 12000 | Furnishings | | |
| 13000 | Special Construction | | |
| 14200 | Elevators | | |
| 15400 | Plumbing | | |
| 15500 | Sprinklers | | |
| 15800 | HVAC | | |
| 16000 | Electrical | | |
| | | | |

*Bid depository

**Figure 13.1**    Computer Spreadsheet of a Subtrade Checklist

### Nominated Subcontractors

Occasionally the project designers or even the owner may make a deal with a subtrade sometime before bids are called from general contractors. A clause is then inserted into the bid documents informing all bidders that the subtrade named is to be awarded the subcontract for the work of a certain trade. The subtrade named is usually referred to as a "nominated subcontractor" and the bidders may also be instructed to include a specified sum of money in their bid price to cover the cost of this subtrade's work. This is not a common practice as owners and designers are reluctant to bear the risk of possible responsibility for a "nominated subcontractor" that may result from this procedure. Yet the practice has sometimes been adopted in order to secure the employment of subtrades who offer unique services or products that the owner feels are indispensable for the project.

### Unsolicited Subtrade Bids

More commonly, a contractor will be faced with an unfamiliar subtrade when a low price is received for a trade from a company who is not known to the contractor. This can put the contractor in a difficult position: On one hand the contractor may need to use the low price to maintain a competitive bid and, on the other hand, the contractor may be hesitant to accept the risk of using an unknown subcontractor. Legal and ethical considerations also impact on this situation. One course of action that has been adopted by some contractors faced with this decision is to use the low price in the bid but not commit to using the subtrade who submitted the price. This solution, as well as being ethically questionable, may also be contrary to bid conditions. In most major contract bids the contractor will be required to name its subcontractors on the bid document. Failure to properly comply with such bid requirements can put the contractor into a vulnerable legal position.

The problem a contractor faces when an unsolicited low bid is received can be further aggravated by time constraints. While there is no sure recipe that can be followed to avoid all the potential problems a contractor can have with a subcontractor, there are certain precautions, outlined in the following, that the prudent contractor can take to minimize the risk. But when a contractor is confronted with a low price from an unknown subtrade just hours or even minutes before the bid to the owner is due to be submitted, there is little or no time available to make effective inquiries about the subtrade.

This problem of lack of time can be avoided to some extent if the estimator takes some precautions earlier in the estimating process. First the estimator must try to learn which subtrades are planning to bid on a project. Information can sometimes be obtained from construction trade associations disclosing the names of their members who are bidding on a project. However, a better source of this kind of information is usually the contractor's own subtrades. They frequently hear from their own industry contacts details about which of their competitors intends to bid on a job. If new subtrade names come to the attention of the estimator during the course of these inquiries, investigation and evaluation of the unfamiliar subtrades can begin in a timely fashion.

## Evaluating Subcontractors

The contractor's two main concerns when considering an unfamiliar subtrade are:

1.  Is the subtrade capable of meeting the quality and schedule requirements of its work?
2.  Does the subtrade have the financial capability required?

Potential subcontractors should be evaluated in these two areas by the estimator or another member of the contractor's team so that an assessment of risk can be made before a subtrade is named in a bid. Carrying an unknown subtrade in a bid without carefully considering the consequences is asking for trouble.

An effective evaluation of subtrade companies would involve an investigation of a number of factors including:

1. The general reputation of the company
2. The quality of previous work performed by the company
3. The quality of the company's management
4. The company's disposition in terms of its general cooperativeness and the ease of dealing with company site and office personnel
5. The financial status of the company

This kind of information can be obtainable from several sources. The company itself may be willing to supply general details about the number of years it has been in business and the number of people it employs, and it might offer references especially if it is eager to secure contract work. An assessment of the stature of a subtrade can sometimes be made by examining a list of the equipment that is owned by the company. A company's status may also be gauged from whether, and how long, it has been a member of a trade association. Construction companies that have previously hired the subtrade may be willing to comment on their experience with the subtrade, and other industry contacts, such as material suppliers, who have had contact with the subtrade may offer their opinions about the company.

When investigating the financial condition of a subtrade, a useful source of information can be bonding companies. If the subtrade wishes to bid on larger value projects, it must be able to obtain surety bonds. The prime contractor's own bonding company that is usually concerned about the business health of the prime contractor can be approached to share any financial information about the subtrade it may have. The prime contractor's bank can also be a source of financial data on the subtrade, and, last, the prime contractor may be able to pick up a financial report on the subtrade, for a fee, from a credit reporting agency.

The reader should note that subcontracting problems are not exclusive to the subtrades new to the prime contractor. Even subtrades that a contractor has worked with over many years can be the source of problem or two, so some of the inquiries suggested for unfamiliar subtrades may also be appropriately applied from time to time in an effort to ensure that the regular subtrades remain reliable.

## Bonding of Subtrades

Under the terms of the contract it has with the owner, the prime contractor is always held responsible for the performance of the work, including the performance of the work done by subcontractors. If a subtrade fails to perform as required, the prime contractor will have to cope with the problem within the project budget. Certainly there will be no additional funds available from the owner to rectify the defaults of a subtrade.

Just as the owner, in an effort to manage the risk of a nonperforming prime contractor, can require the prime contractor to obtain surety bonds to guarantee contract performance, so, too, can the prime contractor require subtrades to provide such bonds to guarantee the performance of their work.

There has been much discussion in construction management texts about the merits and demerits of surety bonds as a way of managing construction risk. Suffice it to state here that some contractors do consider it worthwhile to require subcontractor surety bonds even where it can add more than 1% to subtrade prices. Contractors who demand bonded subtrades will impose a bid condition on subtrades specifying that all subcontracts over a stipulated value, often set at $100,000, shall be guaranteed by a performance bond that names the prime contractor as obligee. Then, should the subtrade default on its contractual obligations, the prime contractor will be able to look to the bonding company to finance the due performance of the subtrade's work.

Subtrades may also be required to provide **bid bonds**. A subtrade bid bond would give the prime contractor the right to damages from the bonding company in the event that a subtrade fails to execute the subcontract after its bid has been properly accepted by the prime contractor.

When subtrade bonds are requested on a project, subtrades may be instructed to include the cost of bonding in their bid prices or, the bid conditions may call for the price of the bonding to be stated separately from the bid price. The second alternative informs the prime contractor of the cost saving of waivering the bonding requirement, an option that might be considered with a low-risk, well-established subcontractor.

## Pre-Bid Subtrade Proposals

At one time most subtrades submitted their bids to the prime contractor by telephone shortly before the prime contractor's bid was due to be presented to the owner. The main terms of the bid would be stated over the telephone and a written confirmation of the bid would be sent to the prime contractor within a few days of the bid closing. One of the problems with this procedure was that the bid conditions stipulated by one subtrade may not be the same as those stipulated by another subtrade bidding the same work. Compounding the problem was the fact that all of these conditions may not have been known until the written confirmations were received from the bidders after the estimate was completed and the bid submitted.

The method of submitting bids now preferred by most subcontractors is to use the fax, and increasingly e-mail is being used. These methods provide a hard copy of the subtrade's bid before the prime contractor's bid goes in and all conditions attached to the bid are presented at this time, but the problem of varying conditions from bidder to bidder still remains. Also, bids are still being received only a short time before the owner's bid closing. What has aggravated the problem is the way bid conditions have proliferated in recent years. While this increase in the number of bid conditions has probably nothing to do with the change in the method of submitting bids, it has occurred since the fax was introduced and it does make bid analysis far more difficult to pursue.

Subcontractors generally cannot be persuaded to submit their bids earlier because they are concerned that their prices will become public knowledge (a discussion of bid shopping follows in the section on bid depositories) and, subsequently their bids will be undercut by competitors. But some time has to be available if any analysis of subtrade bids is to be possible. A practical solution to this dilemma is to have subtrade bidders submit pre-bid proposals so that the contractor's estimator has the opportunity to examine subtrade bids and make comparisons where a number of bids are received for the same work. A pre-bid proposal is a copy of the subtrade's bid without prices but complete in all other details including the conditions the bidder wishes to attach to the bid. Figure 13.2 shows an example of a pre-bid proposal from a glass and windows subcontractor.

**WWW WINDOWS INC.**

**TENDER**

Date:

To:  XYZ Construction Inc.

Re:  Laboratory Building
     Townsville

We propose to supply and install the following materials for the above-named project:

1.  Aluminum storefront including glass and glazing:

2.  Inferior aluminum partitions incl. glass and glazing: (separate price)

3.  Glazing to wood interior partitions: (separate price)

4.  Bathroom mirrors: (separate price)

5.  Alternative price to substitute aluminum partitions for wood partitions: (extra cost)

Note: If we are awarded items 1, 2, 3 and 4 above we offer a discount of 10% on the
      prices quoted.

**EXCLUSIONS:**

1.  Final cleaning
2.  Breakage by others
3.  Scaffolding and hoisting
4.  Heating and hoarding
5.  Sill flashings
6.  Any other glass or glazing not listed above
7.  Temporary glazing or enclosures

**Figure 13.2** Pre-Bid Proposal

## Analyzing Subtrade Bids

The estimator has to ensure that all of the work required from a trade, as defined by
the project drawings and specifications, is included in the subtrade bid. One problem
is that even though the CSI MasterFormat neatly classifies construction trades, the
scope of work of trade contractors may not exactly match that of the CSI classification. For example, a subtrade bidding masonry may not do all the work described in
the Masonry Section 4000 of the specifications; or the roofing contractor may include for all of Section 7500, Membrane Roofing, plus additional work from another
section of the specifications. The number of possible combinations of trades and part
trades is endless.

When a bid proposal has been obtained from a subtrade, it should list inclusions
that detail all the work that the bidder is offering to do for the price quoted. However, this list may or may not coincide with the inclusions listed by other subcontractors bidding the same trade. Meaningful bid comparisons can only be made
when the bidders are offering to perform the same work. Excluded items should also

be listed on the subtrade's proposal. Most exclusions are items of work that, by standard practice, are always omitted from the work scope of the trade. Exclusions from one trade are sometimes covered elsewhere in the estimate by other trades, or by the prime contractor in his or her work. The estimator has to be constantly on the lookout for items excluded from the work of a trade and not priced anywhere else in the estimate.

The subtrade in Figure 13.2 has listed all the work items it intends to price together with an alternative price for a change in design (alternative prices are discussed in Chapter 15). The sill flashings listed in this quote as an exclusion are typical of the kind of item that is excluded by one trade but may not be included by any other subtrade on the project. If this is the case here, the estimator will have to make sure an additional price is obtained to cover the cost of these flashings.

Subtrade bids can be further complicated by multiple trade bids and discounts that are sometimes offered. Subcontractors frequently offer prices for a number of different trades on a project. They may state on their quote that they are only interested in a contract for the entire group of trades they have included in their bid. Alternatively, the bidder may allow the prime contractor to choose which trades to award contracts for, but, as an inducement, the subtrade may offer a discount if it is awarded a contract for all of the trades it submitted bids on. In Figure 13.2, if all the subtrade's prices are accepted by the prime contractor, a price reduction of 10% is offered.

To facilitate subtrade bid analysis, a spreadsheet like the one shown in Figure 13.3 can be set up from the information obtained in pre-bid proposals for each of the trades required for the estimate. Each spreadsheet is compiled sometime before the bid closing day and lists the names of bidders together with the items of work to

| | | | | | | | | | |
|---|---|---|---|---|---|---|---|---|---|
| | | | | | | | | | |

**SUBTRADE ANALYSIS**

**Project:** **Laboratory Building—Townsville**      Page No:

**TRADES:** **8500 Metal Windows**      Date:
**8800 Glazing**

| SUBTRADES | Storefront & Glazg. | Int. Alum Partitions & Glazg. | Glazg. Wood Partitions | Bathroom Mirrors | Sill Flashgs. | Discounts | BASE BID TOTAL | ALTERNATE Alum. Int. Ptns. ADD: |
|---|---|---|---|---|---|---|---|---|
| WWW WINDOWS INC. | $152,000 | $18,900 | $9,000 | $950 | NIC | <$18,085> | $168,365 | $25,250 |
| | | | | | $5,600 | (SMITH) | | |
| XXX WINDOWS INC. | $169,500 | inc. | inc. | inc. | inc. | no | $169,500 | $21,000 |
| YYY WINDOWS INC. | $148,900 | $23,500 | inc. | inc. | inc. | no | $172,400 | $38,000 |
| ZZZ WINDOWS INC. | $189,000 | $15,850 | inc. | NIC | inc. | <$20,485> | $185,315 | $30,000 |
| | | | | $950 | (WWW) | | | |
| | | | | | | | | |
| | | | | | | | | |

**Figure 13.3** Subtrade Analysis Spreadsheet

be included in their bids so that, when prices are received, they can be easily inserted onto the spreadsheet to provide a speedy bid analysis.

## Bid Depositories

There is always a temptation for those who receive bids to disclose the amount of a bid from company A to a preferred company B on the understanding that if company B can equal or better the price, it will be awarded the contract. While this practice, which is variously known as "bid shopping" or "bid peddling," may not be illegal, it is certainly unethical and there have been many attempts to put an end to such practices in the construction industry. The use of bid depositories to assemble bids from subtrades and distribute them to prime contractors is one of the ways this objective has been pursued over the years.

Bid depositories have been set up at different times in many places by construction and trade associations following a variety of procedures but are generally alike in that they require subtrades to submit their bids at a specified time usually 2 days before the prime bids are due to be submitted. This gives general contractors time to properly analyze the bids and select the subtrades they will use in their estimate from those that submitted bids through the depository. Subtrades are not allowed to change their bids after the subtrade bid closing time, but they may be allowed to withdraw their bids before the prime bids are delivered.

In addition to helping combat the practice of bid shopping and reduce the last minute rush of subcontractor bids to the prime contractor, the bid depository has also been used in some instances to introduce generally accepted definitions of the scope of work for trades that submit their prices through the depository. When trade definitions are in place and a prime contractor receives a price through the bid depository for a designated trade, the rules of the depository define what is included in and what is excluded from the work of the subcontractor bidding this trade. Also, together with the trade scope definitions, a mechanism is usually set up to rule on disputes over whose scope applies in situations in which an item of work is not clearly included in the work of any of the defined trades.

## Scope of Work

If a bid depository using trade scope definitions is operating in the area where a project is to be built, anyone preparing a bid for the project is strongly advised to obtain the list of trade definitions published by the local bid depository. These definitions can be a valuable tool in the process of analyzing subtrade bids. All the subcontractors submitting their bids through the bid depository where trade definitions are in use will base their prices on the same rules regarding the scope of work of their trade. An estimator equipped with the trade definitions will be able to scan the definitions to determine exactly what is and is not included in the bids for each trade received.

Even if the bid depository is not used on a project in the area, the trade scope definitions provide a useful guide to what is accepted locally as the definitions of subtrades' scope of work. This enables the estimator to investigate overlap and underlap of subtrade bids far more easily than he or she would be able to on projects in which there are no trade definitions.

Following are examples of trade scope definitions for the masonry, windows, and glass and glazing trades. These definitions are based on the trade scopes published by the Alberta Construction Tendering System, which operates a thriving bid depository in Alberta, Canada. The reader is cautioned that these definitions are provided for illustrative purposes only; they are not intended to be complete and specific trade rules may vary from location to location.

## Masonry Scope of Work

The following items shall be *included* in the work of the masonry trade:

- Brickwork
- Concrete blocks, glazed blocks and glass blocks
- Building-in of miscellaneous metals (supplied by others) in masonry
- Built-in metal flashing—supplied by others
- Caulking (other than firestopping), masonry to masonry
- Cavity wall flashings behind masonry
- Wall insulation complete with adhesives, anchors and suchlike behind masonry, when masonry is the last component to be installed
- Clay flue lining
- Clay tiles
- Cleaning masonry
- Clear waterproofing of exterior masonry
- Concrete and grout fill in masonry only
- Masonry anchors and ties
- Damp-proof course in masonry
- Expansion and control joints in masonry or masonry to other materials, including caulking where it forms part of the expansion or control joint
- Guarantee
- Hoisting
- Inspection
- Installation only of precast concrete band courses, copings, sills and heads that fall within masonry cladding, including supply and installation of insulation and vapor barrier behind these items
- Loose fill or foam insulation in masonry
- Mortar
- Nonrigid and fabric flashing when built into masonry
- Parging of cavity masonry walls only when specified
- Protection of other trades' work from damage by this trade
- Reinforcing steel placed only in masonry—supplied by others to the job site, clearly tagged and identified as to end use
- Scaffolding
- Shop drawings
- Stone (natural and artificial), granite, terra cotta facing over 2" thick, including anchorages
- Testing of material of this trade when specifically called for in the masonry specifications
- Trade cleanup
- Weep hole and venting devices
- Wire reinforcing to mortar joints

The following items shall be *excluded* from the work of the masonry trade.

- Anchor slot
- Demolition
- Fabrication and tying of reinforcing steel
- Firestopping

- General and final clean up
- Hoarding
- Installation of frames
- Lateral support devices on top of masonry walls
- Temporary heat (including mortar shack), light, power, sanitation and water
- Temporary support and shoring for lintels
- Welding
- Wind-bracing

## Windows Scope of Work

The following items shall be *included* in the work of the windows trade.

- Anchoring devices:
  (a)  Supply and installation of anchors that are part of the framing
  (b)  Supply only anchors or inserts that are part of the structure
- Architectural glass and metal doors including sliding, swinging, revolving and balanced doors, complete with hardware (including plastic or plastic laminate push pulls when an integral part of the door); all components for complete installation of automatic doors (including any interconnecting wiring), when they form an integral part of the door
- Architectural metal patio or balcony doors, complete with hardware
- Architectural metal sills, flashing and trims forming part of entrances, storefronts and windows
- Caulking (other than firestopping), joint backing and perimeter sealing
- Demolition of existing installations when the materials are to be reused or retained as specified
- Darkroom revolving doors of architectural metals
- Electrical security and monitoring devices, mounted on doors and door frames within this scope, for example, locks, card readers, door monitors, switches, strike releases, panic devices, magnets and suchlike supplied and installed; wiring by electrical
- Installation only of components for trade name security and monitoring systems specified to be furnished by one supplier to the electrical contractor and that must be mounted on doors and frames within this scope (these are supplied by the electrical contractor for installation by this contractor).
- Extruded aluminum wall cladding
- Factory preassembled translucent panels
- Fly screens to venting windows
- Architectural metal framing
- Glass, glazing and glazing materials
- Glazing that is sloped in excess of 15 degrees beyond vertical, or curved
- Guardrails including architectural metal and wood and plastic laminate when they are an integral part of and built into metal entrances and storefronts and special windows (installation only of rails other than metal attached to special windows)
- Hoisting
- Industrial steel sash
- Infill panels, including insulation when it forms an integral part of a panel
- Initial cleaning between removable/nonremovable multiple glazing
- Interior architectural metal glazed partitions, including recreational arena and court glass and glazing

- Manufacturer's warranty only beyond the normal 1 year maintenance warranty
- Entrances and storefronts (interior or exterior)
- Cash allowances or contingency funds when a specific dollar amount is listed in the entrances, storefronts and windows specification
- Protection of other trades' work from damage by this trade
- Scaffolding
- Shop drawings and engineering of entrances, storefronts and windows to the specified design criteria
- Single glazing, multiple glazing and sealed units, plastic, fiberglass and similar materials
- Spandrel and infill panels, for example, extruded aluminum wall cladding, asbestos cement, porcelain and suchlike including insulation when it forms an integral part of a panel
- Supply only of architectural framing to be embedded in concrete
- Supply and installation of operating hardware
- Testing when specifically required by the specifications
- Thermally formed wall domes or pyramids that are not part of or connected to the roofing system
- Trade cleanup
- Venetian blinds between glass supplied under this scope

The following items shall be *excluded* from the work of the windows trade:

- Concrete
- Doors, door frames and entrances, storefronts and window frames of steel, wood or plastic laminate
- Firestopping
- General and final clean up
- Glass, plastic or asbestos panels when supplied and installed under Architectural Woodwork
- Hoarding
- Installation of membranes or flashing, included under this scope, that cannot be installed or fastened during entrances, storefronts and windows installation
- Insulation, support or furring members, within entrance and storefront systems and windows, that are normally supplied and erected by others
- Interior wood sills or surrounds
- Insulation within framing supplied and erected by others
- Pressed steel frames
- Movable, demountable partitions
- Shower doors
- Structural steel
- Sun control devices not between glass, for example, Venetian blinds, drapes and suchlike
- Sun screens, bird screens, guard screens
- Supply of flashing, sills or trim prefinished to match any materials not included in this scope
- Supply and installation of nonrigid membrane air/vapor barriers that form part of the structure and connects to the entrances, storefronts and windows
- Temporary heat, light, power, sanitation and water
- Weather louvers, mechanical and architectural
- Windows of wood, clad wood, polyvinyl chloride or fire glass

### Glass and Glazing Scope of Work

The following items shall be *included* in the work of the glass and glazing trade:

- Caulking, joint backing and perimeter sealing to glass and glazing
- Demolition of existing installations when the materials are to be reused or retained as specified
- Glass doors (except in architectural woodwork) and sliding hardware, including locks, for site fabricated display cases
- Leaded stained glass, glass, plastics, x-ray and asbestos infill panels to:
  (a) Balustrades (other than escalators)
  (b) Hollow metal
  (c) Metal clad wood doors
  (d) Metal display cases
  (e) Metal frames other than those falling under
      —curtainwall scope
      —entrances, storefronts and windows scope
  (f) Pressed steel (frames and applied stops to be predrilled, fitted and installed by supplier prior to glazing by this trade)
  (g) Prefinished demountable partitioning
- Glass shelving (except in architectural woodwork) including shelf standards, clips and brackets
- Glass smoke baffles and sprinkler baffles including shimming, fasteners and caulking for fastening of the glass to the supporting system (supporting system supplied and installed by others)
- Hoisting
- Manufacturer's warranty only beyond the normal 1 year maintenance warranty
- Mirrors, including glass, plastics, stainless steel, chrome plated in standard and tilting tamper-proof frames, J-molds and clips, unless specified to be part of washroom accessories or Architectural Woodwork
- **PC sums** or contingency funds when a specific dollar amount is listed in the glass and glazing specifications
- Protection of other trades' work from damage by this trade
- Shop drawings and engineering of glass and glazing to the specified design criteria
- Scaffolding
- Trade cleanup

The following items shall be *excluded* from the work of the glass and glazing trade:

- Concrete
- Firestopping
- General and final clean up
- Glass to architectural woodwork
- Glass to overhead doors
- Glass to curtainwall and entrances, storefronts and windows
- Hoarding
- Mirrors and glass in wood frames or standard manufactured cabinets, display cases, furniture, fixtures, casework, countertops and tabletops
- Railing, capping and supports to balustrades and wall rails
- Recreational arena and court glass and glazing
- Shower doors
- Temporary heat, light, power, sanitation and water.

## SUMMARY

- Subcontractors often perform 80% or more of the work on building construction projects
- Subtrades will normally be responsible for the takeoff and pricing of their own portions of the work, which would appear to make it easy for the general contractor to price subtrade work—just call a subtrade for a price—but in reality it never seems to be quite that simple.
- Problems with subtrade bids can result from:
  1. Unreliable subcontractors
  2. Errors in subtrade bids
  3. Overlap or underlap in the scope of work subcontracted
  4. Conditional subtrade bids
  5. Bid closing congestion
  6. Positive action programs
- Effective management practices are required to deal with subcontractors in the estimating process. This includes listing required subtrades, contacting prospective subtrades and ensuring subtrade requirements are covered.
- Problems can also develop when the subtrade with the low price is unknown to the general contractor.
- Contractors need to evaluate subtrades by asking:
  1. Is the subtrade capable of meeting the quality and schedule requirements of its work?
  2. Does the subtrade have the financial capability required?
- The evaluation of a subcontractor needs to consider:
  1. The general reputation of the company
  2. The quality of previous work performed by the company
  3. The quality of the company's management
  4. The company's disposition in terms of its general cooperativeness and the ease of dealing with company site and office personnel
  5. The financial status of the company
- One way to manage the risk of nonperforming subtrades is to have subtrades perform performance bonds.
- Another tool in subtrade management is the pre-bid subtrade proposal. This outlines for the contractor's estimator the terms of the subtrade bid in good time before the subtrade releases its price at the last minute.
- The contractor's estimator needs to carefully analyze subtrade bids to ensure that all of the work required from a trade, as defined by the project drawings and specifications, is included for in the subtrade bid price.
- Bid depositories have been set up by contractors and trade associations to help combat the practice of "bid shopping" and reduce the last-minute rush of subtrade bids to the prime contractor.
- The bid depository has also been used in some instances to introduce generally accepted definitions of the scope of work for trades that submit their prices through the depository.

## REVIEW QUESTIONS

1. What should an estimator do when a bid is received from a subcontractor with a very low price suggesting that the subtrade has made an estimating error?
2. Discuss overlap and underlap of subcontracted work and the problems associated with each condition.

3. Why are conditional bids from subtrades undesirable?
4. How can the problem of bid closing congestion associated with receiving subtrade bids be reduced?
5. What are the general contractor's two main concerns when a low bid has been received from a subcontractor who is new to the general contractor?
6. What are the benefits and disadvantages to a general contractor of having subtrades provide surety bonds?
7. What is a pre-bid subtrade proposal and of what benefit is it to the contractor's estimator?
8. What are the advantages and disadvantages of using a bid depository to (a) a general contractor and (b) a subcontractor?

# 14

# Pricing General Expenses

## OBJECTIVES

*After reading this chapter and completing the review questions, you should be able to do the following:*

■ Define general expenses.
■ Use a checklist to identify general expense requirements for a project.
■ Explain how general expense requirements are assessed.
■ Describe how general expenses are calculated.
■ Calculate general expenses for the following items:
    a. Site personnel
    b. Safety and first aid
    c. Travel and accommodation
    d. Temporary site offices
    e. Temporary site services
    f. Hoardings and temporary enclosures
    g. Temporary heating
    h. Site access and storage space
    i. Site security
    j. Site equipment
    k. Trucking
    l. Dewatering
    m. Site cleanup
■ Identify general expense items that are calculated as add-ons.
■ Complete the pricing of project general expenses using manual methods.
■ Use the Precision Estimating software to price general expenses for a project.

## Introduction

The cost of labor, material and equipment expended on the items that were measured in the quantity takeoffs is usually referred to as the "direct costs" of the work. The "general expenses" of a project comprise all of the additional, indirect costs that are also necessary to facilitate the construction of the project. These indirect costs are sometimes titled General Requirements of the Project or Project Overheads.

There are an enormous number of possible general expense items on a project. The General Expense sheet presented as Figure 14.1 provides a summary of some of

PROJECT:
LOCATION:
DATE:                                    **GENERAL EXPENSES**
ESTIMATOR:
JOB DURATION:

| DESCRIPTION | QUANTITY UNIT | U/RATE LABOR | LABOR $ | U/RATE MATL. | MATL. $ | U/RATE EQUIP. | EQUIP. $ | U/RATE SUB. | SUB. $ | TOTAL $ |
|---|---|---|---|---|---|---|---|---|---|---|
| Superintendent | | | | | | | | | | |
| Site Engineer | | | | | | | | | | |
| Safety Coordinator | | | | | | | | | | |
| Timekeeper | | | | | | | | | | |
| Watchman | | | | | | | | | | |
| Accountant | | | | | | | | | | |
| Surveyor | | | | | | | | | | |
| Layout | | | | | | | | | | |
| First Aid | | | | | | | | | | |
| | | | | | | | | | | |
| Travel-Field | | | | | | | | | | |
| Travel-Head Office | | | | | | | | | | |
| Room & Board | | | | | | | | | | |
| | | | | | | | | | | |
| Tel: Hook-up | | | | | | | | | | |
| Tel: Charges | | | | | | | | | | |
| Elect: Hook-up | | | | | | | | | | |
| Elect: Charges | | | | | | | | | | |
| Temp. Water | | | | | | | | | | |
| Temp. Hoardings | | | | | | | | | | |
| Temp. Heat | | | | | | | | | | |
| | | | | | | | | | | |
| Temp. Office | | | | | | | | | | |
| Office Supp. | | | | | | | | | | |
| Temp. Toilet | | | | | | | | | | |
| | | | | | | | | | | |
| Temp. Roads | | | | | | | | | | |
| Storage | | | | | | | | | | |
| Security | | | | | | | | | | |
| | | | | | | | | | | |
| Site Equipment | | | | | | | | | | |
| Trucking | | | | | | | | | | |
| Dewatering | | | | | | | | | | |
| Saw Setup | | | | | | | | | | |
| | | | | | | | | | | |
| Photographs | | | | | | | | | | |
| Job Signs | | | | | | | | | | |
| Testing | | | | | | | | | | |
| Current Cleanup | | | | | | | | | | |
| Final Cleanup | | | | | | | | | | |
| Warranty | | | | | | | | | | |
| **TOTALS TO SUMMARY:** | | | | | | | | | | |

**Figure 14.1**   General Expense Sheet

the more common items of general expenses. On a small to medium size project, this General Expense sheet can be used as a checklist that allows the estimator to choose those items that are required on the project under consideration. The items selected can then be priced either by inserting the prices and calculating the total costs manually or by setting up the general expenses summary on a computer spreadsheet and allowing the computer to perform the applicable calculations. On larger projects a more comprehensive list of items may be required.

Some general expense items are mentioned in Division 0, Bidding & Contract Requirements, and others in Division 1, General Requirements, of the CSI Master-Format, but many items are not found anywhere in the specifications. Nonetheless, these unlisted items can still be indispensable to the completion of the project. Also, general expense items are not usually detailed on drawings even though items such as scaffolding and temporary enclosures can be substantial structures in their own right. The assessment of the extent of general expense needs for a project, together with the design of any temporary structures required is normally the general contractor's responsibility alone.

Because most general expense items are not referred to in the project specifications and because of the lack of details on contract documents regarding temporary structures, the estimator needs considerable construction knowledge and experience to first of all determine which general expense items are required for the project and, second, to accurately price these requirements. The price of general expenses can be a substantial component of the bid, amounting to 15% and more of the total price in some cases. Accurate assessment of general expenses can be critical to the competitiveness and financial success of the venture, so many estimators, including highly experienced veterans, consider it beneficial to consult with other staff about the general expense demands of a project.

Job superintendents, project managers and other senior company personnel can usually help with the appraisal and pricing of general expenses, especially when site facilities and equipment are being considered. There are often several alternative ways of providing general expense facilities on a project, and discussions can sometimes yield solutions that are more competitive or more effective than those elicited by the estimator working alone. Some companies go so far as to make it a policy that a team of senior staff meet and consider the general expense needs of all major projects.

General expense requirements are assessed by examining the plans and specifications and considering the amount of work involved in the project as measured in the takeoff process, so the task of estimating general expense prices can begin only when the quantity takeoffs are complete. However, some general expenses cannot be priced until the total price of the work is determined. These items are separated from the other general expense items and are priced at the end of the estimate as "add-ons." Figure 14.2 contains a list of some general expense add-on items.

## Project Schedule

A realistic schedule of the project work is necessary at the time of the estimate because one of the principal variables that determines the cost of general expense items is the duration of the project. Also, the project duration specified by the contractor in the bid is often considered by owners when they select the contractor. Sometimes the owner is looking for the shortest possible project duration, in which case the contractor with a better schedule may win over a lower-priced bidder.

When the quantity takeoff is complete, a reliable schedule of the contractor's work can be prepared based on anticipated crew productivities, but this would address only a small portion of the project work since the greater part of the work is usually subcontracted. The contractor does not normally measure the subtrade work,

| DESCRIPTION | METHOD OF CALCULATION | BASED ON $ |
|---|---|---|
| | | |
| Small tools and consumables | Percentage | Part labor or total labor |
| Wage increase | Percentage | Part labor or total labor |
| Overtime premium | Percentage | Part labor or total labor |
| Payroll additive | Percentage | Total labor |
| Data processing charges | Percentage | Total labor |
| Sales tax | Percentage | Total materials |
| Building permit | Percentage or step rate | Total bid |
| Surety bonds | Percentage or step rate | Total bid |
| Insurance—builder's risk | Percentage | Total bid |
| —liability | Percentage | Total bid |
| Contractor's fee | Percentage or lump sum | Total bid |
| Value added tax | Percentage | Total bid |

**Figure 14.2**   Selection of General Expense Add-Ons

so information about subtrade quantities and prices will not be available until sub-contractor bids are received. As we have seen in Chapter 13, subtrade quotes are generally not obtained until shortly before the prime contractor's bid closes. Because of this, and the time constraints of the estimating process, detailed scheduling techniques such as critical path method are not a practical means of preparing schedules at the time of the estimate; more approximate methods have to be resorted to.

One approach taken by many contractors is to simply search previous job records for a project of size and type similar to the one being estimated and use the actual duration of the past project as a guide to the duration of the new project. When a contractor repeatedly builds projects of a similar description, this technique can produce quite accurate results. If detailed information has been kept about the time taken to complete the work of each of the various trades involved in previous projects, a fairly sophisticated schedule of the future project can be produced.

Even when the new project is of a different type or size from the contractor's previous projects, an approximate schedule in the form of a bar chart of the major activities can often still be drafted. From the information contained in the plans and specifications together with details from the takeoffs of the amount of work involved in the project, veteran supervisors can usually put together at least a rough schedule based on their experience on past jobs.

An example of a **bar-chart** schedule prepared for an estimate is shown as Figure 14.3. This kind of schedule can be roughed out in just a few minutes by using a computer spreadsheet template that has been previously set up for the purpose. The first draft of the estimate schedule can be based on mere guesswork but the durations of

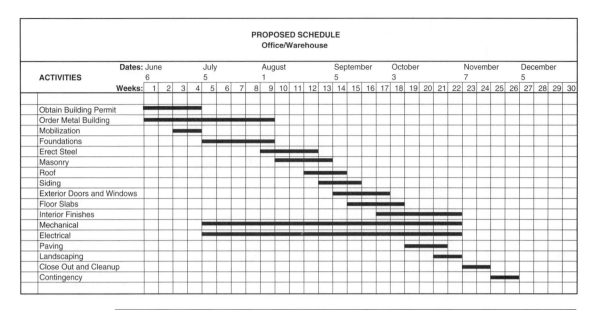

**Figure 14.3**   Sample Bar-Chart Schedule for an Estimate

key trades, such as steel work and masonry work in the example shown, can be updated as information becomes available from subcontractors when they complete their estimates.

## Site Personnel

The estimated cost of site personnel is one of the numerous general expense items that is directly proportional to the duration of the project. After a list of the site personnel required on the project is compiled, the price of this item is determined by multiplying the monthly payroll cost of the personnel on the list by the number of months they are required to be on the project.

The site personnel needs of a project can vary enormously. A part-time supervisor may be all that is called for on a small project, while a staff of 20 people or more may be required on a large project.

The most common members of a construction company's site staff are listed at the top of Figure 14.1. The estimator should be able to price the needs of most small- and medium-size projects by selecting from this list. On larger projects additional specialized site staff may be required. Figure 14.4 provides a more comprehensive list of possible site staff that should be adequate for almost any size of project.

The wages used to price site staff may or may not include payroll burden (the cost of fringe benefits). Estimators should follow a standard practice regarding payroll burden. If it is not firmly established whether net wages or gross wages are used in the pricing process, it can be very easy to make an overpricing or underpricing error. In the examples that follow, net wage amounts are used to price site personnel and the total amount for site personnel is added to the rest of the labor cost of the estimate. An allowance for payroll burden is subsequently added to the total labor amount at the end of the estimate.

| | | | PROJECT: | | |
| LOCATION: | | | | | |
| DATE: | | | **SITE PERSONNEL** | | |
| ESTIMATOR: | | | | | |
| JOB DURATION: | | | | | |

| DESCRIPTION | QUANTITY | UNIT | U/RATE LABOR | LABOR $ |
|---|---|---|---|---|
| Project Manager | | | | |
| Superintendent | | | | |
| Assistant Superintendent | | | | |
| General Foreman | | | | |
| Safety Coordinator | | | | |
| Surveyor | | | | |
| Survey Helper | | | | |
| Site Engineer | | | | |
| Quantity Surveyor | | | | |
| Estimator | | | | |
| Expeditor | | | | |
| Inspector | | | | |
| Accountant | | | | |
| Purchasing Agent | | | | |
| Timekeeper | | | | |
| Administrator | | | | |
| Watchman | | | | |
| Secretary | | | | |
| Driver | | | | |
| Janitor | | | | |
| **TOTAL TO SUMMARY** | | | | |

**Figure 14.4**   Comprehensive List of Possible Site Staff

## Safety and First Aid

In this item we consider the cost of meeting general safety and first aid requirements at the site. Safety requirements for a project may include one or more of the following expenses:

1. The cost of workers' time when attending safety meetings
2. The cost of workers' time when engaged in other safety-related activities
3. The provision of safety awards on the project
4. The provision of safety supplies at the site

The work required to be executed in order to comply with safety regulations on a project is usually measured in the takeoff process together with the rest of the work of the project. Examples include cutting back excavation embankments and providing shoring to the sides of deep trenches, which are taken off and priced with the excavation work, and installing safety guard rails on the edge of a slab forming system, which will normally be included in the cost of the formwork system. But there may be activities on a job that are purely safety related and are not part of any particular work item. Workers' time spent attending safety meetings or taking safety courses are examples of this kind of safety activity. The estimator, generally by reviewing records of previous projects, has to anticipate how much labor time will be spent on these safety matters.

Safety supplies such as hard hats, lifelines, flotation devices, special safety clothing and suchlike are expended during the course of construction activities. Some companies include the cost of these items in the "Small Tools and Consumables" item in the add-ons; other companies prefer to estimate these costs separately. The estimator should determine company policy and work accordingly.

First aid expenses can be divided into two basic categories:

1. The cost of providing on-site first aid supplies
2. The cost of personnel engaged in administering first aid services at the site

Occupational Health and Safety Regulations require construction sites to have first aid kits sufficient to meet the needs of the project. It is necessary to estimate the cost of supplies used up over the course of the work. This is usually done by investigating the amounts expended on first aid supplies on past jobs from project records.

Regulations also call for persons qualified to administer first aid to be present on construction sites. This requirement is met, on projects up to a certain size, by having a number of the trades people and site staff obtain first aid qualifications. But when a carpenter with a first aid ticket spends an hour providing first aid to a fellow worker, the cost of that hour will rightly be charged to the project first aid budget, so an estimate of the total number of hours administering first aid is required in order to establish the budget. On larger jobs this task is easier for the estimator (but more expensive for the contractor) because regulations often require that a full-time paramedic or nurse be employed on site. The estimator simply calculates the total wage bill for the time that person is needed on site.

In order to properly estimate the cost of these first aid provisions, the estimator needs to be familiar with the particular Occupational Health and Safety Regulations that apply to the project. For example, some regulations specify that a registered nurse shall be in full-time attendance at an industrial construction site where the number of on-site workers exceeds 200. This type of requirement has to be carefully considered because the cost of paramedics or nurses would be a significant expense on a long-term project.

## Travel and Accommodation

Travel and accommodation expenses for work crews and staff employed on out-of-town projects can amount to substantial sums on projects that cannot be staffed by local residents. The following charges are included in this category.

1. Travel expenses for site staff
2. Room and board expenses for site staff
3. Travel expenses plus possible travel time allowances for the project work force
4. Room and board or camp costs for the project work force
5. Travel and other expenses incurred by head office staff on visits to the project site

The amounts paid for travel and accommodation depend on the employment agreements that are in place. When unionized labor is employed, the collective agreement with each union will generally include details of the specific methods of calculating travel expenses that have been agreed on. Details vary from one union contract to another, so the estimator has to be prepared to investigate the requirement of each union agreement that applies to the different trades employed on the project. This may result in a separate calculation of travel expenses for each trade involved.

With nonunion labor there are usually company policies in place that deal with travel allowances paid to employees. Here again allowances may vary from one employee grade to another. Whether union labor or nonunion labor is employed, there are often alternative methods of allowing for travel expenses. Figure 14.5 illustrates the possible travel expense calculations for staff and workers on a project located 100 miles out of town. In this example, paying the crew travel expenses and travel time for a weekly round-trip from the city to the site plus an allowance for room and board per week is the least expensive alternative. But clearly, using local labor offers considerable savings over paying travel expenses to city workers. Even the lower price alternative in the example has the effect of almost doubling the labor cost of the project.

## Temporary Site Offices

The requirements for temporary site office accommodation will vary a great deal depending mostly on the size of the project to be constructed. On a very small job there may be no need for any temporary buildings on site. The project supervisor is said to run the job from his or her pickup truck in these situations. In contrast, large projects can call for a multitude of buildings to be set up on or near the site for the duration of the project. The temporary site office cost can include any or all of the following items:

1. Rentals: office trailers, lunch rooms, tool lockups, toilets
2. Costs to move trailers in and out
3. Office furniture
4. Stationery supplies
5. Fax machines
6. Copying machines
7. Computers, printers and related equipment
8. Office heating costs
9. Office utility costs
10. Janitorial services
11. Catering facilities

In addition to offices and other trailers set up to meet the needs of the contractor on site, the project contract documents may call for the contractor to provide office facilities for the use of the design consultants and their project site staff. The consultants' offices are usually required to be fully equipped and serviced by a phone line that is separate from the contractor's line.

Office accommodation for the project staff can be provided by alternatives other than the usual site trailers. In fact on some projects limited space on site may necessitate the use of alternative arrangements. There might be existing buildings on or near the site of the work that could provide office or storage space. Rental rates for floor space in some buildings can be quite comparable to the cost of trailers and, although there might be a cost to set up offices in these spaces, the expense of hauling trailers will be avoided.

On some projects the owner is able to provide quarters that meet the office and storage space needs of the contractor. These might be available in a building near to

### TRAVEL EXPENSES

**NOTES:**
1. The site of the project is located 100 miles from the contractor's home city.
2. The duration of the project is expected to be 50 weeks.
3. Labor agreements call for payment of travel expenses of $0.25 per mile.
4. Travel time also has to be paid to workers at the rate of 1-minute per mile traveled at their regular wage rate.
5. There are to be two site staff on the project who are paid travel expenses but not travel time.

**Preliminary Calculations**

| | | |
|---|---|---|
| Total Project Labor: | This is obtained from the estimate: | $384,000.00 |
| Average Labor Wage: | .......... | $20.00 per hour |
| Total Project Labor-hours: | $384,000/$20 = | 19,2000 labor-hours |
| Total Project Labor-days: | 19,2000 hours/8 hours/day = | 2,400 labor-days |
| Project Duration in Days: | 50 weeks × 5 days/week less 10 holidays = | 240 days |
| Average Labor Force: | 2.400 Labor-days/240 days = | 10 workers |

**Alternative 1:** Workers are pad travel expenses and travel time to drive to the site daily

| | | |
|---|---|---|
| Total Miles Traveled: | 10 workers × 240 days × 200 miles = | 480,000 miles |
| Travel Expenses: | 480,000 miles × $0.25 = | $120,000.00 |
| Travel Time: | 480,000 miles × 1 mins./60 mins. × $20.00 = | $160,000.00 |
| Payroll Additive: | 45% of $160,000.00 = | $ 72,000.00 |
| | **Total Alternative 1:** | **$352,000.00** |

**Alternative 2:** Workers are paid travel time and expenses for a round-trip once a week to the site, plus a weekly allowance for accommodation near the site.

| | | |
|---|---|---|
| Total Miles Traveled: | 10 workers × 50 weeks × 200 miles = | 100,000 miles |
| Travel Expenses: | 100,000 miles × $0.25 = | $ 25,000.00 |
| Travel Time: | 100,000 miles × 1 mins./60 mins. × $20.00 = | $ 33,333.33 |
| Payroll Additive: | 45% of $33,333.33 = | $ 15,000.00 |
| Room and Board Allowance: | 10 workers × 50 weeks × $400.00/week = | $200,000.00 |
| | **Total Alternative 2:** | **$273,333.33** |

**Expenses for Site Staff**

| | | |
|---|---|---|
| Travel Expenses: | 2 staff × 50 trips × 200 miles × $0.25 = | $ 5,000.00 |
| Room and Board Allowance: | 2 staff × 50 weeks × $400.00/week = | $40,000.00 |
| | **Total Staff Expenses:** | **$45,000.00** |

**Expenses for Head Office Staff**

This is based on an average of one head office staff member visiting the site each week:

| | | |
|---|---|---|
| Travel Expenses: | 50 weeks × 200 miles × $0.25 = | $2,500.00 |
| Meals: | 50 trips × $35.00/trip = | $1,750.00 |
| | **Total Head Office Staff Expenses:** | **$4,250.00** |

**Figure 14.5**  Travel Expenses Calculations

the project site or, when the project involves the construction of a building, the project building itself may provide the required accommodation when it reaches a certain stage. It is important to determine whether the contractor or the owner will pay heating, utilities and other such expenses in connection with the use of these premises.

Regarding the provision of catering facilities, on small- to medium-size projects the contractor often relies on coffee trucks to supply snacks and refreshments to the site. Vending machines can be set up in a lunch room; these may provide a small source of revenue to the general contractor that can offset the cost of providing the lunch room. On large projects a cafeteria facility could be established on-site. Again, if this is efficiently organized, there should be some revenues generated to compensate for the cost of setting up and running the service.

Catering facilities may be required by the owner under the provisions of the contract, or the contractor may see some benefits in providing this service for the site work force. Whichever is the case, the estimator needs to investigate the particulars and cost of any licenses and fees required to set up a site catering facility. The requirements of health regulations in connection with catering facilities especially need to be considered.

Often there will also be health department rules to comply with regarding temporary site sanitary facilities. Contractors often use portable chemical toilets to meet these needs. In some situations, toilets that are connected to the sewer system and equipped with adequate plumbing may be necessary.

## Temporary Site Services

A number of site services are usually required in order to pursue construction operations on a project. Site services include the following items:

1. Water services
2. Electrical services for power and lighting
3. Sanitary and storm sewer services
4. Telephone services

In each case there is usually an installation cost to extend the particular service to the site, and a cost for the resources consumed during the course of construction operations. Installation costs vary widely according to how far the services have to be brought to reach the site. The information about which services are available on site and which have to be brought in from afar may be provided in the bid documents, but often this is only available from a thorough investigation of the site. See Chapter 2 for a discussion of site investigations

The largest consumption cost of all of these items will probably be for electrical power used on the project. The cost of power for heavy users of electricity, such as tower cranes, temporary hoists and elevators, should be calculated separately form the general electrical consumption on the project. The example of the pricing of a project tower crane illustrated in Figure 14.9 on page 336 includes a price for the estimated amount of energy consumed by the crane.

The cost of the electricity used to meet the general power and lighting needs of the job is most often estimated by once again using the historic costs of the item on previous projects as a basis for making a prediction about future expenses.

This method is also frequently used to estimate the project phone charges after the cost of installing a telephone service is calculated from cost information obtained from the phone company and project information gathered in the site investigation.

### Hoardings and Temporary Enclosures

Hoardings and temporary enclosures may be constructed on a project in response to specific instructions of the contract documents or they may be required as a result of the contractor's own assessment of what is needed for the project. This is another item that can often be better evaluated if some of the contractor's senior personnel take the time to discuss the requirements of the job with the estimator. There is a wide variety of different temporary structures that could possibly be needed on a project, including:

1. Fences or barricades set up around the perimeter of work areas
2. Covered walkways and supports for overhead office and storage trailers
3. Access stairs and railings up to overhead offices and down to bottoms of deep excavations
4. Temporary enclosures built over work areas to provide protection or to allow the areas to be heated
5. Temporary covers and barricades to slab openings, stairways and elevator shaft openings
6. Temporary covers to the exterior openings of buildings to improve security or to allow the building to be heated
7. Temporary partitions or dust curtains erected in or around a building to confine the debris and noise of construction operations
8. General protection of existing structures on or adjacent to the work site

In each case the total cost of work includes the cost of materials, labor and any necessary equipment required to erect the structures involved, plus the cost of taking down and removing the materials from the site after their use.

Figure 14.6 shows the sketch of a temporary enclosure erected over a building under construction to provide a heated work area in winter conditions. Lightweight joists are supported on 2 × 6 stud walls on the exterior and 4 × 4 posts supporting double 2 × 10 beams on the inside. All of the lumber will be used many times over so its unit cost will be low, and the joists will be rented only for the time they are required for use on the enclosure.

A takeoff of the work involved in this enclosure together with prices for materials and labor to erect and dismantle the items involved is shown as Figure 14.7.

Air support structures rented for the time needed on-site have proved to be a viable alternative on some projects to the kind temporary structures described here.

### Temporary Heating

Temporary heating can be divided into three categories:

1. Heating provided in localized areas to allow operations like masonry to continue in cold weather
2. Heating a temporary enclosure
3. Heating the building to allow the finishing trades to complete their work

Items 1 and 2 are usually achieved by the use of portable space heaters that are oil, gas or propane fueled. This type of heater may also be used for item 3, or the contractor may be allowed to use the building's permanent heating system, if it is operational.

Two costs need to be estimated when pricing these items: (1) the rental rate or ownership costs of the portable space heaters used and (2) the cost of fuel consumed, both of which depend on the duration of the heating period and the amount of heat

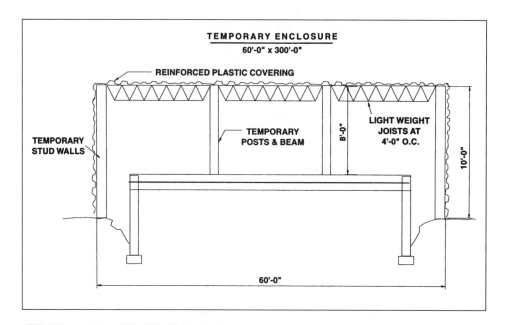

**Figure 14.6**   Temporary Enclosure Details

**TEMPORARY ENCLOSURE**

**Takeoff**

| | | | |
|---|---|---|---|
| Lightweight Joists: | (300.0/4.0) + 1 = 76 Joists × 3 rows | = | 228 No. |
| Posts & Beam: | 2 × 300.0 | = | 600 LF |
| Stud Walls: | 2 × 300.0 × 10.0 = 6,000 | | |
| | 2 × 60.0 × 10.0 = 1,200 | = | 7,200 SF |
| | | | |
| Plastic Covering: | Walls: 7,200 | | |
| | Roof: 300.0 × 60.0 = 18,000 | = | 25,200 |
| | Add 10% for laps: | | 2,520 |
| | | | 27,720 SF |

| ITEMS | QUANTITY | UNIT | U/RATE LABOR | LABOR $ | U/RATE MATL. | MATERIALS $ | TOTAL $ |
|---|---|---|---|---|---|---|---|
| Lightweight Joists | 228 | No. | $20.00 | 4,560 | $15.00 | 3,420 | 7,980 |
| Posts & Beam | 600 | LF | $ 4.00 | 2,400 | $ 1.20 | 720 | 3,120 |
| Stud Walls | 7,200 | SF | $ 0.65 | 4,680 | $ 0.12 | 864 | 5,544 |
| Plastic Covering | 27,720 | SF | $ 0.04 | 1,109 | $ 0.07 | 1,940 | 3,049 |
| | | | | | | | |
| **TOTAL PRICE:** | | | | 12,749 | | 6,944 | 19,693 |

**Figure 14.7**   Temporary Enclosure Materials and Labor Costs

required. There will also be labor costs for maintenance and operation of heaters but this cost should be considered with the other site personnel expenses considered previously.

If the contractor proposes to use the permanent heating system of a building as a source of temporary heating during construction, the estimator should contact the mechanical subtrade responsible for installing the heating system since there may be costs in addition to fuel costs to take into account.

Figure 14.8 shows the calculation of the temporary heating expenses to provide heat for a period of 30 days to the temporary enclosure detailed on Figure 14.7. This estimate is based on a heat consumption rate of 100 BTUs per cubic yard of space heated per hour, which is an estimate of the amount of heat required to maintain acceptable working conditions inside the enclosure. The actual heating rate encountered on the project can vary over a wide range of possibilities depending on the quality of the enclosure, the air temperature outside the enclosure and wind conditions; these are all difficult to accurately predict. Because of this uncertainty, there is a large amount of risk attached to estimates of heating costs; the estimator and the contractor's management should alert themselves to this risk.

## Site Access and Storage Space

Considerations of site access and storage space are grouped together because the two are often interconnected. For instance, a city center site may have excellent accessibility but little or no storage space on-site, whereas more than adequate storage space may be available on an out-of-town job but access to the site may be limited to a dirt road. These items need to receive careful attention on the site visit if an accurate assessment of access and storage problems is to be made.

---

**TEMPORARY HEATING**

**Notes:**
1. A temporary enclosure of size: 300'0" × 60'0" × 8'0" is to be heated for a period of 30 days.
2. Heating is required to be continuous 24-hours per day at the rate of 100 BTUs per CY per hour.
3. Heating is to be provided by propane heaters with an output of 200,000 BTUs per hour.
4. Heaters are rented for $75.00 per week and propane is available for $1.05 per gallon.
5. A gallon of propane provides 90,000 BTUs of heat.

**Calculation:**                                                                                          **Cost**

| | | |
|---|---|---|
| Volume to be Heated: | 300'0" × 60'0" × 8'0" | = 144,000 cu. ft. |
| | | = 5,333 cu. yd. |
| Heat Required: | 5,333 cu. yd. × 100 BTUs per hr.: | 533,333 BTUs/hr. |

Heater Rental Cost:   $\dfrac{533,333}{200,000}$ = 3 Heaters × \$75.00/week × 4 weeks    = \$ 900.00

Propane Consumption:   533,333 × 30 days × 24 hr. = 384,000,000 BTUs

Propane cost:   384,000,000 × $\dfrac{\$1.05}{90,000}$    = \$4,480.00

**Total Heating Cost:** \$5,380.00

---

**Figure 14.8**   Cost of Heating a Temporary Enclosure

In some cases roads or storage space may have to be cleared and constructed for the project. These requirements will usually be detailed in the contract specifications. On other jobs access roads may be in place but the cost of maintaining these roads may need to be estimated. Gravel roads must be graded from time to time, and paved roads may require cleaning or even repairs if they are seriously soiled or damaged by traffic to and from the project.

Municipal bylaws in some localities have imposed liability on contractors regarding cleanup and restoration of city pavements affected by construction operations. The estimator should check into the regulations that apply to the project as the cost of cleaning up roads can be expensive after trucks have left long trails of mud from a wet site.

The problem of lack of space on congested sites for trailers, storage and parking can be difficult to solve. Contractors have sometimes been able to utilize space on the streets next to the site with municipal approval and on payment of the necessary fees. Alternative solutions have included renting space on vacant lots adjacent to or, at least, near to the site; building temporary storage platforms over streets and sidewalks; and, where a waterway is near to a site, renting barges to use for storage. The problem has to be confronted at the time of the estimate so that the price of the chosen remedy can be determined and included in the bid price.

## Site Security

Project specifications may spell out security requirements for the job that can vary anywhere from the provision of a site fence to an elaborate continuous surveillance system incorporating personal I.D. cards for site personnel and security checks on all persons and vehicles passing through the site gates. Clearly, the more sophisticated security systems are the province of organizations that specialize in this work. Such companies would be hired as subcontractors to provide the necessary services where they are required on the project. When this is the case, the estimator should investigate what the security company requires from the contractor if this is not clearly defined in the specifications. Requirements can include special site fencing, gates, and, possibly, guard houses at the gates in some cases.

When specifications are silent on security requirements, the contractor will have to decide what measures are to be taken in this regard for the project. In some cases there is a trade-off between the cost of losses and the cost of security. Installation of an inexpensive alarm system on offices and trailers may be sufficient, or just commonsense precautions taken by supervisors and workers during the course of the work may be all that is necessary on certain projects.

Aside from specification conditions, the contractor may have to consider requirements of its insurance company and the recommendations of local law enforcement agencies when reviewing site security needs. But whatever the situation is, decisions have to be in place before the estimator can to put a price on them.

## Site Equipment

The equipment used on the project is priced in three different sections of the estimate using three different pricing methods:

1. Major items of equipment used in excavation and concreting operations are priced under the recaps of that work. This category of equipment includes items such as excavators, bulldozers, graders, rollers, concrete buggies and concrete pumps. As we have seen in Chapters 9 and 10, these items are priced

using rental rates or ownership costs against the hours the equipment is used on site.

2.  Hand tools and other items of equipment valued at less than $1000 each are priced at the end of the estimate as a percentage of the total labor price of the estimate. See "Small Tools and Consumables" following.

3.  Items of equipment kept at the site, sometimes for extended periods, and used only intermittently during its time on site compose a category of equipment priced with the general expenses, and it is this equipment that we will consider here.

Some of the more common items of site equipment include:

- Table saws
- Radial-arm saws
- Plate compactors
- Jumping jack compactors
- Air compressors
- Pneumatic tools and attachments
- Generators
- Concrete vibrators
- Surveying instruments
- Welders
- Hoists and cranes
- Cement and mortar mixers
- Forklifts

From the contractor's list of site equipment, the estimator selects the pieces required for the job and prices each item by multiplying its price rate by the length of time the item is to be assigned to the project. An amount may also be added to cover the cost of fuel and maintenance expenses on this equipment.

### Tower Crane

Major items of equipment such as hoists and cranes can have many other costs associated with them such as rigging, setup and dismantling expenses that also have to be accounted for. Figure 14.9 illustrates the calculation of the estimated full cost of providing a tower crane on a project.

For this example the project consists of a 10-story building and the tower crane is to be of the climbing type set up in an elevator shaft. As shown on the example estimate, there are many prices to consider in addition to the basic rental and operating costs of the crane.

The crane rises as the building rises under construction, and the loads on the crane are transferred to the floors adjacent to the crane. As a result, these adjacent floors have to be shored to safely carry the loads applied to them. The cost of the shores and labor to set up and take them down is added into the total cost. Mobile cranes are utilized in the erecting and dismantling of this crane. The cost of this work can be substantial as a very large mobile crane is often needed for the dismantling operation.

Electrical consumption costs may be calculated by applying a rate per horsepower-hour to the length of time the crane is in use on the project; however, by simply utilizing the historic record of crane electrical consumption, an estimate that is probably equally as accurate can be obtained.

Note that the labor prices in this example are net of payroll additive, which would be applied to the total labor content of the bid price at the end of the estimate.

PROJECT:  
LOCATION:  
DATE:  
ESTIMATOR:  
DURATION ON SITE:

**TOWER CRANE COSTS**

| DESCRIPTION | QUANTITY UNIT | U/RATE LABOR | LABOR $ | U/RATE MATL. | MATL. $ | U/RATE EQUIP. | EQUIP. $ | TOTAL $ |
|---|---|---|---|---|---|---|---|---|
| 1. Move-in Cost | 1 No. | $500.00 | 500 | — | — | $1,000.00 | 1,000 | 1,500 |
| 2. Unloading at Site | 1 No. | $200.00 | 200 | — | — | $500.00 | 500 | 700 |
| 3. Base for Crane/Rigging | 1 No. | $1,500.00 | 1,500 | $1,000.00 | 1,000 | — | — | 2,500 |
| 4. Erecting Costs | 1 No. | $3,000.00 | 3,000 | — | — | $4,000.00 | 4,000 | 7,000 |
| 5. Rental | 12 Mo. | — | — | — | — | $12,000.00 | 144,000 | 144,000 |
| 6. Jacking | 5 Lifts | $240.00 | 1,200 | — | — | — | — | 1,200 |
| 7. Post Shores | 10 Floors | $100.00 | 1,000 | $50.00 | 500 | — | — | 1,500 |
| 8. Crane Operator | 12 Mo. | $5,000.00 | 60,000 | — | — | — | — | 60,000 |
| 9. Floor Openings | N/A—Crane erected in elevator shaft | | | — | — | — | — | — |
| 10. Work Platforms | 5 No. | $200.00 | 1,000 | $200.00 | 1,000 | — | — | 2,000 |
| 11. Crane Repairs | Nil | — | — | — | — | — | — | — |
| 12. Dismantle and Remove | 1 No. | $4,000.00 | 4,000 | — | — | $8,000.00 | 8,000 | 12,000 |
| 13. Oil and Grease | Nil | — | — | — | — | — | — | — |
| 14. Electrical Costs: | | | | | | | | |
| —Install Service | 1 No. | $500.00 | 500 | — | — | — | 500 | |
| —Consumption | 12 Mo. | — | — | $400.00 | 4,800 | — | 4,800 | |
| 15. Testing | Nil | — | — | — | — | — | — | — |
| **TOTALS TO SUMMARY:** | | | 72,900 | | 7,300 | | 157,500 | 237,700 |

**Figure 14.9** Tower Crane Costs Calculation

## Trucking

This item concerns the cost of providing the incidental trucking on a project associated with minor items of equipment and small material orders that need to be transported to or from the job site. However, the transportation cost for shipping large items of equipment is best dealt with when the cost of the equipment is estimated. This was done in the case of major excavation equipment and the town crane that have been previously considered in the text.

Trucking requirements vary with the size and type of project under construction. On small to medium size projects, the trucking requirements may call for no more than use of the project supervisor's pickup truck. On projects that are spread over a large area, there may be a need for a larger truck or multiple trucks for the purpose of transporting materials and equipment from one area of the site to another.

## Dewatering

Dewatering expenses refer to the cost of removing water from excavations and in some cases from the basements of projects under construction. The usual method of dewatering makes use of submersible electric pumps and hoses, but more elaborate techniques are available including the well-point system of dewatering.

Sometimes when the need for dewatering is quite evident to the project designers, the specifications and drawings will provide details of a particular system to be used in the work. More often, however, the topic of dewatering is not discussed other than in very general terms in the specifications, and the contractor is simply left to allow for what is considered necessary to control moisture conditions in the work areas.

This is yet another region of great uncertainty and, consequently, high risk in the estimate. On projects where a large amount of pumping is expected, very little may ultimately be needed, and where little or no moisture was anticipated on a job, extensive dewatering can turn out to be necessary. These errors in judgment are possible even after diligent investigations of the site, previous conditions encountered in the vicinity of the work and careful study of all available soils reports.

## Site Cleanup

Site cleanup is generally divided into two categories: the ongoing daily cleanup expenses and the final cleanup expenses. The daily cleanup expenses include the cost of renting large garbage containers and the cost of constructing and later taking down temporary garbage chutes on the project. This expense item does not include the labor cost of the cleanup required in connection with the work of the project. The cost of cleaning up after the concrete work, for instance, should be included in the total estimated cost of the concrete work and, under the terms of subcontracts, subtrades are made responsible for any cleanup necessary after their own work and the cost is to be included in that subtrade's price for its work.

The final cleanup expenses include all the costs involved in getting the work site ready for handing over the finished project to the owner. This will involve cleaning floors, fixtures and windows when the project is a building. The work is usually subcontracted to a company that provides this cleaning service on the basis of a price per square foot of gross floor area.

## Miscellaneous Expenses

There are many types of general expenses that an estimator may not encounter as often as those just considered but that, nonetheless, do need to be considered on some projects. While it is impossible to make an exhaustive list of these miscellaneous expenses some of the more common items are:

1.  Layout—This is an allowance for the cost of helpers assisting a surveyor, engineer or the project superintendent with survey work involved in setting out the lines and levels of the project.

2.  Snow Removal—Contractors operating in northern areas often find that this can be a significant expense on the project.

3.  Saw Setup—When formwork and carpentry work is fabricated at the site, the contractor often builds a temporary enclosure for the main table saw or radial-arm saw. The cost to build and later remove the enclosure together with any other expanse to set up the saw is priced in this item.

4.  Photographs—Photographs may be called for in the specifications, in which case it may be necessary to include the price of professional photographers who perform this work. If not specified, an allowance for film and processing can be included when the contractor wishes to assemble a photographic record of progress of the work.

5.  Project Signs—The signs displayed outside the site of a project can vary from a small advertisement of the name of the general contractor to large, elaborate placards depicting artist's renderings of the finished project and listings of the names of all the major participants in the project. Site signs will normally be specified in the contract documents outlining in detailed or general terms the owner's requirements. Sign painters are usually hired to produce the signs, which are then installed by the prime contractor.

6.  Materials Testing—The specifications may state that the contractor shall pay for concrete testing, soils testing and other tests required, in which case the cost of providing the testing services is obtained from a testing agency. Often, however, the testing is to be paid for directly by the owner, but, even when this is so, the estimator may still need to consider the costs of providing samples and assistance to the owner's testing agencies.

7.  Water Leakage Tests—On projects such as the construction of reservoirs and swimming pools, the contract may call for leakage tests. In some cases a testing agency may be required to conduct the tests or the test may consist merely of a visual examination by the design consultants. One item that has been overlooked by contractors in the past is the cost of the water necessary for these tests.

8.  Warranty—Standard construction contracts include warranty provisions that call for the contractor, at no extra expense, to perform all remedial or maintenance work required on the project over a defined time period, often the 12 months following project completion. The average cost to the contractor of providing this service can usually be estimated from cost records of previous jobs.

9.  Temporary Fire Protection—This is normally the cost of providing fire extinguishers on-site during the construction period. The estimator should be careful to note any more elaborate protection requirements such as the provision of fire hydrants as this can be an expensive item.

10.  Wind Bracing Masonry—In many places the masonry trade does not include in the scope of its work the temporary wind bracing that is often necessary to support masonry work while it is under construction. When wind bracing is required, the contractor allows for the cost under this item.

11.  Traffic Control—This may be needed on downtown jobs where traffic adjacent to the site has to be controlled and on highway maintenance or construction jobs.

12.  Municipal Charges—There are many possible municipal charges that a contractor may be liable to pay in connection with any construction project. In addition to building permit fees (discussed later), fees may be payable for the

use of city streets for unloading, storage or parking; the construction of city sidewalk crossovers; municipal business taxes; and contractor license fees, to name but a few. Estimators should be familiar with the requirements of their own city, and on out-of-town jobs they must investigate the municipal charges in vicinities that are new to them.

13. Financing Charges—With an accurate forecast of the cash flow of a project, an estimator would be able to calculate the interest charges on the contractor's running overdraft that is required to finance the project. But precise forecasts of the inflow and outflow of cash on a construction project, even when detailed schedules are available, are extremely difficult to produce. Here again the estimator usually has to fall back on previous job records, with necessary adjustments for changes in interest rates, to determine the probable financing charges on a future project.

14. Taxes—There are as many ways of accounting for taxes in an estimate as there are ways of applying taxes to the price of construction work, and because the substance of a tax can vary widely from place to place, there is no single formula for dealing with taxes. How a tax is calculated depends on the particulars of how the tax is assessed. For instance:

    a. Where a sales tax applies to materials but varies from one material to another, the tax has to be calculated separately on each individual material price.

    b. Where a uniform sales tax applies to all project materials, an add-on can be set up at the end of the estimate to add the necessary percentage to the value of the total material content of the estimate.

    c. Payroll taxes that apply to the total labor price of the estimate can also be handled with an add-on. This is the case with the payroll additive add-on discussed in a following section.

    d. A European- and Canadian-style goods and services tax or value added tax that applies to the prices of all the components of an estimate can be set up as a percentage add-on calculated on the very last line of the estimate so that the applicable rate applies to the final complete estimate price.

## Labor Add-Ons

The following general expenses items are proportional to the total labor cost of a project, so the dollar value of these items can only be calculated when the total estimated labor cost has been determined. This point is reached in the estimating process when the contractor's own work, the subtrade work and the other general expenses have been priced. See Figure 15.1 for the location of these add-ons on a typical Estimate Summary sheet.

### Electronic Data Processing (E.D.P.)

This add-on accounts for the cost of using computer payroll or computer cost reporting systems. This expense is not incurred as frequently as it was in the past because many contractors have switched to the use of personal computers for this purpose, the cost of which is generally considered to be a company overhead (thus part of the fee) rather than a project overhead cost.

### Small Tools and Consumables

In this expense item the cost of hand tools and equipment valued at less than $1000 per piece is priced together with miscellaneous supplies used in the construction process. These supplies would include such items as replacement drill bits, saw blades, chalk lines, batteries and all the other incidental components used up in connection to the work of a construction project.

### Wage Increases

Most estimators price the work of the project using current wage rates then allow, under this item, for any wage increases anticipated during the course of the work by adding a percentage to the total wage bill of the project.

A more precise estimate of the wage increase can be obtained when the date and amount of wage raise is known at the time of the estimate. This data may be accessible when unionized labor is employed since the amounts of wage rates and the dates of increases are usually defined in collective agreements. In this situation the estimator needs to identify, from a reliable schedule of the work, those activities that will be completed before the wage raise and those that will proceed after the raise. The increase in wages is then applied to the latter category and the amount of the additional wages can be included in the estimate under this general expense item.

### Overtime Premium

The overtime premium is the extra payment over and above the basic wage rate that is paid to employees who work time in excess of the standard work day or work week. The premium may be applied in response to government regulations, company policy or union agreements, and the pay rate for overtime work may be set at time and a quarter, time and a half or double time. Wage premiums are also paid when workers are required to work shifts on a project. The premium in this case is often 10% above the standard pay rate.

The overtime premium is usually accounted for by adding a percentage or a fixed dollar amount to the estimate labor component that is calculated using standard rates of pay.

### Payroll Additive

The labor component of the estimate has so far been priced using the base wage rates for the trades involved. Every employer also makes a variety of contributions in relation to its total payroll in response to federal, state and municipal requirements and company policies. Payroll additive may account for any or all of the following items:

1. Social Security Tax—This is a federal government requirement to provide retirement benefits to persons who become eligible.
2. Unemployment Compensation Tax—This is a state tax that is gathered to provide funds to compensate workers in periods of unemployment.
3. Worker's Compensation and Employer's Liability Insurance—As required by state regulations, contributions are collected to provide a fund from which workers who are injured while working or as a result of working are compensated.
4. Public Liability and Property Damage Insurance—This insurance protects the employer against claims by the public for injuries or damage sustained as a result of the work activities of an employee.
5. Fringe Benefits—This includes a number of items that could be contained in a contract of employment or a union collective agreement. Examples include health care insurance, pensions and disability insurance.

The sum of these contributions can vary enormously from place to place, from company to company, and even from trade to trade within a company. The task of the estimator is to ascertain the rates of contributions that apply to the project labor force and allow for them in the estimate. Different rates of payroll additive may have to be applied to each separate trade employed on the project, or the estimator may be able to use a single average rate against the total labor component of the estimate.

## Bid Total Add-Ons

The final group of add-ons are assessed as a proportion of the total bid price so they are calculated at the end of the estimate after all other items, including the project markup or fee, have been added in.

The simple way to calculate these add-ons is to apply percentage rates to the final subtotal of the estimate, but this produces a slight error since the add-on rates ought to be calculated as a percentage of the total bid price rather than the final subtotal before the bid price.

Consider the case in which there is a single bid total add-on for the cost of a building permit for the project. Let us assume that the price charged for building permits is 1% of the bid price and the final subtotal of the estimate is $10,000,000. Calculating the add-on as a percentage of the final subtotal produces this outcome:

| | | | |
|---|---|---|---|
| Subtotal: | | | $10,000,000 |
| Add-on for building permit: | 1% | = | 100,000 |
| Total bid price: | | | $10,100,000 |

But when the contractor comes to purchase the building permit, the fee will be 1% of the total bid price, which is 1% of $10,100,000 = $101,000, which is $1,000 more than was allowed in the estimate.

To estimate the correct amount, the rate of the add-on is calculated by using the following formula:

$$\text{Add-on Rate} = \frac{\text{Actual Rate}}{1 - \text{Actual Rate}}$$

So, in the preceding example, the correct add-on rate $= \dfrac{0.01}{1 - 0.01}$

$$= .010101$$

With the corrected add-on rate, the last lines of the estimate will be revised to this:

| | | | |
|---|---|---|---|
| Subtotal: | | | $10,000,000 |
| Add-on for building permit: | 1.0101% | = | 101,010 |
| Total bid price: | | | $10,101,010 |

The fee for the permit would now be 1% of $10,101,010, which amounts to $101,010—precisely the amount estimated.

Where there are a number of bid total add-ons, the following formula can be used to determine the combined add-on rate:

$$\text{The combined add-on rate} = \frac{\text{Sum of Actual Rates}}{1 - \text{Sum of Actual Rates}}$$

At the time when the estimating process was completed by manual methods using calculators only, the prospect of handling this arithmetic right at the close of the estimate was not always attractive to estimators; many found it easier to accept the relatively small error that results from using the simpler method of calculation. However, now that computers are widely used to compile the estimate, using the correct add-on rates is not difficult since computer programs can be preset to make the correct calculations automatically.

### General Expenses Pricing Notes Example 1—House

#### General Expenses Figure 14.10

1. This project is expected to take 4 months to complete based on the duration of previous projects of a similar type and size.

2. Because this is such a small project, a number of the general expenses are shared with other projects so that only a portion of their total cost is included in this estimate. These items are marked with asterisks on the General Expenses sheet.

3. The $1750 amount against the Surveyor item is an allowance for employing a certified land surveying company to provide a legal survey of the site showing the accurate dimensions and layout of the lot including the location of the house on the property.

4. The 2 days of labor allowed against the Layout item is the price of a helper to assist the superintendent surveying and setting out the building.

5. The amount shown against First Aid is an allowance for first aid supplies consumed during the course of the work. This is based on the average amount spent on this item on previous projects.

6. There are no Travel or Room & Board expenses anticipated on this project when it is to be constructed in the builder's home city.

7. The office trailer, telephone and electrical charges reflect the cost of a single serviced trailer set up to be used by the superintendent as a base to supervise a number of small projects.

8. The allowance against security is the contribution of this project to the cost of providing a security patrol at the location of the work.

9. The Site Equipment to be used on this project would be listed and priced on the back of the General Expenses sheet. The note: "See Backup Sheet" is directing the reviewer of this estimate to consult a supplementary calculation sheet usually located on the back of the General Expenses sheet. The equipment list for this project is as follows:

| | | | | |
|---|---|---|---|---|
| 1 Compactor | 2 weeks | × $120.00 | = | $ 240.00 |
| 1 Compressor | 2 months | × $305.00 | = | 610.00 |
| 2 Concrete vibrators | 3 days | × $22.00 | = | 132.00 |
| 1 Crane (hoisting steps) | 1 day | × $250.00 | = | 250.00 |
| | | | | $1232.00 |
| Allowance for fuel and maintenance (25% × $1100.00) | | | | 275.00 |
| | | | Total: | $1507.00 |

10. The fuel and maintenance allowance is calculated as 25% of the items that require fuel (nonelectrical items).

11. The Trucking amount allowed is the portion of the cost of the superintendent's pickup truck that applies to this project.

12. The Current Cleanup amount is a contribution to the cost of renting a garbage container for the site.

13. The Final Cleanup amount is the price charged by a subtrade to clean the carpets and windows and generally get the house cleaned up and ready to hand over to the owner.

14. Add-on amounts are not included on this General Expense sheet since add-ons cannot be calculated until the summary is completed. Add-ons that apply to this job can be found at the end of the Summary sheet shown in Figure 15.10.

PROJECT: House Example
LOCATION:
DATE:
ESTIMATOR: ABF
JOB DURATION: 4 Months

**GENERAL EXPENSES**

| DESCRIPTION | QUANTITY | UNIT | U/RATE LABOR | LABOR $ | U/RATE MATL. | MATL. $ | U/RATE EQUIP. | EQUIP. $ | U/RATE SUB. | SUB. $ | TOTAL $ |
|---|---|---|---|---|---|---|---|---|---|---|---|
| Superintendent | * 4 | Mo. | $1,250 | 5,000 | | — | | — | | — | 5,000 |
| Site Engineer | — | | — | — | | — | | — | | — | — |
| Safety Coordinator | — | | — | — | | — | | — | | — | — |
| Timekeeper | * 4 | Mo. | $900 | 3,600 | | — | | — | | — | 3,600 |
| Watchman | — | | — | — | | — | | — | | — | — |
| Accountant | — | | — | — | | — | | — | | — | — |
| Surveyor | Legal survey by subtrade | | | | | — | | — | | 1,750 | 1,750 |
| Layout | 2 | Days | $168 | 336 | | — | | — | | — | 336 |
| First Aid | Allow | | — | — | | 50 | | — | | — | 50 |
| | | | | | | | | | | | |
| Travel-Field | — | | — | — | | — | | — | | — | — |
| Travel-Head Office | — | | — | — | | — | | — | | — | — |
| Room & Board | — | | — | — | | — | | — | | — | — |
| | | | | | | | | | | | |
| Tel: Hook-up | * 1 | No. | — | — | | — | | — | $150 | 150 | 150 |
| Tel: Charges | * 4 | Mo. | — | — | $100 | 400 | | — | | — | 400 |
| Elect: Hook-up | * 1 | No. | — | — | | — | | — | $350 | 350 | 350 |
| Elect: Charges | 4 | Mo. | — | — | $150 | 600 | | — | | — | 600 |
| Temp. Water | — | | — | — | | — | | — | | — | — |
| Temp. Hoardings | — | | — | — | | — | | — | | — | — |
| Temp. Heat | — | | — | — | | — | | — | | — | — |
| | | | | | | | | | | | |
| Temp. Office | * 4 | Mo. | — | — | — | — | $150 | 600 | | — | 600 |
| Office Supp. | 4 | Mo. | — | — | $75 | 300 | | — | | — | 300 |
| Temp. Toilet | 4 | Mo. | — | — | — | — | $120 | 480 | | — | 480 |
| | | | | | | | | | | | |
| Temp. Roads | — | | — | — | | — | | — | | — | — |
| Storage | — | | — | — | | — | | — | | — | — |
| Security | * 4 | Mo. | $500 | 2,000 | | — | | — | | — | 2,000 |
| | | | | | | | | | | | |
| Site Equipment | See Backup Sheet | | — | | | — | | 1,507 | | — | 1,507 |
| Trucking | * 4 | Mo. | — | — | | — | $300 | 1,200 | | — | 1,200 |
| Dewatering | — | | — | — | | — | | — | | — | — |
| Saw Setup | — | | — | — | | — | | — | | — | — |
| | | | | | | | | | | | |
| Photographs | — | | — | — | | — | | — | | — | — |
| Job Signs | 1 | No. | $350 | 350 | $850 | 850 | | — | | — | 1,200 |
| Testing | — | | — | — | | — | | — | | — | — |
| Current Cleanup | 4 | Mo. | $150 | 600 | | — | $250 | 1,000 | | — | 1,600 |
| Final Cleanup | 1 | No. | — | — | | — | | — | | 400 | 400 |
| Warranty | — | | — | — | | — | | — | | — | — |
| **TOTALS TO SUMMARY:** | | | | 11,866 | | 2,200 | | 4,787 | | 2,650 | 21,523 |

**Figure 14.10**   House Example—General Expense

### General Expenses Pricing Notes Example 2—Office/Warehouse Building

#### Precision Estimating Takeoff

In practice when you are estimating one job after another, Precision Estimating allows you to speed up the process of taking off and pricing the general expenses and add-ons in a number of ways. For instance, if you are using the Standard Edition or the Extended Edition of Precision Estimating, you can utilize an *assembly*. An assembly is a collection of items that can be accessed in one step in the takeoff process. So, because general expense items do not vary a great deal from one project to another, you can compile a list of the general expenses items that are encountered on a typical project and save them in a single assembly. With this assembly in place, on each new estimate you will be able to take off all of these general expense items in a single step.

Another way to increase the efficiency of your takeoff, even when using the Basic Edition as we are doing here, is to set up a template that contains the items required for a typical project. This template is an estimate that consists of a takeoff of all of the general expense items you encounter on a typical job. You can also set up on the totals page of this estimate template a list of the usual job add-ons required on your projects. (See add-ons in Chapter 15.) After you have completed such a template, you begin each new estimate by copying the template onto the new estimate. This will provide you with all the more common general expense items and add-ons in place ready to adjust prices to reflect the specific job circumstances. Other general expenses and add-ons required for a particular project can be added during the estimating process.

Because we are unable to use assembly takeoff and we lack an estimate template for the office/warehouse example, we are going to take off the general expenses from scratch. There are very few items in the sample database, so we begin by setting up new phases and new items. Figure 14.11 lists the new phases needed for the takeoff and Figure 14.12 is a list of all the general expense items and prices.

| PHASE CODE | DESCRIPTION | UNIT |
|---|---|---|
| *1000.00 | GENERAL EXPENSES | Mo. |
| *Note: this is a Group Phase. | | |
| 1304.01 | Site Layout | No. |
| 1310.01 | First Aid | No. |
| 1400.01 | Travel & Accommodation | Mo. |
| **1520.03 | Temp: Site Office | No. |
| **There is an existing Phase with this number; just modify it. | | |
| 1530.01 | Site Access | No. |
| 1532.01 | Storage | Mo. |
| 1536.01 | Site Security | Mo. |
| 1542.01 | Trucking | Mo. |
| 1550.01 | Dewatering | Mo. |
| 1600.01 | Photographs | No. |
| 1630.01 | Testing | No. |
| 1650.01 | Cleanup | Mo. |

**Figure 14.11** New Phases—General Expenses

| PHASE CODE | ITEM CODE | ITEM DESCRIPTION | UNIT | COST CATEGORIES | UNIT PRICES $ | | | | |
|---|---|---|---|---|---|---|---|---|---|
| | | | | | LABOR | MATL. | EQUIP. | SUB. | OTHER |
| 1300.01 | 15 | Project Manager | Mo. | L | 6,000.00 | — | — | — | — |
| 1300.01 | 20 | Site Engineer | Mo. | L | 4,000.00 | — | — | — | — |
| 1300.01 | 30 | Safety Coordinator | Mo. | L | 2,000.00 | — | — | — | — |
| 1300.01 | 40 | Timekeeper | Mo. | L | 3,000.00 | — | — | — | — |
| 1300.01 | 70 | Surveyor | Mo. | S | — | — | — | 3,000.00 | — |
| 1304.01 | 10 | Layout | Mo. | L | 3,640.00 | — | — | — | — |
| 1310.01 | 10 | First Aid | Mo. | L, M | 100.00 | 50.00 | — | — | — |
| 1510.01 | 05 | Elect: Hook-up | No. | O | — | — | — | — | 950.00 |
| 1510.01 | 10 | Temp. Electricity | Mo. | O | — | — | — | — | 500.00 |
| 1510.01 | 20 | Temp. Heat | Mo. | O | — | — | — | — | 250.00 |
| 1510.01 | 30 | Tel: Hook-up | No. | O | — | — | — | — | 750.00 |
| 1510.01 | 50 | Temp. Hoardings | No. | L, M | 6,825.00 | 4,725.00 | — | — | — |
| 1510.01 | 60 | Temp. Water | Mo. | O | — | — | — | — | 60.00 |
| 1510.01 | 80 | Temp. Toilet | Mo. | M | — | 240.00 | — | — | — |
| 1520.03 | 05 | Temp. Office | Mo. | O | — | — | — | — | 750.00 |
| 1520.03 | 10 | Job Sign | No. | L, M | 200.00 | 600.00 | — | — | — |
| 1520.03 | 15 | Office Supplies | Mo. | O | — | — | — | — | 150.00 |
| 1536.01 | 10 | Security | Mo. | S | — | — | — | 3,000.00 | — |
| 1540.01 | 10 | Tools & Equipment | No. | E | — | — | 16,456.25 | — | — |
| 1540.01 | 20 | Saw Setup | No. | L, M | 400.00 | 250.00 | — | — | — |
| 1542.01 | 10 | Trucking | Mo. | E | — | — | 800.00 | — | — |
| 1550.01 | 10 | Dewatering | Mo. | L, E | 100.00 | — | 200.00 | — | — |
| 1600.01 | 10 | Photographs | Mo. | O | — | — | — | — | 25.00 |
| 1650.01 | 10 | Current Cleanup | Mo. | L, E | 150.00 | — | 150.00 | — | — |
| 1650.01 | 20 | Final Cleanup | No. | S | — | — | — | 2,700.00 | — |

**Figure 14.12**   New Items—General Expenses

### General Expenses—Pricing Notes

1. This project is estimated to take 6 months to complete based on the preliminary schedule shown on Figure 14.3.
2. A site engineer is required only for the layout of the building, which is expected to be completed over a period of 1 month.
3. The safety coordinator amount allows for a part-time safety person for this size of job.

4.  The Surveying amount is the price of a registered land surveying company to provide certified layout and benchmark information on-site.

5.  The Layout amount reflects the cost of a helper for the site engineer.

6.  The material amount shown against First Aid is an allowance for first aid supplies consumed during the course of the work. The labor amount is for the time spent by first aid qualified workers administering first aid on-site. These allowances are based on the average cost of these items on previous projects.

7.  The Temporary Hoardings amounts are the estimated prices to construct a temporary fence around the site and remove it on completion of the work. The materials price is based on the maximum use of recycled materials.

8.  The Temporary Heat allowance is for the cost of fuel heating the building in the last few weeks of the project. This is based on the use of the permanent heating system, which the mechanical subtrade informs us will be operational at that time. The project owner has agreed to this.

9.  Under the item for Security we have allowed the price of a subtrade providing a security patrol at the site.

10. The Site Equipment comprises the following items:

| | | | | |
|---|---|---|---|---:|
| 1 Radial-arm saw | 4 Months × $400.00 | | = | $1,600.00 |
| 2 Compactors* | 1 Month × 700.00 | | = | 1,400.00 |
| 1 Compressor* | 4 Months × 1200.00 | | = | 4,800.00 |
| 3 Concrete vibrators | 3 Months × 350.00 | | = | 3,150.00 |
| 1 Generator* | 3 Months × 375.00 | | = | 1,125.00 |
| 1 Electrical panel | 5 Months × 150.00 | | = | 750.00 |
| 1 Mortar mixer* | 1 Month × 600.00 | | = | 600.00 |
| 1 Forklift* | 1 Month × 1850.00 | | = | 1,850.00 |
| | | | | $15,275.00 |
| *Fuel and Maintenance on These Items: 25% of | $4,725.00 = | | | 1,181.25 |
| | | | Total: | $16,456.25 |

11. The amount against Dewatering is a contingency allowance to cover the cost of renting pumps and of labor setting up and maintaining these pumps should some dewatering of excavations be necessary.

12. The Saw Setup is an estimate of the cost of building and removing a temporary enclosure and lockup for the radial-arm saw that is to be used on-site in connection with the formwork and carpentry work.

13. The allowance against Photographs is for film and processing expenses incurred taking progress photographs of the work.

14. The Current Cleanup amount is for the cost of renting a garbage container at the site.

15. The Final Cleanup amount is the price charged by a subtrade to clean the floors and windows and generally get the project cleaned up ready for occupancy by the owner.

16. The add-on amounts are included at the end of the office/warehouse summary shown in Figure 15.11.

17. Figure 14.13 shows the General Expenses as they appear on the Precision Estimating spreadsheet.

PEWin - C:\Sample Commercial GC
File Edit View Database Takeoff Pricing Reports Interface Window Help

Office-Warehouse

| Group | Phase | Description | Takeoff Quantity | Labor Cost/Unit | Labor Amount | Material Price | Material Amount | Sub Amount | Equip Price | Equip Amount | Other Price |
|---|---|---|---|---|---|---|---|---|---|---|---|
| 1000.000 | | GENERAL EXPENSES | | | | | | | | | |
| | 1300.010 | Personnel: Supervision | | | | | | | | | |
| | | Project Manager | 6.00 Mo. | 6,000.00 /Mo. | 36,000 | - | - | - | - | - | - |
| | | Site Engineer | 1.00 Mo. | 4,000.00 /Mo. | 4,000 | - | - | - | - | - | - |
| | | Safety Coordinator | 6.00 Mo. | 2,000.00 /Mo. | 12,000 | - | - | - | - | - | - |
| | | Timekeeper | 6.00 Mo. | 3,000.00 /Mo. | 18,000 | - | - | - | - | - | - |
| | | Surveyor | 1.00 Mo. | - | - | - | - | 3,000 | - | - | - |
| | 1304.010 | Site layout | | | | | | | | | |
| | | Layout | 1.00 Mo. | 3,640.00 /Mo. | 3,640 | - | - | - | - | - | - |
| | 1310.010 | First Aid | | | | | | | | | |
| | | First Aid | 6.00 Mo. | 100.00 /Mo. | 600 | 50.00 /Mo. | 300 | - | - | - | - |
| | 1510.010 | Utilities: Temporary | | | | | | | | | |
| | | Elect: Hook-up | 1.00 No. | - | - | - | - | - | - | - | 950.00 /No. |
| | | Temp Electricity | 6.00 mo | - | - | - | - | - | - | - | 500.00 /mo |
| | | Temp Heat | 2.00 mo | - | - | - | - | - | - | - | 250.00 /mo |
| | | Tel: Hook-up | 1.00 No. | - | - | - | - | - | - | - | 750.00 /No. |
| | | Temp Phone | 6.00 mo | - | - | - | - | - | - | - | 120.00 /mo |
| | | Temp. Hoardings | 1.00 No. | 6,825.00 /No. | 6,825 | 4,725.00 /No. | 4,725 | - | - | - | /No. |
| | | Temp Water | 6.00 mo | - | - | - | - | - | - | - | 60.00 /mo |
| | | Temp Toilet | 6.00 mo | - | - | 240.00 /mo | 1,440 | - | - | - | - |
| | 1520.030 | Temp: Site Office | | | | | | | | | |
| | | Temp. Office | 6.00 Mo. | - | - | - | - | - | - | - | 750.00 /Mo. |
| | | Job Sign | 1.00 ea | 200.00 /ea | 200 | 600.00 /ea | 600 | - | - | - | - |
| | | Office Supplies | 6.00 Mo. | - | - | - | - | - | - | - | 150.00 /Mo. |
| | 1536.010 | Site Security | | | | | | | | | |
| | | Security | 6.00 Mo. | - | - | - | - | 18,000 | - | - | - |

Phase/Item / Takeoff Order

For Help, press F1          7/12/2002

PEWin - C:\Sample Commercial GC
File Edit View Database Takeoff Pricing Reports Interface Window Help

Office-Warehouse

| Group | Phase | Description | Takeoff Quantity | Labor Cost/Unit | Labor Amount | Material Price | Material Amount | Sub Amount | Equip Price | Equip Amount | Other Price |
|---|---|---|---|---|---|---|---|---|---|---|---|
| | 1540.010 | Temp: Tools & Equipment | | | | | | | | | |
| | | Tools & Equipment | 1.00 No. | 0.00 /No. | 0 | 0.00 /No. | 0 | - | 16,456.25 /No. | 16,456 | - |
| | | Saw Setup | 1.00 No. | 400.00 /No. | 400 | 250.00 /No. | 250 | - | - | - | - |
| | 1542.010 | Trucking | | | | | | | | | |
| | | Trucking | 6.00 Mo. | - | - | - | - | - | 800.00 /Mo. | 4,800 | - |
| | 1550.010 | Dewatering | | | | | | | | | |
| | | Dewatering | 2.00 Mo. | 100.00 /Mo. | 200 | - | - | - | 200.00 /Mo. | 400 | - |
| | 1600.010 | Photographs | | | | | | | | | |
| | | Photographs | 6.00 Mo. | - | - | - | - | - | - | - | 25.00 /Mo. |
| | 1650.010 | Cleanup | | | | | | | | | |
| | | Current Cleanup | 6.00 Mo. | 150.00 /Mo. | 900 | - | - | - | 150.00 /Mo. | 900 | - |
| | | Final Cleanup | 1.00 No. | - | - | - | - | 2,700 | - | - | - |
| 2000.000 | | SITEWORK | | | | | | | | | |
| | 2144.000 | Gravel @ Slab | | | | | | | | | |
| | | Gravel under SOG | 402.90 CY | 6.20 /CY | 2,498 | 24.38 /CY | 11,787 | - | 3.94 /CY | 1,587 | - |
| | 2201.000 | Earthwk: Remove Topsoil | | | | | | | | | |
| | | Remove Topsoil | 1,087.04 CY | 0.60 /CY | 652 | - | - | - | 1.63 /CY | 1,772 | - |
| | 2205.000 | Earthwk: Excav Trench | | | | | | | | | |
| | | Excavate Trench | 248.82 CY | 0.81 /CY | 202 | - | - | - | 1.68 /CY | 418 | - |
| | 2210.000 | Earthwk: Excav Pits | | | | | | | | | |
| | | Excavate Pits | 13.22 CY | 1.62 /CY | 21 | - | - | - | 1.93 /CY | 26 | - |
| | 2220.000 | Earthwk: Bulk Cut | | | | | | | | | |
| | | Bulk Cut | 149.67 CY | 0.75 /CY | 112 | - | - | - | 2.03 /CY | 304 | - |
| | 2220.400 | Earthwk: Grade & Trim | | | | | | | | | |
| | | Grade & Trim Bottoms of Excav | 2,944.00 sf | 0.50 /sf | 1,472 | - | - | - | /sf | | - |
| | 2221.500 | Earthwk: Gravel Fill | | | | | | | | | |

Phase/Item / Takeoff Order

For Help, press F1          7/12/2002

**Figure 14.13** General Expenses Spreadsheet

## SUMMARY

- General expenses or project overheads comprise the indirect costs expended on a construction project.
- It is usually the general contractor's responsibility to assess and price the general expenses on a construction project.
- Because there are so many possible general expense items, estimators usually make use of a checklist when general expenses are priced.
- General expense requirements are assessed by examining the plans and specifications and considering the amount of work involved in the project as measured in the takeoff process. Job superintendents and project managers can help the estimator define general expense requirements for a project.
- Some general expense items are calculated as a percentage of the estimate or as a percentage of some part of the estimate; these items are included as add-ons at the end of the estimate.
- Many general expense items are calculated as a function of the duration of the project, so at least a rough schedule is required before these items can be evaluated.
- Major items to consider when pricing general expenses include:

  1. Site personnel
  2. Safety and first aid
  3. Travel and accommodation
  4. Temporary site offices
  5. Temporary site services
  6. Hoardings and temporary enclosures
  7. Temporary heating
  8. Site access and storage space
  9. Site security
  10. Site equipment
  11. Trucking
  12. Dewatering
  13. Site cleanup

- General expenses calculated as add-ons include:

  1. Electronic data processing
  2. Small tools and consumables
  3. Wage increases
  4. Overtime premium
  5. Payroll additive
  6. Building permits
  7. Surety bonds
  8. Insurance

- Pricing general expenses can be done manually or via computer

## REVIEW QUESTIONS

1. Explain what the general expenses of a project are.
2. How can the total duration of the construction work of a project be determined in the short time available in the estimating process?
3. Estimate the cost for the heating of a temporary enclosure of this size: 250 feet × 300 feet × 12 feet high. The space is to be heated at the rate of 75 BTUs per cubic yard per hour for a period of 2 weeks (14 days), 24 hours per day. Use 150,000 BTU propane heaters at a rental of $25.00 per week. The cost of propane is $0.90 per gallon and assume 1 gallon provides 90,000 BTUs of heat.
4. List the four temporary site services that are usually required on a construction project and describe how each service is priced.
5. List three possible ways of dealing with the problem of limited storage space on a downtown site.
6. List the two categories of site cleanup that have to be considered in an estimate and describe how each category is priced.
7. What are the contractor's financing charges on a project and how is the cost of financing charges estimated?
8. With the aid of examples, describe the two types of add-ons considered in an estimate and explain how they are calculated.

# 15

# CLOSING THE BID

## OBJECTIVES

*After reading this chapter and completing the review questions, you should be able to do the following:*

- Describe the last steps in the estimating process, including the use of summary sheets.
- Identify the three stages of a construction contract bid and explain why the processes involved have to be well planned and organized.
- Describe the items that are addressed in the advance stage of the bid process including:
  a. Completion of bid forms
  b. Calculation and listing of unit prices
  c. Alternative prices (alternates)
  d. Separate prices
  e. Allowances
  f. Bid security—bid bonds and deposits
  g. Bid conditions
- Explain the three main objectives of a pre-bid review meeting.
- Identify the two components of bid markup and describe how markup is assessed.
- Explain how the contractor's risk on a project is identified and how the risk factor influences the bid markup.
- Identify the problems often associated with bid-closings and explain how these problems can be dealt with.
- Distinguish a unit-price bid from a lump-sum bid and describe the additional requirements of a unit-price bid.
- Outline the tasks that should be addressed following a bid submission.
- Finalize a unit-price bid using manual methods.
- Finalize a lump-sum estimate and bid using manual methods.
- Use the Precision Estimating software to finalize a lump-sum estimate and bid for a project.

## Introduction

In this chapter we consider the last stages of the estimate when the bid clock, which was started when the bid documents were received, ticks off the final hours to the bid-closing time. In this period the estimate prices are summarized, the components of the bid are assembled and the bid documents are submitted to the place designated for the bid-closing.

Most contractors use a bid summary sheet to gather together all of the constituent prices of the estimate. An example of the format of a summary sheet is shown in Figure 15.1. The summary not only provides a means of recording prices and presenting the total amounts, but it also allows the estimator, before pricing begins, to distinguish the items that have to be priced in order to complete the estimate. At the very start of the estimating process, the estimator can set up the summary sheet by checking off the trades that will be completed by "own forces" and those that will be subcontracted. When the prices of the trades are obtained, they can be written onto preprinted summary sheets like the one shown in the example, or a summary sheet in the form of a computer spreadsheet can be used to store and manipulate the estimate prices.

The prices of the contractor's own work are entered onto the top of the summary sheet so that "Subtotal 1" provides the estimated direct cost of the contractor's work. Subtrade prices and the price of general expenses are recorded as they become available until the bid total at the foot of the summary is finally established. This total can then be transferred to the bid documents and the bid submitted.

To successfully prepare a bid that is free from errors and complete in all essentials, the process has to be carefully planned and organized from the point when the bid documents arrive at the contractor's office up to the last act in the process, the submission of the bid. The busiest period, and the period when the process is most error prone, is the last few hours before the bid is delivered. Every process that can possibly be accomplished before bid-closing day has to be out of the way so that these last few hours can be devoted to those activities that can only occur at this time.

## In Advance of the Bid-Closing

A construction contract bid is compiled in three stages: the advance stage, the review stage, and the closing stage. The following processes are performed in the advance stage, which should be concluded at least 2 workdays before the day of the bid-closing:

1. The quantity takeoff, the recaps, the general expenses and the summary sheet, except for the subcontractor prices, are completed.
2. Unit prices and alternative prices relating to the contractor's work are calculated.
3. Bid bonds, consent of surety forms and any other documents required to accompany the bid are obtained and completed.
4. Bid forms are signed by company officers, witnessed, sealed and completed as far as possible.
5. Duplicate copies of all bid documents are made for retention by the contractor.
6. An envelope to contain all bid documents is prepared.
7. All bid documents and their envelope are retained by the chief estimator until the closing day.
8. A plan is formulated for the delivery of the bid. The **bid runner** (the person who is to deliver the bid) is carefully briefed about his or her role and how he or she will complete the bid documents if this is required.
9. The office staff members are briefed regarding their roles on bid-closing day.
10. All fax machines, computers and telephones are checked to ensure they will be operational for closing day. Backup fax machines and computers are made available.

PROJECT:
GROSS FLOOR AREA:
LOCATION:
DATE:
JOB ESTIMATOR:

**ESTIMATE SUMMARY**

| DESCRIPTION | LABOR $ | MATL. $ | EQUIP. $ | SUBS. $ | OTHER $ | TOTAL $ |
|---|---|---|---|---|---|---|
| **OWN WORK:** | | | | | | |
| EXCAVATION & FILL | | | | | | |
| CONCRETE WORK | | | | | | |
| FORMWORK | | | | | | |
| MISCELLANEOUS | | | | | | |
| MASONRY | | | | | | |
| ROUGH CARPENTRY | | | | | | |
| FINISH CARPENTRY | | | | | | |
| EXTERIOR FINISHES | | | | | | |
| INTERIOR FINISHES | | | | | | |
| | | | | | | |
| **Subtotal 1:** | | | | | | |
| **SUBCONTRACTORS:** | | | | | | |
| DEMOLITION | | | | | | |
| LANDSCAPING | | | | | | |
| PILING | | | | | | |
| REINFORCING STEEL | | | | | | |
| PRECAST | | | | | | |
| MASONRY | | | | | | |
| STRUCTURAL AND MISC. STEEL | | | | | | |
| CARPENTRY | | | | | | |
| MILLWORK | | | | | | |
| ROOFING | | | | | | |
| CAULKING AND DAMP-PROOFING | | | | | | |
| DOORS AND FRAMES | | | | | | |
| WINDOWS | | | | | | |
| RESILIENT FLOORING | | | | | | |
| CARPET | | | | | | |
| DRYWALL | | | | | | |
| ACOUSTIC CEILING | | | | | | |
| PAINTING | | | | | | |
| SPECIALTIES | | | | | | |
| PLUMBING | | | | | | |
| HEATING AND VENTILATING | | | | | | |
| ELECTRICAL | | | | | | |
| CASH ALLOWANCES | | | | | | |
| | | | | | | |
| **Subtotal 2:** | | | | | | |
| | | | | | | |
| **GENERAL EXPENSES** | | | | | | |
| | | | | | | |
| **Subtotal 3:** | | | | | | |
| | | | | | | |
| **ADD-ONS** | | | | | | |
| E.D.P. | | | | | | |
| SMALL TOOLS | | | | | | |
| PAYROLL ADDITIVE | | | | | | |
| | | | | | | |
| **Subtotal 4:** | | | | | | |
| | | | | | | |
| BUILDING PERMIT | | | | | | |
| *50% PERF. BOND | | | | | | |
| *100% PERF. BOND | | | | | | |
| INSURANCE | | | | | | |
| FEE | | | | | | |
| **Subtotal 5:** | | | | | | |
| ADJUSTMENT | | | | | | |
| | | | BID TOTAL: | | | |

*Choose one bond
Price per S.F. =

**Figure 15.1** Sample Estimate Summary Sheet Format

Many estimators prepare a checklist of the items that have to be completed during the entire estimating process to ensure that all needs are taken care of in good time. There are few things more frustrating and more liable to trigger panic among the estimators closing the bid than learning, moments before the bid is to be submitted, that an item required for the bid, such as a bid bond or a signature on a key document, is missing. Those of us who have been involved in numerous closings can attest to the truth of Murphy's Law: "Whatever can go wrong, will go wrong" in these situations.

### The Bid Form

The bid, which may sometimes be called a "tender" or a "proposal," can be presented in a form entirely of the contractor's choosing or it may be written on a standard bid form such as the type available from many construction associations. A third option that is quite common, especially when the owner is a public agency, is to stipulate in the instructions to bidders that all bidding contractors shall use the bid form provided with the contract documents. In all cases the bid form should contain certain basic information:

1. The name and address of the project
2. The identity of the owner
3. The identity of the bidding contractor
4. A description of the work to be done
5. A list of the bid documents
6. A list of addenda
7. The bid price
8. The duration or completion date of the project
9. A declaration that the bidder agrees to leave the bid open for acceptance by the owner for a specified period of time and to comply with all other bid requirements
10. The signatures of the bidders if they are not incorporated, or the corporate seal if the bidder is a corporation, all duly witnessed
11. The time, date and place of the bid

These items are generally considered to be the essential ingredients of a bid, but many bids have been written in an informal style omitting one or more of those items. The lack of some detail in a bid may have no effect on the course of the contract, but owners and contractors alike have sometimes wished the bid had been more explicit so that a problem might have been avoided.

### Additional Bid Form Information

In addition to the basic bid information, the bid form may also have a number of other items attached to it, including:

1. A list of unit prices for pricing changes in the work
2. Alternative prices
3. Separate prices
4. Allowances
5. Bid security
6. Bid conditions

*Unit Prices*

Bidders may be requested to indicate against each item on a schedule of work items, furnished with the bid documents, a unit price that is to be used for pricing added or

---

**UNIT PRICES**

Unit prices for added or subtracted material are to include all labor, equipment, materials and incidentals necessary to perform the work, and all as described and required in Divisions 2 to 16 inclusive of the Detailed Specifications.

| | | UNIT | ADD $ | DEDUCT $ |
|---|---|---|---|---|
| 1. | Reinforced Concrete Grade Beams in Place | cu. yd. | | |
| 2. | Reinforced Concrete Columns in Place | cu. yd. | | |
| 3. | Reinforced Concrete Slab in Place | cu. yd. | | |
| 4. | Structural Steel in Place | ton | | |
| 5. | 8" Concrete Block Wall in Place | sq. ft. | | |
| 6. | Excavation Common Dry | cu. yd. | | |
| 7. | Excavation Common Wet | cu. yd. | | |
| 8. | Rock Excavation | cu. yd. | | |
| 9. | Compacted Granular Backfill | cu. yd. | | |
| 10. | Reinforced Concrete Piles | lin. foot | | |

**Figure 15.2**   Sample Unit-Price Schedule

deleted work on the project. Sometimes a single unit price for each work item will be called for on the schedule of work items. This single price will be used as the basis for pricing both added and deleted work, but the contractor should be allowed to add an amount for overheads and profit to this price when it is used for pricing additional work. At other times there will be provision for the contractor to list two prices against each item, a price for added work and a second for deleted work. The process of applying these unit prices varies from contract to contract, but details about how the unit prices are to be used on the project will usually be found in the supplementary conditions of the contract documents.

Figure 15.2 shows an example of a unit-price schedule for the prime contractor's work of a project only. While concrete work, excavation work and, possibly, masonry work are the more frequent subjects of unit-price schedules, the bid documents can contain a schedule that lists hundreds of items encompassing the work of every trade on the project. When subtrade unit prices are requested, bidding subcontractors need to be forewarned that unit prices are required with their bid because many subtrades will not be aware of the details of the prime contractor's bid form that calls for unit prices.

The following example illustrates how the estimator calculates unit prices for inclusion in the schedule of unit prices. The prices used in this example are based on prices taken from the recaps of the office/warehouse example detailed in the text.

**EXAMPLE**

---

For reinforced concrete to grade beams, the price of a 100-foot long section of grade beam is estimated. Then the total price obtained for this is converted to a price per cubic yard:

| WORK ITEM | QUANTITY | UNIT | UNIT PRICE | LABOR $ | UNIT PRICE | MATL. $ | UNIT PRICE | EQUIP. $ | TOTAL $ |
|---|---|---|---|---|---|---|---|---|---|
| Concrete Grade Beams | 5 | cu. yd. | 9.00 | 45.00 | 80.00 | 400.00 | 7.50 | 37.50 | 482.50 |
| Form Sides of Grade Beam | 400 | sq. ft. | 3.50 | 1,400.00 | 0.75 | 300.00 | — | — | 1,700.00 |
| Void Forms | 100 | ft. | 1.50 | 150.00 | 0.75 | 75.00 | — | — | 225.00 |
| Strip, Clean & Oil Forms | 400 | sq. ft. | 0.75 | 300.00 | 0.08 | 32.00 | — | — | 332.00 |
| Rough Hardware | 15% Form Matls. | | — | — | — | 61.05 | — | — | 61.05 |
| Reinforcing Steel | 804 | lbs | 0.18 | 144.72 | 0.30 | 241.20 | — | — | 385.92 |
| Subtotal: | | | | 2,039.72 | | 1,109.25 | | 37.50 | 3,186.47 |
| Add-ons—Small Tools & Payroll Additive: | 50% of Labor: | | | 1,019.86 | | — | | — | 1,019.86 |
| Net Price per 100 ft. (5 cu. yds.): | | | | 3,059.58 | | 1,109.25 | | 37.50 | 4,206.33 |
| **Net price per cu. yd. (for deductions):** | | | | | | | | | 841.27 |
| Overheads: (20%) | | | | | | | | | 168.25 |
| Fee: (10%) | | | | | | | | | 84.13 |
| **Price per cu. yd. (for additions):** | | | | | | | | | **1,093.65** |

It can be appreciated, even from this single example, that a large amount of time may be needed to calculate the unit prices required on a bid. This task can be scheduled well before the bid-closing time in which the unit prices concern only the prime contractor's work, but when subtrade prices are involved, it usually has to wait until the subtrade bids are received on bid-closing day, resulting in more work for that busy day.

*Alternative Prices*

Alternative prices are quoted for proposed changes in the specifications. The contractor normally bids a price for the project as specified and then quotes alternative prices to perform the work in accordance with the different requirements called for in the alternates. For instance, if the specifications indicate that 26-ounce carpet is required, this type of carpet will be priced in the base bid. Then, should an alternative price be requested to substitute 32-ounce carpet, the extra cost of supplying the heavier carpet will be calculated and quoted in the bid. Alternative prices are usually expressed as additions to or deletions from the base bid and are often referred to as "alternates." In this example, the alternate submitted with the bid may state: "To substitute 32-ounce carpet for 26-ounce carpet, add $45,000."

Bid documents call for alternative prices when owners or their design consultants wish to provide some price flexibility in the bid. This allows them to substitute lower price alternates if the bid price exceeds the owner's budget and vice versa. A few alternative prices can be reasonably expected on a bid, but in some cases several dozen prices have been required. This puts the estimator under added pressure on bid-closing day since many of the alternatives involve subtrades and so alternates cannot be priced until information is received with the subtrade bids that arrive in the last few hours of the estimating period.

Alternates can be extremely complex, indeed alternative prices have been requested for the deletion of entire sections of a building comprising several floors in some cases. This kind of alternate will impact the prices of most of the subtrades on the project. A more common situation, but one that still involves a large number of subtrades, is when an alternate is requested for deleting the finishes from one or more floors of a building. When an alternate involves a significant number of trades, the estimator is advised to set up a spreadsheet in advance to streamline the process of pricing the alternate when subtrade bids are received.

Figure 15.3 shows a template for pricing some example alternates on the office/warehouse project. This kind of spreadsheet can be set up early in the esti-

| | BID ALTERNATIVES | | |
|---|---|---|---|
| | FILE: | | |
| | PROJECT: Office/Warehouse Example | | |
| | LOCATION: | | |
| | DATE: | | |
| | ESTIMATOR: DJP | | |
| **NUMBER** | **ALTERNATIVE PRICES** | **ADD $** | **DEDUCT $** |
| ALT. 1 | To substitute the Single-Ply Roofing System as Specified: | 10,000 | — |
| ALT. 2 | To Delete Insulation from Overhead Doors: | — | 5,000 |
| ALT. 3 | To Add Air Conditioning Unit As Specified: | 15,674 | — |
| ALT. 4 | To Delete Finishes from 2nd Floor Offices: | | |
| | - Millwork | — | 4,500 |
| | - Installation | — | 2,000 |
| | - Resil Flooring | — | 3,005 |
| | - Drywall | — | 8,256 |
| | - Acoustic Ceiling | — | 4,000 |
| | - Painting | — | 4,982 |
| | - Toilet Partitions | — | 1,150 |
| | - Washroom Accessories | — | 350 |
| | - Installation | — | 100 |
| | - Plumbing and Heating | — | 23,850 |
| | - Electrical | — | 12,900 |
| | **TOTAL DEDUCTION** | — | 65,093 |

**Figure 15.3**   Template Showing Sample Alternates Pricing

mating process with the spaces for prices left blank until the pricing information is obtained for the subtrades.

Alternative prices may also be introduced into the bid on the contractor's own initiative. These alternatives are invariably offers to perform the work for a reduced price on the condition that certain changes be made to the design or specifications. They can be in the form of a simple substitution of a product of equal quality but lower cost for the one specified, or the alternative may consist of offering a completely different design for the project. This second case is more common with engineering works than with building construction.

### Separate Prices

Bidders are sometimes required to state separately from the base bid the prices of some specific items of work. There can be several different reasons for requesting this information from bidders. In some cases the owner needs this information to facilitate a value analysis of some items for the work. More commonly, a part of work may be considered to be an optional component of the project; in which case it is priced separately so that it can be included or excluded from the scope of the contract. The decision usually depends on the owner's budget situation.

The estimator has to be especially careful with regard to separate prices because most contracts that include separate prices state that the separate prices and the work they relate to are in addition to the work of the contract and shall not be included in the base bid. However, some contracts indicate that separate prices and the work they relate to are part of the work of the contract and shall be included in the base bid. Clearly, the correct interpretation of the contract conditions relating to separate prices has to be ascertained; otherwise serious bid errors can be made.

Work of substantial proportions can be the subject of separate prices in a bid. In these situations estimators should consider the effect that deducting or adding this large quantity of work will have on the project schedule. While there is usually space provided on the bid form to insert the separate prices, there is seldom provision made for indicating the schedule changes associated with the work involved. A note about the effect on the schedule may have to be added to the bid under "bid clarifications," which are discussed in the section on the pre-bid review.

### Allowances

Contractors can be instructed to include in their bids a stipulated sum of money to be expended on a specific work item on the project. These sums, which are usually referred to as cash allowances or prime cost sums (PC sums), are not normally required to be listed on the bid documents as are alternative or separate prices, but they are to be included in the bidder's price.

The instructions to bidders could say, for example, "Allow $10,000 for the finish hardware as specified in section 08705 of the specifications." Contract conditions will normally go on to describe how the cash allowances are to be disbursed. Instructions usually state that the cash allowance shall include the actual supply cost of the goods required for the work but shall not include the contractor's markup. The contractor is required to include the profit that applies to the cash allowance work elsewhere in its bid price.

A second type of allowance, often called a "**contingency sum,**" differs from that previously described in that the funds are to be made available for a broader purpose. A bidder may be instructed to include a contingency sum of $15,000 in its bid price for extra work authorized on the project, in which case the cost of extra work will be paid out of this sum.

When the estimator first reviews the bid documents at the start of the estimating process, any allowances specified should be highlighted. Many estimators set up the summary sheet for the estimate at this time and enter details of the bid cash allowances on the sheet at once so that they are not later forgotten.

### Bid Security

On just about all public bids, and on many bids to private owners these days, bid security is required to accompany the contractor's bid. Bid security is usually in the form of a bid bond or may consist of a certified check or other form of negotiable instrument. It is used to guarantee that if the bid is accepted within the time specified for acceptance of bids, the contractor will enter into a contract and furnish satisfactory contract bonds to the owner; otherwise the bid security is forfeited.

Bid bonds are issued by surety companies and are favored by contractors over other forms of bid security since their use does not require large sums of money to be tied up for extended time periods. Certain owners, especially government bodies, require surety bonds to be drafted in a particular form, in which case the estimator has to ensure that the bonds obtained do comply with these specifications. Requests for bonds should be placed as soon as the bonding provisions of a project are determined.

The face value of a bid bond is usually set at 10% of the bid price, but a 5% value has been specified with regard to some multimillion-dollar contracts. Denoting the dollar value of a bid bond on its face can present a problem when the bid price will not be established until only minutes before the bid is presented to the owner. Sometimes simply defining the value as "10% of the bid price" may be acceptable to the owner; when it is not, writing in the dollar value becomes yet another chore assigned to the bid runner who delivers the bid.

A second form of bid security, the Consent of Surety, is also frequently required to accompany bids. A Consent of Surety is a written statement made by a bonding company indicating that it will provide surety bonds as required by the contract to the contractor named in the statement. The estimator should be able to collect this document at the same time and from the same source as the bid bond.

### Bid Conditions

The owner often lists on the bid forms a number of conditions to which the bidder expresses agreement by signing or sealing the bid. The list of owner's conditions has been known to run to many pages on some bid forms, but some of the more usual examples include declarations that:

1.  The bidder has examined all of the contract documents and addenda as listed by the bidder on the bid
2.  The bidder has inspected the site of the work and is familiar with all of the conditions relating to the construction of the project
3.  The bidder agrees to provide all labor, materials, equipment and any other requirements necessary to construct the complete project in strict accordance with the contract documents.
4.  If the bidder whose bid is accepted within the specified acceptance period refuses or fails, within 15 days after the bid is accepted, (1) to enter into a contract with the owner or (2) to provide contract performance security, or both, the bidder shall be liable to the owner for the difference in money between the amount of the bid and the amount for which the contract is entered into with some other person. (Note that the tender security is in place to guarantee this last condition.)

Many of the conditions found in the bid documents are placed there to shift a risk factor from the owner to the contractor. These conditions should be highlighted and brought to the attention of company management so that the consequence of assuming the risk involved can be assessed. The estimator should ensure that the risks imposed by both bid and contract conditions have all been addressed in the estimate.

## The Pre-Bid Review

The objectives of the pre-bid review are threefold: (1) to ensure that there are no obvious omissions from the bid, (2) to consider ways of improving the competitiveness of the bid price and (3) to consider an appropriate fee for the project. While the estimators will have previously considered all three of these elements during the estimating process, the additional experience and different viewpoints contributed by other people at the review meeting can often be of great benefit.

Ideally a pre-bid review meeting will take place at least 2 workdays before the bid-closing day. This gives the estimator time to take action on the decisions that arise from the meeting and allows sufficient time for a brief follow-up meeting when it is necessary.

The pre-bid meeting will be attended by at least the project estimator and his or her supervisor who would normally be the chief estimator. The presence of other participants would depend on both the size of the construction company and the size of the project to be bid. On a large project closing for a substantial contractor, a company vice president or even the president may attend, together with a regional general manager, an operations manager, a project manager, the project estimating team and the company's chief estimator.

Before the review, the takeoff of the contractor's work should be complete and priced, the general expenses should have been priced, as far as possible, and pre-bid subtrade proposals should have been obtained from the major subcontractors on the project. By the time of the meeting, the chief estimator should have looked over the bid documents and discussed the estimate generally, and the prices specifically, with the estimator. It is also advisable for everybody else attending the review to examine the drawings and, at least, glance over the other bid documents before the meeting commences; otherwise much of the meeting time will be taken up "enlightening" these attendees about the basic characteristics of the project.

The principal topics discussed at the pre-bid review together with typical questions raised about these subjects are:

1. The nature and scope of the project work: Is it clear what is involved in the project? Does the estimate address all of the constituent parts that it should?
2. The construction methods proposed for the project: Can more effective or more efficient methods be used? If so, what will be the effect on the price of the work?
3. The supervisory needs of the project: What key positions will need to be filled on the project? Will there be company personnel available when these positions have to be filled?
4. The labor resources required for the project: What will be the project labor demands? Are any problems anticipated in meeting these demands?
5. The equipment needs of the project and availability of key items: What are the main items of equipment required? Will there be any problems meeting these requirements?
6. The prices used in the estimate: Is there general agreement regarding the prices of key labor and equipment items used in the estimate?
7. The project risks: Which items of the takeoff are considered to be risky and how much has been allowed to cover the risk involved?

8. The general expense prices: What has been allowed for in the key items of general expenses? Can anything be done to improve the competitiveness of the prices used?

9. The response from subcontractors: Will prices be forthcoming for all the trades involved in the work? What problems have been identified by subcontractors?

10. The markup fee to be added to the estimate: What is the exposure to risk contained in this bid? What is an appropriate fee to include in the bid price? (See "Bid Markup" in the following.)

11. Clarifications to be submitted with the bid: Is it necessary to include a statement about the basis of the contractor's price with the bid? (See "Bid Clarifications" in the following.)

All of these items and any others that affect the bid should be discussed in an open and mutually supportive climate at the bid review meeting. It should be made clear that the object of the meeting is not to criticize the work of the estimators, but, rather, to discuss what can be done to maximize the quality of the bid. Good bid reviews are infused by a team spirit that can restore the confidence of an estimator who is suffering from doubts and the feeling of aloneness that can torment even the best of the profession.

Important decisions can be reached in many of the areas discussed at the meeting but two topics in particular—the bid markup and the need for bid clarifications—require careful consideration.

## Bid Markup

A markup or fee is added to the estimated price of a project to contribute to the company overheads and profit margin in the bid. Company overheads encompass all the expenses, apart from the project costs, incurred in operating a construction company. Some expenses are difficult to distinguish as either company overheads or project expenses, but any company expenditure that is not the direct cost of project work or a part of the general expenses of a project is normally considered to be company overhead. Company overhead usually includes the following expenses:

1. The cost of providing single or multiple company office premises including mortgage or rent costs
2. Office utilities
3. Office furnishings and equipment
4. Office maintenance and cleaning
5. Executive and office personnel salaries and benefits
6. Company travel and entertainment costs
7. Accounting and legal consultant's fees
8. Advertising
9. Business taxes and licenses
10. Interest and bank charges

The amount of company overhead establishes the minimum fee a contractor needs to obtain from each project. In the case in which a construction company has an overhead of \$1,000,000 and the volume of project work completed per year is estimated to be \$20 million, the minimum fee is calculated as:

$$\frac{1,000,000}{20,000,000} = 5\%$$

Profit, defined as the amount of money remaining after all project expenses and company expenses have been paid, can be assessed in two ways in the construction business: as a return on the investment or as compensation for the risk assumed by the construction company owners. Owners who have extensive investments in their company, usually in the form of construction equipment, may wish to analyze their profits in terms of a percentage of the amount they have invested in the business. An option these owners always have, at least in theory, is to sell all the assets of the company and invest the funds elsewhere. Many construction company owners, however, have relatively small amounts invested in the operation; for these owners the risk assessment is far more relevant.

Right from the beginning of the pricing process, and even before in the takeoff section of the estimate, the question of risk is considered, and it is difficult to deny the claim that every price in the estimate has an element of risk attached to it. But some items in the estimate are more risky than others. As outlined in the discussion on risk in Chapter 8, the items that hold the most risk for the contractor are those in which the price depends on the productivity of the contractor's work force or on the duration of the project. The value of these items is equal to the sum of the labor and equipment prices of the estimate plus the price of time-dependent general expense items. This amount is considered to be the risk component of the estimate.

Once the value of the risk component is established, the fee for the project can be calculated as a percentage of this risk component. The percentage amount used in the fee calculation is determined by a large number of factors, many of which were touched on in items discussed in the pre-bid review listed earlier, but the factors affecting the fee markup can be summarized into three main categories:

1. The construction market and the expected competition for the project
2. The contractor's desire for and capacity to handle more work
3. The desirability of the project under consideration

However, calculating the project fee as a percentage of the value of risk in a bid has a certain danger attached to it. The reasons for applying a low fee to a bid—high competition, the contractor's need for work, a desirable project—can also influence the estimator into pricing the estimate optimistically. This can then produce a compounding effect since the low prices used in the estimate will lead to a low risk component that will have a low percentage fee applied to it. The complete reverse of this is also possible: high prices, high risk factor resulting in an excessively high bid price. To avoid these extremes the estimator has to be impartial in the pricing process and try to avoid both undue optimism or pessimism.

### Bid Clarifications

The instructions to bidders, especially on public projects, will frequently state that qualified bids or bids that have conditions attached to them will be considered invalid and these bids may be rejected. One of the reasons why it is important to obtain clear answers to the questions raised in the estimating process is that it is not advisable to state on the bid that it is based on this or that interpretation since the presence of such a statement can cause the bid to be rejected.

However, occasionally a situation arises in which some aspect of the bid is still not perfectly clear even after explanations are received from the designers, and the contractor wishes to avoid the consequences of conflicting interpretations of the contract documents. In such a case the group gathered together at the pre-bid review has to weigh the risk of these consequences against the risk of having the bid rejected. Sometimes the anticipated dangers are large enough to justify attaching the condition to the bid and risking rejection.

## The Closing

The process of closing a bid is a time of great excitement in the contractor's office. From the first day the project drawings and specifications arrived in the office, the estimator has been keenly aware of the bid deadline; first the days are seen to slip by, then the hours are noted and finally the minutes tick away to bid-closing time. Those estimators who enjoy the thrill of a bid-closing may approach the final day with eager anticipation, but many estimators approach it with dread.

After the bid has been reviewed, the amount of fee set and the changes to the bid decided on in the review process made, all that remains to be done is to add in the subtrade prices and complete the bid documents. This is what some estimators refer to as the crazy part of the process.

On a project with more than 30 subtrades when, over a period of just a few hours, as many as ten bids can be received for each separate subtrade, the activity level in the prime contractor's office can reach frenzied proportions. Under these conditions in which each subtrade price can range up to $1 million or more and a single mistake can cost hundreds of thousands of dollars, the stress on estimators can be enormous. Cool heads and total control is essential if panic is to be avoided.

The computer can be a useful tool in the bid-closing process and, if used effectively, can reduce the strain on the estimators considerably. In this last stage of the bid, successful bidding, more than ever, depends on good organization.

### Staffing the Bid-Closing

On small building projects or civil engineering jobs with few subtrades involved, the bid-closing may be handled in the office by no more than the estimator and the chief estimator. When many subtrade prices are expected for a project, however, additional people will be required to assist in the process. The number of estimators and assistants engaged in a bid-closing has to be sufficient to deal with all telephone calls, e-mail, and to gather up all fax sheets promptly, leaving the project estimator and the chief estimator free to analyze the incoming bids and complete the pricing of the estimate. A team of six to eight people is usually sufficient to cope with the task, but on very large projects, or in situations in which several projects close at the same time, this number may need to be further increased.

Where the bid is to be hand delivered, which is the case with most bids, one more person has to be added to the bid-closing team for any size of project. This is the bid runner, the person who is to deliver the bid. Whether the place of the bid-closing is just across town or in a city many hundreds of miles away, the bid runner will usually have to write the final bid information onto the bid documents. This may consist of merely the final bid price, but many other items of information can be undetermined when the bid runner leaves the office with the bid documents. Subtrade names and alternative, separate and breakdown prices are just a few of the possible items that may need to be completed by the bid runner on the documents.

The bid runner will be equipped with a cellular phone so that, once he or she reaches the location of the bid-closing and finds a place nearby to complete the documents, contact can be established with the office and the final bid details received. Certain subtrades, being aware of this arrangement, will deliberately withhold its price until the last moment to avoid the possibility of a contractor disclosing its bid to other subtrades. Sometimes, therefore, bids can still be received by the contractor in the last few minutes before the closing. If care is not taken at this point, a disastrous situation can develop in which the bid is delayed until just after the closing time and is disqualified for being too late.

Some contractors have found it necessary to set a policy and inform all subtrades that no bid will be accepted after a stipulated time. This can be set at one hour or even half an hour before the owner's bid-closing time. The contractor then has time to handle the last-minute bids and the bid runner has time to complete the bid and hand it in before the time for accepting bids expires.

### Telephones and Fax Machines

As we have previously stated, the subcontractor's preferred method of submitting a bid is by means of the fax machine, but many bid prices are still presented over the telephone and e-mail is now sometimes used to submit subtrade bids. Apart from the bid price, the person in the contractor's office receiving the bid has to make careful note of several items of information from the bidder including:

1. The name of the subtrade bidding
2. The trade being bid
3. The name of the person submitting the bid
4. The telephone number where this person can be contacted
5. The base bid price
6. Whether or not the bid is based on plans and specifications, and whether everything required of this trade is covered by the bid price
7. Alternative, unit and separate prices if required
8. The taxes applicable to this item and whether they have or have not been included in the price
9. Any conditions attached to the bid
10. The time and date the bid was received
11. The name of the person receiving the bid

Figure 15.4 shows a standard form used for logging telephone bids. This provides a checklist that the person receiving the bid can follow to ensure that all the information required is recorded when the bid is received.

While it is unusual for a telephone to malfunction, fax machines are far more prone to problems if only because they have more mechanical parts that can break down or jam up. A bid-closing is a very effective test of a fax machine as it can put the machine to almost continuous use for hours on end. Backup machines are desirable, not only to hedge against mechanical failure, but also to provide added capacity at peak periods when the bulk of bids are arriving since a busy fax can result in the loss of an incoming price. An operator who is sending up to 20 faxes manually may forget to return to a busy line or, if an auto-dial system is being used, the system may not function properly and fail to complete transmission to a number that is constantly busy.

### Subtrade Conditions

In Chapter 13 we discussed the problems that result from subcontractors submitting conditional bids; the estimator closing a bid should be reminded to examine the terms of subtrade bids carefully. When a subcontractor's bid has conditions attached, the estimator needs to evaluate the consequences of accepting those conditions and make allowances, if necessary, in the bid price for them.

Often the subtrade's condition is in the form of an exclusion from the usual work performed by that trade, in which case the estimator has to calculate the cost of the exclusion and add it to the subtrade's bid price. If the subtrade's adjusted price is still the lowest received for that trade, the subtrade is used in the bid and the excluded work is added to the prime contractor's work.

Sometimes the condition imposed by the subtrade is something the contractor needs to discuss with the owner. For instance, the subtrade may offer a price on the

---

**TELEPHONE BID**

*Trade Classification:* _____

                                                    *Date:* _____

*Project:* _____

*Bidder:* _____ *By:* _____

*Address:* _____ *Phone:* _____

====================================================================

**BASE BID:** _____

| | INCLUDED | EXCLUDED |
|---|---|---|
| *TAXES:*   *State:* _____ | | |
| *Federal:* _____ | | |

*REMARKS:*_____

_____

_____

_____

_____

_____

_____

_____

_____

====================================================================

*ALTERNATES:*

_____

_____

_____

_____

_____

_____

_____

_____

====================================================================

Taken By: _____

**Figure 15.4**    Standard Form for Logging Telephone Bids

condition that it have the use of a certain size storage space at the site. If the contractor is aware of the condition early enough, the question can be raised with the owner's design consultant and confirmation of whether or not the required space is available to be obtained before the bid-closing. But when the contractor is made aware of this condition only an hour or so before the bid-closing time, there is no time to discuss the matter. This is the kind of situation in which the benefit of pre-bid subtrade proposals is most appreciated.

### Naming Subcontractors

Compilation of the list of proposed subcontractors can be yet another problem for the bidder. Because of time constraints, subtrade bid conditions and the overwhelming need to put together the most competitive price, the final list of subcontractors may not be available until moments before the bid is to be handed in. Often having just a few more minutes after the bid is submitted would be useful to the contractor for sorting out exactly whose prices were used in the last-minute scramble to finish off the bid.

While some owners and their design consultants are sympathetic about the contractor's difficulties, they point out, with some grounds, that to allow even a short time after the bid for contractors to make their list of subtrades, would be an invitation to the unscrupulous contractor to engage in bid shopping of subtrade prices. So the required list of proposed subcontractors can be expected to remain as a pre-bid condition on most contracts.

### Bid Breakdown

To further complicate the task of the estimator who is trying to close out a bid, some owners call for a cost breakdown to be attached to the bid documents. This breakdown is often required in a form similar to the estimate summary in which a price is indicated for each of the trades involved in the work. The information is generally requested to provide a basis for evaluating monthly progress payments on the project. More reasonable contract requirements usually state that the bidder awarded the contract shall submit this breakdown to the owner within a stated time after the contract award date.

Preparation of this information at the time of the bid is a major chore for the estimator because the prices stated have to include profit and overhead amounts and many of these prices will not be available until the day of the bid-closing, sometimes only minutes before the closing. In some cases bidders just "rough out" the breakdown and submit the approximate values with the bid. In a few cases contractors have risked having their bids disqualified for failing to include the breakdown with their bid. These contractors may state on their bid that the breakdown is not available at this time but will be forwarded to the owner within 24 hours of the bid-closing time.

### Computerized Bid-Closing

Computer programs are available to perform all the calculations and price compilations that are necessary in the closing stages of an estimate. Merely having the bid summary set up on a computer spreadsheet provides a great advantage over manual methods. When a subtrade price is changed on the summary the computer can be preprogrammed to automatically recalculate all consequential amounts and immediately display the revised bid price. But there are software packages designed especially for bid-closing situations that offer a number of useful additional features.

Estimating staff may still be required to handle the telephones and collect the subtrade faxes but, with bid-closing software operating on a computer network, prices can be constantly inputted into the system from any of the computers on the network. The estimator compiling the bid is then able to continually analyze the incoming prices and make selections from the combined input from all the estimators situated at their own terminals.

Bid-closing software is normally compatible with other estimating software so that prices estimated on one system can also be used as input into the bid-closing system. Changes made in material, labor or equipment prices in one part of the estimate will be automatically reflected in the bid price.

Bells and whistles (literally) are also available with some systems to warn the estimator in charge about the time remaining before bid-closing time.

## Unit-Price Bids

Unit-price bids differ from lump-sum bids in that the contractor is not bidding a price to complete the project but is offering to perform the work of the project for the unit prices quoted. At the conclusion of a lump-sum contract, in the absence of any changes in the work, the contractor will have been paid exactly the price quoted in the bid. In contrast, when a contractor has performed a unit-price contract, the total amount paid to the contractor will depend on the quantity of work completed by the contractor during the course of the project.

With a unit-price bid the primary bid document is a "Schedule of Prices" that comprises an itemized breakdown of the work of the project. The contractor completes the bid by inserting a price per unit of work measured against each work item listed on the schedule and then uses the quantities shown on the schedule to calculate the cost of the work. An example of a Schedule of Prices, before the contractor's unit prices are inserted, is shown as Figure 15.5.

The estimate of a unit-price bid begins in the same way as the estimate of a lump-sum bid: with the takeoff of the quantities of work from the drawings that are furnished with the bid documents. Unit-price contract terms usually state that the design depicted on the bid drawings is subject to change; actual quantities of work required for the project may be more or less than those shown on the drawings. However, the estimator has no choice but to base the bid price on these drawings since there is nothing else to go by. If the actual work performed on the project is significantly different from that shown on the drawings or indicated in the Schedule of Prices provided with the bid, most unit-price contracts allow for the adjustment of the contract unit prices.

### The Recap

The recap format for the takeoff items of a unit-price bid is quite different from the format used with a lump-sum bid. The unit-price recap has to reflect the list of items provided on the bid Schedule of Prices. Sorting work items according to trade, which is done for convenience of pricing on a lump-sum bid, is not possible on a unit-price bid in which a separate recap is required for each item listed on the Schedule of Prices.

The recap for item number 201, "Excavation Common Dry," of the First Avenue Overpass Bid is shown as Figure 15.6. On this recap all the work associated with this work item is listed. The estimator has to be sure that all the costs involved in the work of a bid item are taken off and priced because, if they are not and the work is not listed elsewhere as a separate item on the Schedule of Prices, the contractor will not be paid for it. For instance, in this example the contractor knows that the excavated soil will have to be trucked away from the site. There is no item for removing excavated material on the Schedule of Prices, so the cost of this operation has to be added to the cost of excavating the material. When this has been done, and the contractor is later paid for excavation work on the project, the amount received will be sufficient to pay for the cost of excavation and for the removal of surplus soil.

After each recap is priced, the pay quantity of that item has to be calculated. If the contractor performs the amount of work taken off for this item, what quantity of work will be paid for in accordance with the terms of the contract? This is the

PROJECT: FIRST AVENUE OVERPASS

LOCATION:                                    **SCHEDULE OF PRICES**

DATE:

CONTRACTOR:

| ITEM | QUANTITY | UNIT | UNIT PRICE | AMOUNT $ |
|---|---|---|---|---|
| 100 Site Clearance | PC | SUM | — | 3000.00 |
|  |  |  |  |  |
| 200    EXCAVATION |  |  |  |  |
| 201 Common Dry | 6100 | cu. yd. |  |  |
|  |  |  |  |  |
| 300    BACKFILL |  |  |  |  |
| 301 Common | 1500 | cu. yd. |  |  |
| 302 Pit-run | 3100 | cu. yd |  |  |
|  |  |  |  |  |
| 400    CONCRETE |  |  |  |  |
| 401 Class "A" | 37 | cu. yd. |  |  |
| 402 Class "B" | 1030 | cu. yd. |  |  |
| 403 Class "D" | 17 | cu. yd. |  |  |
| 404 Slope protection | 250 | cu. yd. |  |  |
| 405 Wearing surface | 27 | cu. yd. |  |  |
| 406 Nonshrink grout | 25 | cu. ft. |  |  |
|  |  |  |  |  |
| 500    REINFORCING STEEL | 195 | ton |  |  |
|  |  |  |  |  |
| 600    GUARDRAIL | 180 | foot |  |  |
|  |  |  |  |  |
| 700    EXPANSION JOINTS | 201 | foot |  |  |
|  |  |  |  |  |
| 800    PRECAST STRINGERS | 26 | no. |  |  |
|  |  |  |  |  |
| 900    MISC. FORCE ACCOUNT | PC | SUM | — | 7500.00 |
|  |  |  |  |  |
|  |  |  |  |  |
| SUBTOTAL |  |  |  |  |
|  |  |  |  |  |
| 10% Allowance for extra work |  |  |  |  |
|  |  |  |  |  |
| BID TOTAL |  |  |  |  |

**Figure 15.5**  Schedule of Prices Example

PRICING SHEET                    SHEET No.    1 of 13

JOB: FIRST AVENUE OVERPASS              DATE:

ESTIMATOR: DJP

| NO. | DESCRIPTION | QUANTITY | UNIT | UNIT PRICE | LABOR $ | UNIT PRICE | MATL. $ | UNIT PRICE | EQUIP. $ | TOTAL $ |
|-----|-------------|----------|------|------------|---------|------------|---------|------------|----------|---------|
| | | | | | | | | | | |
| 201 | EXCAVATION COMMON DRY | | | | | | | | | |
| | | | | | | | | | | |
| 201.1 | Excavate trench for footings | 6,168 | cu. yd. | 0.95 | 5,860 | — | — | 1.55 | 9,560 | 15,420 |
| | | | | | | | | | | |
| 201.2 | Hand trim bottoms of excavation | 3,600 | sq. ft. | 1.05 | 3,780 | — | — | — | — | 3,780 |
| | | | | | | | | | | |
| 201.3 | Remove surplus material | 1,636 | cu. yd. | 2.50 | 4,090 | — | — | 5.24 | 8,573 | 12,663 |
| | | | | | | | | | | |
| | | | | | | | | | | |
| | TRANSPORT EQUIPMENT | | | | | | | | | |
| | | | | | | | | | | |
| 201.4 | 1.5 cu. yd. Loader | | | | 250 | | | | | 250 |
| | | | | | | | | | | |
| 201.5 | 0.75 cu. yd. Backhoe | | | | 250 | | | | | 250 |
| | | | | | | | | | | |
| | TOTALS: | | | | 14,230 | | — | | 18,133 | 32,363 |
| | | | | | | | | | | |
| | | | | | | | | | | |
| | PAY QUANTITY = 6,168 cu. yd. | | | | | | | | | |
| | | | | | | | | | | |
| | | | | | | | | | | |

**Figure 15.6**   Item 201—Recap

question that needs to be answered in order to determine the pay quantity, because the quantity of work done as measured by the method defined in the contract will normally be quite different from the actual amount of work done. Every unit-price contract has to provide definitions of how the work done is to be measured for payment; otherwise the owner may end up paying for a lot of unnecessary work.

With some work the contract terms may reflect the same kind of measurement rules used in the takeoff section of this book: net in-place quantities are measured that conform with the sizes shown on the drawings. But where sizes are not shown on the drawings, as is the case with excavation work, another way to measure quantities has to be employed. The method of measurement of excavation work defined in a unit-price contract usually results in a theoretical quantity that is much lower in value than the actual quantity excavated. In this case two quantities are measured by the estimator: the quantity that reflects the actual amount of work expected and the (theoretical) quantity measured in accordance with the contract defined methods. It is this second theoretical quantity that we refer to as the "pay quantity."

In the example recap on Figure 15.6, the pay quantity has been made equal to the actual quantity but, in practice, the pay quantity will normally be less than the actual quantity.

### The Summary

When the recap for each of the Schedule of Prices items has been completed, the overheads and fee for the project have to be distributed against the recap prices obtained. This distribution can be achieved with the use of a wide spreadsheet as shown in Figure 15.7. The item descriptions together with the labor, material, equipment and subtotal 1 prices are copied directly from the recaps.

The general expenses for the project are calculated in the usual way on a separate General Expenses sheet, then the total is inserted at the bottom of the General Expenses column on the spreadsheet. This total is then distributed against the work items in proportion to the value of the labor content of each work item. The labor value is chosen as the basis of distribution because the overheads represented by the general expenses will vary in accordance with the labor cost of the work rather than the material or equipment costs.

In the example, the total general expenses were calculated to be \$77,849 so the ratio as a proportion of the total labor cost of the work will be:

$$\frac{\$77,849}{\$321,696} = 0.242$$

Then the amount of general expenses assigned to item 201, Common Dry Excavation, will be: $\$14,230 \times 0.242 = \$3,444$, and so on.

The amount for add-ons 1, which includes a small tools allowance and the payroll additive, is calculated as a percentage of the labor cost of each item. Adding subtotal 1, the general expenses, and add-ons 1 together gives subtotal 2. Add-ons 2, for surety bonds and insurance, are then calculated as a percentage of subtotal 2 and added in to give subtotal 3. The fee is distributed as a percentage of subtotal 3. Notice that no fee can be added to the PC sums for site clearing and miscellaneous force account work. A column is provided so that any necessary adjustments can be made at this stage and, finally, the total amount for each item is determined.

The unit price of each work item is calculated by dividing the total amount for the item by the pay quantity in each case. The pay quantity, as previously discussed, is copied here directly from the recap of each work item. Finally we return to the bid form. See Figure 15.8.

### The Unit-Price Bid Form

The last stage in the preparation of a unit-price bid is to complete the bid form. This is achieved by inserting the unit prices calculated on the spreadsheet against each of the work items listed on the Schedule of Prices bid form. Then the amounts for each item are calculated by multiplying the unit price by the contract quantity in each case. The contract quantity is the quantity that is indicated against each work item on the Schedule of Prices; it is determined by the design consultants who prepare the contract documents. These contract quantities are usually preprinted onto the Sched-

PROJECT: FIRST AVENUE OVERPASS  
LOCATION:  
DATE:  
ESTIMATOR: DJP

SUMMARY

Add-ons 1:  
Small tools: 5.00%  
Payroll Add.: 25.00%

Add-ons 2:  
50% Bond: 0.73%  
Insurance: 1.50%

| DESCRIPTION | LABOR | MATL. | EQUIP. | SUBTOTAL 1 | GENERAL EXPENSES | ADD-ONS 1 | SUBTOTAL 2 | ADD-ONS 2 | SUBTOTAL 3 | FEE | ADJUST | TOTAL | PAY QUANTITY | UNIT | UNIT PRICE |
|---|---|---|---|---|---|---|---|---|---|---|---|---|---|---|---|
| 100 Clear Site (PC Sum) | — | — | 3,000 | 3,000 | — | — | 3,000 | — | 3,000 | — | — | 3,000 | — | — | 3,000.00 |
| 201 Common dry | 14,230 | — | 18,133 | 32,363 | 3,444 | 4,269 | 40,076 | 892 | 40,967 | 2,048 | — | 43,016 | 6,168 | cu. yd. | 6.97 |
| 301 Common B/fill | 13,392 | — | 9,698 | 23,090 | 3,241 | 4,018 | 30,348 | 675 | 31,024 | 1,551 | — | 32,575 | 2,309 | cu. yd. | 14.11 |
| 302 Pit-run | 9,856 | 22,800 | 35,840 | 68,496 | 2,385 | 2,957 | 73,838 | 1,643 | 75,481 | 3,774 | — | 79,255 | 1,792 | cu. yd. | 44.23 |
| 401 Class "A" | 338 | 2,797 | 0 | 3,135 | 82 | 101 | 3,318 | 74 | 3,392 | 170 | — | 3,562 | 39 | cu. yd. | 91.32 |
| 402 Class "B" | 108,402 | 140,175 | 11,751 | 260,328 | 26,233 | 32,521 | 319,081 | 7,100 | 326,181 | 16,309 | — | 342,490 | 1,022 | cu. yd. | 335.12 |
| 403 Class "D" | 939 | 1,716 | 200 | 2,855 | 227 | 282 | 3,364 | 75 | 3,439 | 172 | — | 3,611 | 15 | cu. yd. | 240.71 |
| 404 Slope protection | 48,383 | 32,074 | 2,005 | 82,462 | 11,708 | 14,515 | 108,685 | 2,418 | 111,104 | 5,555 | — | 116,659 | 273 | cu. yd. | 427.32 |
| 405 Wearing surface | 1,388 | 2,109 | 380 | 3,877 | 336 | 416 | 4,629 | 103 | 4,732 | 237 | — | 4,969 | 25 | cu. yd. | 198.76 |
| 406 Nonshrink grout | 262 | 1,044 | — | 1,306 | 63 | 79 | 1,448 | 32 | 1,480 | 74 | — | 1,554 | 8 | cu. ft. | 194.28 |
| 500 REBAR | 67,850 | 213,095 | — | 280,945 | 16,419 | 20,355 | 317,719 | 7,069 | 324,789 | 16,239 | — | 341,028 | 212 | ton | 1,608.62 |
| 600 GUARDRAIL | 44,446 | 25,016 | — | 69,462 | 10,756 | 13,334 | 93,552 | 2,082 | 95,633 | 4,782 | — | 100,415 | 558 | foot | 179.95 |
| 700 EXPN JOINTS | 12,210 | 25,410 | — | 37,620 | 2,955 | 3,663 | 44,238 | 984 | 45,222 | 2,261 | — | 47,483 | 198 | foot | 239.81 |
| 800 P.C. STRINGERS | — | 26,560 | 185,600 | 212,160 | — | — | 212,160 | 4,721 | 216,881 | 10,844 | — | 227,725 | 52 | no. | 4,379.32 |
| 900 Misc. Force Acct. | — | 3,750 | 3,750 | 7,500 | — | — | 7,500 | — | 7,500 | — | — | 7,500 | — | — | 7,500.00 |
| | 321,696 | 496,546 | 270,357 | 1,088,599 | 77,849 | 96,509 | 1,262,957 | 27,867 | 1,290,824 | 64,016 | — | 1,354,840 | — | — | — |

**Figure 15.7** Unit-Price Bid Summary Spreadsheet

PROJECT: FIRST AVENUE OVERPASS

LOCATION:                                    **SCHEDULE OF PRICES**

DATE:

CONTRACTOR:

| ITEM | QUANTITY | UNIT | UNIT PRICE | AMOUNT $ |
|---|---|---|---|---|
| 100 Site Clearance | PC | SUM | — | 3,000.00 |
| | | | | |
| 200    EXCAVATION | | | | |
| 201 Common dry | 6,100 | cu. yd. | $6.97 | 42,517.00 |
| | | | | |
| 300    BACKFILL | | | | |
| 301 Common | 1,500 | cu. yd. | $14.11 | 21,165.00 |
| 302 Pit-run | 3,100 | cu. yd | $44.23 | 137,113.00 |
| | | | | |
| 400    CONCRETE | | | | |
| 401 Class "A" | 37 | cu. yd. | $91.32 | 3,378.84 |
| 402 Class "B" | 1,030 | cu. yd. | $335.12 | 345,173.60 |
| 403 Class "D" | 17 | cu. yd. | $240.71 | 4,092.07 |
| 404 Slope protection | 250 | cu. yd. | $427.32 | 106,830.00 |
| 405 Wearing surface | 27 | cu. yd. | $198.76 | 5,366.52 |
| 406 Nonshrink grout | 25 | cu. ft. | $194.28 | 4,857.00 |
| | | | | |
| 500    REINFORCING STEEL | 195 | ton | $1,608.62 | 313,680.90 |
| | | | | |
| 600    GUARDRAIL | 180 | foot | $179.95 | 32,391.00 |
| | | | | |
| 700    EXPANSION JOINTS | 201 | foot | $239.81 | 48,201.81 |
| | | | | |
| 800    PRECAST STRINGERS | 26 | no. | $4,379.32 | 113,862.32 |
| | | | | |
| 900    MISC. FORCE ACCOUNT | PC | SUM | — | 7,500.00 |
| | | | | |
| | | | | |
| SUBTOTAL | | | | 1,189,129.06 |
| | | | | |
| 10% Allowance for extra work | | | | 118,912.91 |
| | | | | |
| BID TOTAL | | | | 1,308,041.97 |

**Figure 15.8**   Schedule of Prices—Bid Form

ule of Prices bid document so that all bidders base their total price on the same set of quantities. The contract quantities are what the designers consider to be roughly the amount of work to be done on the project but may bear little resemblance to the pay quantities used by the contractor's estimator to compute the unit prices in the bid.

### Unbalanced Bids

In the example of the aforementioned unit price estimate, the overheads and fee for the project were distributed evenly against the cost of the work, but many contractors see an advantage in allowing a higher proportion of overheads and fee against the work being completed early in the project. The contractor who indulges in this price manipulation expects to obtain payment for all of the project overheads and fee in the first few months of the project duration. However, there is a risk attached to this practice because, if the actual quantities of the items containing overheads and fee decrease and the quantity of the other work increases, the contractor will be left with considerably less payment for overheads and fee than expected.

Often contract documents will state that unbalanced bids will be disqualified. This is because owners do not want to assume the risk of overpaying a contractor in the early stages of a project. In the case in which a contractor has manipulated prices so that some unit prices are obviously overvalued, this contractor may find that its bid is simply rejected by the owner.

## After the Bid

Once the bid is submitted and the bid-closing time passes, those who were involved in the estimate, after perhaps sharing a moment of euphoria in the knowledge that the process is finally at an end, turn their attention to the bid results. A different excitement from the one felt in the final stages of the bid now grips the team. The office scene is now calm after the chaos of the bid-closing but there is an underlying tension, first, about the position of the bid: Was it low? Was it high? And, second, about the bid itself: Was everything covered? Were there any mistakes?

Government agencies and some private owners open the bids in public shortly after the closing, in which case the anxieties of the estimators will soon be relieved or, in the case of a very low bid, they may be multiplied. Whatever the results a review of the bid should be undertaken and all the loose ends left by the estimating process need to be tidied up.

After the bid the items that need to be attended to include:

1.  Tabulate the bid results—Figure 15.9 shows a typical bid results table. This table should be appended to the Bid Report Document that was written up at the start of the estimate. Copies of the bid results table are usually distributed to senior management personnel and to all of the people who were present at the bid review meeting.
2.  Compile the bid documents—Drawings, specifications and other documents provided by the design consultants for the bid are gathered together.
3.  Return bid documents—If the bid is not one of the three low prices, the bid documents are returned to the design consultants and the deposit left for these documents is claimed.
4.  Store bid documents—If the bid results have not been made public or if the bid is one of the three low bids, the bid documents are retained pending the award of contract. Should a contract be awarded to this contractor, these documents will provide evidence of the basis of the bid so they should

**BID RESULTS**

FILE: BR94526

PROJECT: Parkade Structure

LOCATION: Townsville

DATE: 1-SEPT-94

ESTIMATOR: DJP

| | Bid Price $ | |
|---|---|---|
| 1. Construction Company AAAA | 11,007,694 | 1.00 |
| 2. Construction Company BBBB | 11,388,279 | 1.03 |
| 3. Construction Company CCCC | 11,397,239 | 1.04 |
| 4. Construction Company DDDD | 11,526,754 | 1.05 |
| 5. Construction Company EEEE | 11,772,156 | 1.07 |
| 6. Construction Company FFFF | 11,820,000 | 1.07 |
| 7. Construction Company GGGG | 13,665,892 | 1.24 |

**Figure 15.9** Bid Results Tabulation

be held in safe storage by the contractor. In no circumstances should these documents be sent back either to the designer or the owner.

5. Compile and store the estimate file—All the price quotations from material suppliers and subcontractors should be collected, together with all the other estimate documents in an estimate file. This will be the principal resource used in the review process that follows. Afterward, whatever the outcome of the bid, this file should be kept in safe storage for future reference.

6. Handle subcontractors—How contractors deal with subcontractors at this stage is a matter of individual company policy. Immediately after the bid-closing some contractors are willing to disclose whether a subtrade's price was or was not used by the contractor in its bid. Some contractors will disclose this information to subtrades at a later stage when the bid results are made public, while other contractors refuse to even provide this information. These last contractors, if they are awarded the contract, will notify only those subtrades that they intend to award contracts to.

**Post-Bid Review**

If possible the group of people who attended the pre-bid review should return for the post-bid review. Depending on what the bid results were, there will often be many questions that arise after the bid is submitted. In general terms, if the bid was not successful, the prime consideration will be: What needs to be done in the estimating process to improve the success of our bids? Clearly, if mistakes were made, action needs to be taken to try to avoid repeating the same mistakes again. If problems were encountered in the estimating process, the group needs to consider what can be done on future bids to prevent these problems from recurring. If a bid was extremely high or low, the major component prices of the bid should be examined to try to determine why the overall price was so extreme.

Sometimes, after all the prices used in the bid have been poured over and all kinds of alternative approaches and possible lower subtrade prices have been considered, the contractor finds that the price of low bid can be reached only if the estimate is to be priced below cost with no fee included. In this situation the most feasible explanation seems be that the low bid contains an error and this contractor is consoled only by the notion that the successful contractor will lose so much money on the project that it will not be around too much longer to cause any further problems.

## The Estimate–Cost Control Cycle

The bid estimate will be a primary resource in project cost control for the contractor who is awarded the contract to perform the work.

Because construction work is project oriented, construction contractors pursue cost control on a project-by-project basis. Managers need to know how much is being spent on project X so that they can assess whether project X is meeting its budget. This will enable managers to take timely action on project X if a budget problem is found there.

The cost accounting function in a construction company will record all the expenses incurred in the operation of the entire company; the project cost control system will be the component of the overall cost accounting function that deals with individual project costs.

The first step in the cost control process is to use the estimate to set up a budget for each work item on the project. As work advances and expenses are incurred, these costs are allocated to the particular items of work they relate to. Reports are then generated that show the costs expended on each work item next to the budget for that item. This helps the people managing a project identify where problems are occurring so that they can quickly instigate measures, where necessary, to deal with these problems.

There is also a secondary purpose, which is to assemble cost data for use in estimating future jobs. The estimate–cost control process is circular in nature. If we begin with the bid estimate, this defines the budget for the construction work; the actual costs of the work are captured by the cost control system during construction; these costs are then analyzed and retained in databases that provide information for the preparation of subsequent bid estimates. And so the cycle continues.

## Summary Example 1—House

### Notes on Summary Shown as Figure 15.10

1. At the top of the summary sheet, the labor, material, equipment and totals from the recaps of the contractor's work are listed.
2. Because most of the work involved in the construction of this house has been measured in the takeoff, there are few subtrades listed on this summary.
3. The price of plumbing, heating and ventilating work is combined in the amount of $8,500 and the electrical subtrade price is $7815.
4. The price of general expenses is entered from the General Expanse sheet.
5. Two groups of add-ons are included at the foot of the summary. The first group is calculated as a percentage of the total labor amount of the bid, which is $47,403 in this estimate.

PROJECT: House Example
GROSS FLOOR AREA: 1080 SF
LOCATION:
DATE:
JOB ESTIMATOR: ABF

**ESTIMATE SUMMARY**

| DESCRIPTION | LABOR $ | MATL. $ | EQUIP. $ | SUBS. $ | OTHER $ | TOTAL $ |
|---|---|---|---|---|---|---|
| **OWN WORK:** | | | | | | |
| EXCAVATION & FILL | 2,133 | 1,357 | 2,361 | — | — | 5,851 |
| CONCRETE WORK | 731 | 4,748 | 450 | — | — | 5,929 |
| FORMWORK | 7,072 | 1,341 | — | — | — | 8,413 |
| MISCELLANEOUS | 946 | 490 | 59 | — | — | 1,495 |
| MASONRY | 1,013 | 955 | — | — | — | 1,968 |
| ROUGH CARPENTRY | 7,875 | 8,186 | — | — | — | 16,061 |
| FINISH CARPENTRY | 4,015 | 14,151 | — | — | — | 18,166 |
| EXTERIOR FINISHES | 3,767 | 3,968 | — | — | — | 7,735 |
| INTERIOR FINISHES | 7,965 | 7,527 | — | — | — | 15,492 |
| | | | | | | |
| Subtotal 1: | 35,517 | 42,723 | 2,870 | 0 | 0 | 81,110 |
| **SUBCONTRACTORS:** | | | | | | |
| DEMOLITION | — | — | — | — | — | — |
| LANDSCAPING | — | — | — | — | — | — |
| PILING | — | — | — | — | — | — |
| REINFORCING STEEL | — | — | — | — | — | — |
| PRECAST | — | — | — | — | — | — |
| MASONRY | — | — | — | — | — | — |
| STRUCTURAL AND MISC. STEEL | — | — | — | — | — | — |
| CARPENTRY | — | — | — | — | — | — |
| MILLWORK | — | — | — | — | — | — |
| ROOFING | — | — | — | — | — | — |
| CAULKING AND DAMP-PROOFING | — | — | — | — | — | — |
| DOORS AND FRAMES | — | — | — | — | — | — |
| WINDOWS | — | — | — | — | — | — |
| RESILIENT FLOORING | — | — | — | — | — | — |
| CARPET | — | — | — | — | — | — |
| DRYWALL | — | — | — | — | — | — |
| ACOUSTIC CEILING | — | — | — | — | — | — |
| PAINTING | — | — | — | — | — | — |
| SPECIALTIES | — | — | — | — | — | — |
| PLUMBING | — | — | — | 8,500 | — | 8,500 |
| HEATING AND VENTILATING (Included with Plumbing) | — | — | — | — | — | |
| ELECTRICAL | — | — | — | 7,815 | — | 7,815 |
| CASH ALLOWANCES | — | — | — | — | — | — |
| | | | | | | |
| Subtotal 2: | 35,517 | 42,723 | 2,870 | 16,315 | 0 | 97,425 |
| | | | | | | |
| **GENERAL EXPENSES** | 11,886 | 2,200 | 4,787 | 2,650 | — | 21,523 |
| | | | | | | |
| Subtotal 3: | 47,403 | 44,923 | 7,657 | 18,965 | 0 | 118,948 |
| | | | | | | |
| **ADD-ONS** | | | | | | |
| E.D.P.          NIL | — | — | — | — | — | — |
| SMALL TOOLS    5% | — | — | — | — | 2,370 | 2,370 |
| PAYROLL ADDITIVE   25% | — | — | — | — | 11,851 | 11,851 |
| | | | | | | |
| Subtotal 4: | 47,403 | 44,923 | 7,657 | 18,965 | 14,221 | 133,169 |
| | | | | | | |
| BUILDING PERMIT   $6.00/1,000 | — | — | — | — | 960 | 960 |
| *50% PERF. BOND   NIL | — | — | — | — | — | — |
| *100% PERF. BOND   NIL | — | — | — | — | — | — |
| INSURANCE   $5.00/1,000 | — | — | — | — | 800 | 800 |
| FEE   Lump Sum | — | — | — | — | 25,000 | 25,000 |
| Subtotal 5: | 47,403 | 44,923 | 7,657 | 18,965 | 40,980 | 159,928 |
| ADJUSTMENT | | | | | | — |
| | | | | | BID TOTAL: | 159,928 |

*Choose one bond
Price per S.F. =          $148.08

**Figure 15.10**   House Example—Estimate Summary

6.  The second set of add-ons, except for the fee that is a lump sum, is calculated as a percentage of the bid total. When the bid is being compiled by hand, as in this case, this calculation presents something of a problem as these prices are part of the bid total. This problem can be overcome in two ways. A percentage based on the value of subtotal 4 can be calculated using the method shown in Chapter 14 under the heading "Bid Total Add-Ons." Alternatively, a trial and error method can be employed. An approximate value is used for the bid total and the summary completed. The resulting bid total is compared with the value used to calculate the add-ons. If there is a large error, the process is repeated until the margin of error is within acceptable limits.

7.  The cost per square foot is calculated at the end of the summary sheet by dividing the Bid Total by the gross floor area of the project that is listed at the top of the page. This provides a check on the accuracy of the Bid Total; a square foot cost that appears too high or too low would sound a warning to the estimator.

## Summary Example 2—Office/Warehouse Building

### Subcontractors Takeoff

The list of trades to be subcontracted is prepared at the beginning of the estimate so that, at the time of the bid-closing, the estimator is aware of the complete list of trades for which prices are required. When using Precision Estimating, the required subtrades can be taken off at an early stage in the estimating process. Later, when the low subtrade price is received, it is entered against the trade on the spreadsheet and the name of the subtrade is entered beside the price. For the office/warehouse building, we first have to prepare the database by setting up new subtrade phases and items; then we can proceed to the subtrade takeoff.

Here are the phases needed for this estimate:

| Phase | Description | Unit |
|---|---|---|
| 1100.01 | Cash Allowances | Lsum |
| 2100.01 | Division 2 Subtrades | Lsum |
| 3000.01 | Division 3 Subtrades | Lsum |
| 4000.01 | Division 4 Subtrades | Lsum |
| 5000.01 | Division 5 Subtrades | Lsum |
| 6000.01 | Division 6 Subtrades | Lsum |
| 7000.01 | Division 7 Subtrades | Lsum |
| 8000.01 | Division 8 Subtrades | Lsum |
| 9000.01 | Division 9 Subtrades | Lsum |
| 10000.01 | Division 10 Subtrades | Lsum |
| 15000.01 | Division 15 Subtrades | Lsum |
| 16000.01 | Division 16 Subtrades | Lsum |

Following are the subtrade items and prices for this estimate. Prices can be inserted when the item is set up in the database, or the items can be set up without prices and prices entered onto the spreadsheet later. This second alternative is the usual procedure in practice since subtrade prices are not normally known when these database items are set up.

| Phase | Item | Subtrade | Unit | Categories | Sub Price $ |
|-------|------|----------|------|------------|-------------|
| 1100.01 | Fsub | Finish Hardware Allowance | LSUM | Sub | 3,500 |
| 2100.01 | Ssub | Site Paving | LSUM | Sub | 35,460 |
| 2100.01 | Lsub | Landscaping | LSUM | Sub | 9,627 |
| 2100.01 | Psub | Piling | LSUM | Sub | 9,000 |
| 3000.01 | Rsub | Reinforcing Steel | LSUM | Sub | 14,025 |
| 5000.01 | Bsub | Metal Building | LSUM | Sub | 127,400 |
| 5000.01 | Msub | Miscellaneous Steel | LSUM | Sub, L | 8,000 |
| 6000.01 | Csub | Carpentry | LSUM | Sub | 4,500 |
| 6000.01 | Msub | Millwork & Wood Doors | LSUM | Sub, L | 9,500 |
| 7000.01 | Rsub | Roofing | LSUM | Sub | 17,108 |
| 7000.01 | Csub | Caulking & Damp-proofing | LSUM | Sub | 1,250 |
| 8000.01 | Dsub | Hollow Metal Doors & Frames | LSUM | Sub, L | 4,000 |
| 8000.01 | Osub | Overhead Doors | LSUM | Sub | 10,200 |
| 8000.01 | Wsub | Windows | LSUM | Sub | 32,406 |
| 9000.01 | Rsub | Resilient Flooring | LSUM | Sub | 6,405 |
| 9000.01 | Dsub | Drywall | LSUM | Sub | 19,692 |
| 9000.01 | Asub | Acoustic Ceiling | LSUM | Sub | 10,805 |
| 9000.01 | Psub | Painting | LSUM | Sub | 11,240 |
| 10000.01 | Tsub | Toilet Partitions | LSUM | Sub | 3,240 |
| 10000.01 | WSub | Washroom Accessories | LSUM | Sub, L | 750 |
| 15000.01 | Msub | Plumbing & HVAC | LSUM | Sub | 174,550 |
| 16000.01 | Esub | Electrical | LSUM | Sub | 55,200 |

On the following items there is also an additional charge for labor to install the subtrade-supplied items. These prices are entered into the labor column against each particular subtrade:

| Phase | Item | Subtrade | Labor Price $ |
|-------|------|----------|---------------|
| 5000.01 | Msub | Miscellaneous Steel | 2,400 |
| 6000.01 | Msub | Millwork & Wood Doors | 3,800 |
| 8000.01 | Dsub | Hollow Metal Doors & Frames | 1,040 |
| 10000.01 | WSub | Washroom Accessories | 225 |

Once these new phases and items are in place in the database, we are able to proceed with the subtrade takeoff. Note that the takeoff quantity for each subtrade item is 1.

### Inserting Add-Ons

We now finish the estimate for this project by inserting the project add-ons onto the Precision Estimating Totals Window. As we mentioned in Chapter 14, add-ons can be set up at the beginning of an estimate; this then allows you to open the Totals

Window and view the current total amounts at any stage of the estimate. There are no add-ons set up in the sample database we are using so, click on the Database icon on the menu bar, select Add-ons, then click on Add to set up the following add-ons for this estimate:

| Add-On Number | Description | Cost Basis | Rate |
|---|---|---|---|
| 20 | Small Tools | Category—Labor | 3% |
| 30 | Payroll Additive | Category—Labor | 25% |
| 40 | State Taxes | Category—Materials | 12% |
| 50 | Building Permit | Total—Estimate total | $6.00 per 1000 |
| 60 | 50% Bond | Total—Estimate total | $7.25 per 1000 |
| 70 | Insurance | Total—Estimate total | $6.00 per 1000 |
| 100 | Fee | Lump Sum | — |

Once the add-ons are set up in the database, use the following procedure to insert the add-ons into the estimate:

1. On the Toolbar click on the $\Sigma$ icon.
   - The estimate Totals Window will now open showing the status of totals at this stage of the estimate.
2. In the Totals Window click on the Insert Add-on icon, which is the leftmost icon on this Toolbar.
3. Double click on each add-on to insert it onto the Totals Window.
4. Once you have selected all the add-ons you can close the Insert Add-on window.
   - You should now have a list of the add-ons in the Totals Window; at this stage you can check that the cost basis and rate for each add-on is correct according to requirements.
   - Now we can insert a subtotal before the fee on this list.
5. With the cursor on the Fee add-on, click on the Subtotal icon, which is the third from the left on the Toolbar in the Totals Window.
   - This provides an estimate subtotal before the fee is added; a subtotal here can be useful when assessing the amount of fee to enter.
   - Now enter the fee amount for this project.
6. Place the cursor in the Amount column on the Fee add-on line and enter the amount of fee you wish to include in this estimate, say, $30,000, and then press enter.
7. You accept the changes made to the Totals Window by clicking on the OK icon at the bottom of the window, then click on the Close icon to close the window.
   - The estimate is now complete. We will finish off by printing the estimate report.

### Printing the Estimate Report

Now that we have completed the estimate, Precision Estimating allows us to present it in a number of report formats. Here we will select the Standard Estimate Report, which is a good format for reviewing the finished estimate:

1. On the Menu bar click on Reports.
2. Select Standard Estimate from the list.

3. Click on the Report Options button.
4. Switch off the labor and materials Unit Costs.
   - We just need to view the totals on this report.
   - The other default settings in this window can be left as they are.
   - However, we would like to use a landscape printout so that we can view all the columns on one page.
5. Click on the Page Setup tab.
6. Click on the Landscape spot.
7. Click the Fit To spot and insert 1 in the page(s) wide box.
8. Click the Preview button.
9. Click the Print button to start the printout.

See Figure 15.11 for a copy of this printout.

**Basic Edition**

**Standard Estimate Report**
*Office–Warehouse*

Page 1
7/23/2002 2:41 PM

| Item | Description | Takeoff Qty | | Labor Amount | Material Amount | Subcontract Amount | Name | Equipment Amount | Other Amount | Total Amount |
|---|---|---|---|---|---|---|---|---|---|---|
| **1000.000** | **GENERAL EXPENSES** | | | | | | | | | |
| **1100.010** | **Cash Allowances** | | | | | | | | | |
| Fsub | Finish Hardware Allowance | 1.00 | Lsum | - | - | 3,500 | | - | - | 3,500 |
| | **Cash Allowances** | | | | | **3,500** | | | | **3,500** |
| **1300.010** | **Personnel: Supervision** | | | | | | | | | |
| 15 | Project Manager | 6.00 | Mo. | 36,000 | - | - | | - | - | 36,000 |
| 20 | Site Engineer | 1.00 | Mo. | 4,000 | - | - | | - | - | 4,000 |
| 30 | Safety Coordinator | 6.00 | Mo. | 12,000 | - | - | | - | - | 12,000 |
| 40 | Timekeeper | 6.00 | Mo. | 18,000 | - | - | | - | - | 18,000 |
| 70 | Surveyor | 1.00 | Mo. | - | - | 3,000 | | - | - | 3,000 |
| | **Personnel: Supervision** | | | **70,000** | | **3,000** | | | | **73,000** |
| **1304.010** | **Site layout** | | | | | | | | | |
| 10 | Layout | 1.00 | Mo. | 3,640 | - | - | | - | - | 3,640 |
| | **Site layout** | | | **3,640** | | | | | | **3,640** |
| **1310.010** | **First Aid** | | | | | | | | | |
| 10 | First Aid | 6.00 | Mo. | 600 | 300 | - | | - | - | 900 |
| | **First Aid** | | | **600** | **300** | | | | | **900** |
| **1510.010** | **Utilities: Temporary** | | | | | | | | | |
| 05 | Elect: Hook-up | 1.00 | No. | - | - | - | | - | 950 | 950 |
| 10 | Temp Electricity | 6.00 | mo | - | - | - | | - | 3,000 | 3,000 |
| 20 | Temp Heat | 2.00 | mo | - | - | - | | - | 500 | 500 |
| 30 | Tel: Hook-up | 1.00 | No. | - | - | - | | - | 750 | 750 |
| 40 | Temp Phone | 6.00 | mo | - | - | - | | - | 720 | 720 |
| 50 | Temp. Hoardings | 1.00 | No. | 6,825 | 4,725 | - | | - | - | 11,550 |
| 60 | Temp Water | 6.00 | mo | - | - | - | | - | 360 | 360 |
| 80 | Temp Toilet | 6.00 | mo | - | 1,440 | - | | - | - | 1,440 |
| | **Utilities: Temporary** | | | **6,825** | **6,165** | | | | **6,280** | **19,270** |
| **1520.030** | **Temp: Site Office** | | | | | | | | | |
| 05 | Temp. Office | 6.00 | Mo. | 200 | - | - | | - | 4,500 | 4,500 |
| 10 | Job Sign | 1.00 | ea | - | - | - | | - | 800 | 800 |
| 15 | Office Supplies | 6.00 | Mo. | - | 600 | - | | - | 900 | 900 |
| | **Temp: Site Office** | | | **200** | **600** | | | | **5,400** | **6,200** |
| | | | 9.999 Labor hours | | | | | | | |
| **1536.010** | **Site Security** | | | | | | | | | |
| 10 | Security | 6.00 | Mo. | - | - | 18,000 | | - | - | 18,000 |
| | **Site Security** | | | | | **18,000** | | | | **18,000** |
| **1540.010** | **Temp: Tools & Equipment** | | | | | | | | | |
| 10 | Tools & Equipment | 1.00 | No. | 0 | 0 | - | | 16,456 | - | 16,456 |
| 20 | Saw Setup | 1.00 | No. | 400 | 250 | - | | - | - | 650 |

**Figure 15.11**   Office/Warehouse—Estimate Report

**Basic Edition**

**Standard Estimate Report**
*Office-Warehouse*

7/23/2002 2:41 PM
Page 2

| Item | Description | Takeoff Qty | | Labor Amount | Material Amount | Subcontract Amount | Name | Equipment Amount | Other Amount | Total Amount |
|---|---|---|---|---|---|---|---|---|---|---|
| | Temp: Tools & Equipment | | | 400 | 250 | | | 16,456 | | 17,106 |
| | 20.00  Labor hours | | | | | | | | | |
| 1542.010 | Trucking | | | | | | | | | |
| 10 | Trucking | 6.00 | Mo. | - | - | - | | 4,800 | - | 4,800 |
| | **Trucking** | | | | | | | **4,800** | | **4,800** |
| 1550.010 | Dewatering | | | | | | | | | |
| 10 | Dewatering | 2.00 | Mo. | 200 | - | - | | 400 | - | 600 |
| | **Dewatering** | | | **200** | | | | **400** | | **600** |
| 1600.010 | Photographs | | | | | | | | | |
| 10 | Photographs | 6.00 | Mo. | - | - | - | | | 150 | 150 |
| | **Photographs** | | | | | | | | **150** | **150** |
| 1650.010 | Cleanup | | | | | | | | | |
| 10 | Current Cleanuo | 6.00 | Mo. | 900 | - | - | | 900 | - | 1,800 |
| 20 | Final Cleanup | 1.00 | No. | - | - | 2,700 | | - | - | 2,700 |
| | **Cleanup** | | | **900** | | **2,700** | | **900** | | **4,500** |
| | ***GENERAL EXPENSES*** | | | *82,765* | *7,315* | *27,200* | | *22,556* | *11,830* | *151,666* |
| | 29.999  Labor hours | | | | | | | | | |
| ***2000.000*** | ***SITEWORK*** | | | | | | | | | |
| 2100.010 | Division 2 Subtrades | | | | | | | | | |
| Lsub | Landscaping | 1.00 | Lsum | - | - | 9,627 | | - | - | 9,627 |
| Psub | Piling | 1.00 | Lsum | - | - | 9,000 | | - | - | 9,000 |
| Ssub | Site Paving | 1.00 | Lsum | - | - | 35,460 | | - | - | 35,460 |
| | **Division 2 Subtrades** | | | | | **54,087** | | | | **54,087** |
| 2144.000 | Gravel @ Slab | | | | | | | | | |
| 10 | Gravel under SOG | 402.90 | CY | 2,498 | 11,787 | - | | 1,587 | - | 15,873 |
| | **Gravel @ Slab** | | | **2,498** | **11,787** | | | **1,587** | | **15,873** |
| 2201.000 | Earthwk: Remove Topsoil | | | | | | | | | |
| 10 | Remove Topsoil | 1,087.04 | CY | 652 | - | - | | 1,772 | - | 2,424 |
| | **Earthwk: Remove Topsoil** | | | **652** | | | | **1,772** | | **2,424** |
| 2205.000 | Earthwk: Excav Trench | | | | | | | | | |
| 10 | Excavate Trench | 248.82 | CY | 202 | - | - | | 418 | - | 620 |
| | **Earthwk: Excav Trench** | | | **202** | | | | **418** | | **620** |
| 2210.000 | Earthwk: Excav Pits | | | | | | | | | |
| 10 | Excavate Pits | 13.22 | CY | 21 | - | - | | 26 | - | 47 |
| | **Earthwk: Excav Pits** | | | **21** | | | | **26** | | **47** |
| 2220.000 | Earthwk: Bulk Cut | | | | | | | | | |

**Figure 15.11  continued**

**Basic Edition**

## Standard Estimate Report
### Office-Warehouse

Page 3
7/23/2002 2:41 PM

| Item | Description | Takeoff Qty | | Labor Amount | Material Amount | Subcontract Amount | Name | Equipment Amount | Other Amount | Total Amount |
|---|---|---|---|---|---|---|---|---|---|---|
| **2220.000** | **Earthwk: Bulk Cut** | | | | | | | | | |
| 10 | Bulk Cut | 149.670 | CY | 112 | - | - | | 304 | - | 416 |
| | **Earthwk: Bulk Cut** | | | **112** | | | | **304** | | **416** |
| **2220.400** | **Earthwk: Grade & Trim** | | | | | | | | | |
| 10 | Grade & Trim Bottoms of Excav | 2,944.00 | sf | 1,472 | - | - | | | | 1,472 |
| | **Earthwk: Grade & Trim** | | | **1,472** | | | | | | **1,472** |
| **2221.500** | **Earthwk: Gravel Fill** | | | | | | | | | |
| 10 | Gravel Fill | 504.26 | CY | 1,941 | 13,523 | - | | 2,814 | - | 18,278 |
| | **Earthwk: Gravel Fill** | | | **1,941** | **13,523** | | | **2,814** | | **18,278** |
| **2222.000** | **Earthwk: Dispose Surplus** | | | | | | | | | |
| 10 | Dispose Surplus | 211.00 | CY | 551 | - | - | | 800 | - | 1,350 |
| | **Earthwk: Dispose Surplus** | | | **551** | | | | **800** | | **1,350** |
| **2225.000** | **Backfill Trenches** | | | | | | | | | |
| 10 | Bakfill Trenches | 190.82 | CY | 1,183 | - | - | | 752 | - | 1,935 |
| | **Backfill Trenches** | | | **1,183** | | | | **752** | | **1,935** |
| **2230.000** | **Backfill Pits** | | | | | | | | | |
| 10 | Bakfill Pits | 9.89 | CY | 82 | - | - | | 52 | - | 134 |
| | **Backfill Pits** | | | **82** | | | | **52** | | **134** |
| | *SITEWORK* | | | *8,714* | *25,310* | *54,087* | | *8,524* | *0* | *96,636* |
| **3000.000** | **CONCRETE** | | | | | | | | | |
| **3000.010** | **Division 3 Subtrades** | | | | | | | | | |
| Rsub | Reinforcing Steel | 1.00 | Lsum | - | - | 14,025 | | - | | 14,025 |
| | **Division 3 Subtrades** | | | | | **14,025** | | | | **14,025** |
| **3110.100** | **Forms: Footings** | | | | | | | | | |
| 20 | Pad Footing Forms | 143.200 | sf | 345 | 117 | - | | - | - | 462 |
| 30 | Continuous Footing Forms | 329.41 | sf | 609 | 259 | - | | - | - | 869 |
| 50 | Keyway In Footing | 297.33 | lf | 122 | 77 | - | | - | - | 198 |
| | **Forms: Footings** | | | **1,076** | **453** | | | | | **1,530** |
| | 6.095 Labor hours | | | | | | | | | |
| **3112.000** | **Forms: Pile Caps** | | | | | | | | | |
| 10 | Form Pile Caps | 133.300 | sf | 321 | 109 | - | | - | | 430 |
| | **Forms: Pile Caps** | | | **321** | **109** | | | | | **430** |
| **3114.000** | **Forms: Grade Beams** | | | | | | | | | |
| 10 | Form Grade Beams | 1,815.00 | sf | 3,648 | 1,601 | - | | - | | 5,249 |
| | **Forms: Grade Beams** | | | **3,648** | **1,601** | | | | | **5,249** |
| **3114.500** | **Forms: Voids** | | | | | | | | | |

**Figure 15.11** continued

**Basic Edition**

## Standard Estimate Report
### Office-Warehouse

| Item | Description | Takeoff Qty | | Labor Amount | Material Amount | Subcontract Amount | Name | Equipment Amount | Other Amount | Total Amount |
|------|-------------|-------------|---|-------------|-----------------|--------------------|------|-----------------|--------------|--------------|
| 3114.500 | **Forms: Voids** | | | | | | | | | |
| 10 | 4" x 8" Void Forms | 473.67 | lf | 194 | 622 | - | | - | - | 816 |
| | **Forms: Voids** | | | **194** | **622** | | | | | **816** |
| 3115.000 | **Forms: Walls** | | | | | | | | | |
| 10 | Form Walls | 2,310.00 | sf | 5,313 | 1,649 | - | | - | - | 6,962 |
| | **Forms: Walls** | | | **5,313** | **1,649** | | | | | **6,962** |
| 3119.000 | **Forms: Pilasters** | | | | | | | | | |
| 10 | Form Pilasters | 285.00 | sf | 980 | 254 | - | | | | 1,235 |
| | **Forms: Pilasters** | | | **980** | **254** | | | | | **1,235** |
| 3123.000 | **Forms: Pedestals** | | | | | | | | | |
| 10 | Form Pedestals | 48.00 | sf | 165 | 43 | - | | | | 208 |
| | **Forms: Pedestals** | | | **165** | **43** | | | | | **208** |
| 3127.000 | **Forms: Edgeform Slabs** | | | | | | | | | |
| 10 | Form Edge of Slab-On-Grade | 665.00 | sf | 3,205 | 594 | - | | | | 3,799 |
| | **Forms: Edgeform Slabs** | | | **3,205** | **594** | | | | | **3,799** |
| 3135.000 | **Forms: Edgeform Susp Slab** | | | | | | | | | |
| 10 | Form Edge of Susp Slab | 21.00 | sf | 101 | 19 | - | | | | 120 |
| | **Forms: Edgeform Susp Slab** | | | **101** | **19** | | | | | **120** |
| 3139.000 | **Forms: Curbs** | | | | | | | | | |
| 10 | Form Curbs | 1,384.00 | sf | 3,709 | 1,090 | - | | | | 4,799 |
| | **Forms: Curbs** | | | **3,709** | **1,090** | | | | | **4,799** |
| 3159.000 | **Forms: Strip & Oil** | | | | | | | | | |
| 10 | Strip & Oil: Misc | 7,133.90 | sf | 4,209 | 471 | - | | | | 4,680 |
| | **Forms: Strip & Oil** | | | **4,209** | **471** | | | | | **4,680** |
| 3160.000 | **Screeds** | | | | | | | | | |
| 10 | Slab Screeds | 26,539.00 | sf | 1,062 | 1,115 | - | | | | 2,176 |
| | **Screeds** | | | **1,062** | **1,115** | | | | | **2,176** |
| 3214.000 | **Rebar: Slab-On-Grade** | | | | | | | | | |
| 10 | 6 x 6 x 10/10 WWM | 10,564.00 | sf | 739 | 887 | - | | | | 1,627 |
| 20 | 6 x 6 x 6/6 WWM | 16,454.00 | sf | 1,481 | 2,937 | - | | | | 4,418 |
| | **Rebar: Slab-On-Grade** | | | **2,220** | **3,824** | | | | | **6,045** |
| 3305.000 | **Conc: Pile Caps** | | | | | | | | | |
| c30 | 3000psi Pile Caps | 2.33 | CY | 76 | 226 | - | | 0 | - | 301 |
| | **Conc: Pile Caps** | | | **76** | **226** | | | | | **301** |
| 3306.000 | **Conc: Footings** | | | | | | | | | |
| cc30 | 3000psi Continuous Footings | 17.330 | CY | 225 | 1,678 | - | | 0 | - | 1,903 |
| cp30 | 3000psi Pad Footings | 2.00 | CY | 65 | 194 | - | | 0 | - | 259 |

**Figure 15.11 continued**

**Basic Edition**

## Standard Estimate Report
### Office-Warehouse

Page 5
7/23/2002 2:41 PM

| Item | Description | Takeoff Qty | | Labor Amount | Material Amount | Subcontract Amount | Name | Equipment Amount | Other Amount | Total Amount |
|---|---|---|---|---|---|---|---|---|---|---|
| | **Conc: Footings** | | | 290 | 1,872 | | | | | 2,162 |
| **3307.000** | **Conc: Walls** | | | | | | | | | |
| c30 | 3000psi Walls | 30.96 | CY | 560 | 2,998 | - | | 464 | - | 4,022 |
| | **Conc: Walls** | | | 560 | 2,998 | | | 464 | | 4,022 |
| **3309.000** | **Conc: Pedestals** | | | | | | | | | |
| c30 | 3000psi Pedestals | 0.90 | CY | 29 | 87 | - | | 0 | - | 116 |
| | **Conc: Pedestals** | | | 29 | 87 | | | | | 116 |
| **3310.000** | **Conc: Slab-On-Grade** | | | | | | | | | |
| c30 | 3000psi Slab-On-Grade | 360.85 | CY | 3,266 | 33,451 | - | | 2,706 | - | 39,423 |
| | **Conc: Slab-On-Grade** | | | 3,266 | 33,451 | | | 2,706 | | 39,423 |
| **3310.500** | **Conc: Sidewalks** | | | | | | | | | |
| c30 | 3000psi Sidewalks | 35.222 | CY | 319 | 3,410 | - | | 264 | - | 3,993 |
| | **Conc: Sidewalks** | | | 319 | 3,410 | | | 264 | | 3,993 |
| **3314.000** | **Conc: Stairs** | | | | | | | | | |
| c30 | 3000psi Stairs | 0.333 | CY | 92 | 31 | - | | 0 | - | 122 |
| | **Conc: Stairs** | | | 92 | 31 | | | | | 122 |
| **3318.000** | **Conc: Grade Beams** | | | | | | | | | |
| c30 | 3000psi Grade Beams | 27.07 | CY | 327 | 2,621 | - | | 271 | - | 3,218 |
| | **Conc: Grade Beams** | | | 327 | 2,621 | | | 271 | | 3,218 |
| **3319.000** | **Conc:Curbs** | | | | | | | | | |
| c30 | 3000psi Curbs | 14.888 | CY | 135 | 1,441 | - | | 112 | - | 1,688 |
| | **Conc:Curbs** | | | 135 | 1,441 | | | 112 | | 1,688 |
| **3320.000** | **Conc: Suspended Slabs** | | | | | | | | | |
| c30 | 3000psi Suspended Slabs | 44.44 | CY | 447 | 4,120 | - | | 333 | - | 4,900 |
| | **Conc: Suspended Slabs** | | | 447 | 4,120 | | | 333 | | 4,900 |
| **3350.000** | **Conc: Curing** | | | | | | | | | |
| 10 | Curing Slabs | 26,539.00 | sf | 1,062 | 3,503 | - | | - | - | 4,565 |
| | **Conc: Curing** | | | 1,062 | 3,503 | | | | | 4,565 |
| **3380.000** | **Finish: General** | | | | | | | | | |
| 10 | Trowel Finish | 24,289.00 | sf | 8,015 | - | - | | 1,457 | - | 9,473 |
| 15 | Stair Finish | 71.00 | sf | 23 | - | - | | - | - | 23 |
| 20 | Sidewalk Finish | 2,250.00 | sf | 743 | - | - | | 135 | - | 878 |
| 30 | Non-Metalic Hardner | 14,685.00 | sf | 5,140 | 5,815 | - | | - | - | 10,955 |
| 50 | Finish Curbs | 692.00 | LF | 125 | - | - | | - | - | 125 |
| | **Finish: General** | | | 14,046 | 5,815 | | | 1,592 | | 21,453 |
| **3901.000** | **Misc: Grout** | | | | | | | | | |
| 10 | Grout Base Plates | 2.59 | CF | 224 | 271 | - | | - | - | 495 |

**Figure 15.11 continued**

**Basic Edition**

## Standard Estimate Report
### Office-Warehouse

Page 6
7/23/2002 2:41 PM

| Item | Description | Takeoff Qty | | Labor Amount | Material Amount | Subcontract Amount | Subcontract Name | Equipment Amount | Other Amount | Total Amount |
|---|---|---|---|---|---|---|---|---|---|---|
| | **Misc: Grout** | | | 224 | 271 | | | - | | 495 |
| **3950.000** | **Misc: Anchor Bolts** | | | | | | | | | |
| 10 | Install 3/4" Anchor Bolts | 12.00 | No. | 8 | - | - | | - | | 8 |
| 20 | Install 1" Anchor Bolts | 104.00 | No. | 113 | - | - | | - | | 113 |
| | **Misc: Anchor Bolts** | | | **121** | | | | | | **121** |
| **3955.000** | **Misc: Insulation** | | | | | | | | | |
| R2 | 2" Rigid Insulation | 1,243.00 | sf | 311 | 1,750 | - | | | | 2,061 |
| | **Misc: Insulation** | | | **311** | **1,750** | | | | | **2,061** |
| **3960.000** | **Misc: Expansion Joint** | | | | | | | | | |
| 10 | 1/2" x 6" Expn Joint Filler | 979.00 | LF | 715 | 808 | - | | - | | 1,522 |
| | **Misc: Expansion Joint** | | | **715** | **808** | | | | | **1,522** |
| **3961.000** | **Misc: Sawcut Concrete** | | | | | | | | | |
| 10 | 1/4" x 1 1/2" Sawcuts | 960.00 | LF | 0 | - | 1,680 | | - | | 1,680 |
| | **Misc: Sawcut Concrete** | | | **0** | | **1,680** | | | | **1,680** |
| **3965.000** | **Misc: Bollards** | | | | | | | | | |
| 10 | 6" Dia. Conc Filled Steel Pipe Bollards | 19.00 | No. | 0 | - | 2,945 | | - | | 2,945 |
| | **Misc: Bollards** | | | **0** | | **2,945** | | | | **2,945** |
| | *CONCRETE* | | | *48,224* | *74,246* | *18,650* | | *5,743* | *0* | *146,863* |
| | 6.095   Labor hours | | | | | | | | | |
| **4000.000** | *MASONRY* | | | | | | | | | |
| **4060.100** | **Mortar: All Types** | | | | | | | | | |
| 10 | Mortar Type "N" | 21.74 | cy | - | 1,920 | - | | - | | 1,920 |
| | **Mortar: All Types** | | | | **1,920** | | | | | **1,920** |
| **4070.100** | **Mortar: Grout Fill Conc** | | | | | | | | | |
| 10 | 3000 psi Fill Lintel Blocks | 11.56 | CY | 751 | 1,092 | - | | - | | 1,844 |
| 20 | 3000 psi Core Fill Blocks | 15.04 | CY | 978 | 1,421 | - | | - | | 2,399 |
| | **Mortar: Grout Fill Conc** | | | **1,729** | **2,514** | | | | | **4,243** |
| **4080.100** | **Reinforce: Horizontl Wall** | | | | | | | | | |
| ld 8 | Horiz Wall Reinf 8" Ladder | 7,341.75 | LF | 734 | 1,156 | - | | - | | 1,891 |
| | **Reinforce: Horizontl Wall** | | | **734** | **1,156** | | | | | **1,891** |
| **4101.100** | **Control Joints** | | | | | | | | | |
| 08 | PVC Control Joints in 8" Wall | 336.00 | LF | 302 | 1,058 | - | | - | | 1,361 |
| | **Control Joints** | | | **302** | **1,058** | | | | | **1,361** |
| **4220.110** | **Conc. Block: 8"** | | | | | | | | | |
| lw 1 | Blk 8" Standard Face Lt Wt | 11,013.00 | ea | 19,163 | 34,691 | - | | - | | 53,854 |

**Figure 15.11  continued**

**Basic Edition**

## Standard Estimate Report
### Office-Warehouse

Page 7
7/23/2002 2:41 PM

| Item | Description | Takeoff Qty | | Labor Amount | Material Amount | Subcontract Amount | Name | Equipment Amount | Other Amount | Total Amount |
|---|---|---|---|---|---|---|---|---|---|---|
| | **Conc. Block: 8"** | | | | | | | | | |
| | 958.131  Labor hours | | | 19,163 | 34,691 | | | | | 53,854 |
| 4220.170 | **Conc. Block: 8" Lintel** | 1,115.00 | ea | | | | | | | |
| lw 1 | Lintel 8"  Stand Face Lt Wt | | | 1,673 | 1,405 | - | | - | - | 3,077 |
| | **Conc. Block: 8" Lintel** | | | 1,673 | 1,405 | | | | | 3,077 |
| | 83.625  Labor hours | | | | | | | | | |
| 4250.000 | **Misc: Masonry Scaffold** | | | | | | | | | |
| 10 | Masonry Scaffold | 9,799.00 | SF | 490 | - | - | | 980 | - | 1,470 |
| | **Misc: Masonry Scaffold** | | | 490 | | | | 980 | | 1,470 |
| | **MASONRY** | | | 24,091 | 42,745 | 0 | | 980 | 0 | 67,815 |
| | 1,041.756  Labor hours | | | | | | | | | |
| **5000.000** | **METALS** | | | | | | | | | |
| 5000.010 | **Division 5 Subtrades** | 1.00 | Lsum | | | | | | | |
| Bsub | Metal Building | | | - | - | 127,400 | | - | - | 127,400 |
| Msub | Miscellaneous Steel | 1.00 | Lsum | 2,400 | - | 8,000 | | - | - | 10,400 |
| | **Division 5 Subtrades** | | | 2,400 | | 135,400 | | | | 137,800 |
| | **METALS** | | | 2,400 | 0 | 135,400 | | 0 | 0 | 137,800 |
| **6000.000** | **WOOD & PLASTICS** | | | | | | | | | |
| 6000.010 | **Division 6 Subtrades** | 1.00 | Lsum | | | | | | | |
| Csub | Carpentry | | | - | - | 4,500 | | - | - | 4,500 |
| Msub | Millwork & Wood Doors | 1.00 | Lsum | 3,800 | - | 9,500 | | - | - | 13,300 |
| | **Division 6 Subtrades** | | | 3,800 | | 14,000 | | | | 17,800 |
| | **WOOD & PLASTICS** | | | 3,800 | 0 | 14,000 | | 0 | 0 | 17,800 |
| **7000.000** | **THERMAL & MOISTURE PROT.** | | | | | | | | | |
| 7000.010 | **Division 2 Subtrades** | 1.00 | Lsum | | | | | | | |
| Csub | Caulking & Damp-proofing | | | - | - | 1,250 | | - | - | 1,250 |
| Rsub | Roofing | 1.00 | Lsum | - | - | 17,108 | | - | - | 17,108 |
| | **Division 2 Subtrades** | | | | | 18,358 | | | | 18,358 |
| 7100.000 | **Vapor Barrier** | 21,436.00 | sf | | | | | | | |
| 10 | 6Mil Poly Vapor Barrier | | | 857 | 900 | - | | - | - | 1,758 |
| | **Vapor Barrier** | | | 857 | 900 | | | | | 1,758 |
| 7210.050 | **Insulation: Loose Fill** | 78.56 | CY | | | | | | | |
| v10 | Vermiculite Insulation | | | 3,535 | 3,500 | - | | - | - | 7,035 |

**Figure 15.11** continued

*Basic Edition*

## Standard Estimate Report
### Office-Warehouse

Page 8
7/23/2002 2:41 PM

| Item | Description | Takeoff Qty | | Labor Amount | Material Amount | Subcontract Amount | Subcontract Name | Equipment Amount | Other Amount | Total Amount |
|---|---|---|---|---|---|---|---|---|---|---|
| | Insulation: Loose Fill | | | 3,535 | 3,500 | | | | | 7,035 |
| | **THERMAL & MOISTURE PROT.** | | | *4,393* | *4,400* | *18,358* | | *0* | *0* | *27,151* |
| **8000.000** | **DOORS & WINDOWS** | | | | | | | | | |
| **8000.010** | **Division 8 Subtrades** | | | | | | | | | |
| Dsub | Hollow Metal Doors & Frames | 1.00 | Lsum | 1,040 | - | 4,000 | | - | - | 5,040 |
| Osub | Overhead Doors | 1.00 | Lsum | - | - | 10,200 | | - | - | 10,200 |
| Wsub | Windows | 1.00 | Lsum | - | - | 32,406 | | - | - | 32,406 |
| | Division 8 Subtrades | | | 1,040 | | 46,606 | | | | 47,646 |
| | **DOORS & WINDOWS** | | | *1,040* | *0* | *46,606* | | *0* | *0* | *47,646* |
| **9000.000** | **FINISHES** | | | | | | | | | |
| **9000.010** | **Division 9 Subtrades** | | | | | | | | | |
| Asub | Acoustic Ceiling | 1.00 | Lsum | - | - | 10,805 | | - | - | 10,805 |
| Dsub | Drywall | 1.00 | Lsum | - | - | 19,692 | | - | - | 19,692 |
| Psub | Painting | 1.00 | Lsum | - | - | 11,240 | | - | - | 11,240 |
| Rsub | Resilient Flooring | 1.00 | Lsum | - | - | 6,405 | | - | - | 6,405 |
| | Division 9 Subtrades | | | | | 48,142 | | | | 48,142 |
| | **FINISHES** | | | *0* | *0* | *48,142* | | *0* | *0* | *48,142* |
| **10000.000** | **SPECIALTIES** | | | | | | | | | |
| **10000.010** | **Division 10 Subtrades** | | | | | | | | | |
| Tsub | Toilet Partitions | 1.00 | Lsum | - | - | 3,240 | | - | - | 3,240 |
| Wsub | Washroom Accessories | 1.00 | Lsum | 225 | - | 750 | | - | - | 975 |
| | Division 10 Subtrades | | | 225 | | 3,990 | | | | 4,215 |
| | **SPECIALTIES** | | | *225* | *0* | *3,990* | | *0* | *0* | *4,215* |
| **15000.000** | **MECHANICAL** | | | | | | | | | |
| **15000.010** | **Division 15 Subtrades** | | | | | | | | | |
| Msub | Plumbing & HVAC | 1.00 | Lsum | - | - | 174,550 | | - | - | 174,550 |
| | Division 15 Subtrades | | | | | 174,550 | | | | 174,550 |
| | **MECHANICAL** | | | *0* | *0* | *174,550* | | *0* | *0* | *174,550* |
| **16000.000** | **ELECTRICAL** | | | | | | | | | |
| **16000.010** | **Division 16 Subtrades** | | | | | | | | | |

**Figure 15.11 continued**

**Basic Edition**

## Standard Estimate Report
*Office-Warehouse*

Page 9
7/23/2002 2:41 PM

| Item | Description | Takeoff Qty | Labor Amount | Material Amount | Subcontract Amount | Name | Equipment Amount | Other Amount | Total Amount |
|------|-------------|-------------|--------------|-----------------|--------------------|------|-------------------|--------------|--------------|
| 16000.010 | Division 16 Subtrades | | | | | | | | |
| Esub | Electrical | 1.00  Lsum | - | - | 55,200 | | - | - | 55,200 |
| | Division 16 Subtrades | | | | 55,200 | | | | 55,200 |
| | *ELECTRICAL* | | *0* | *0* | *55,200* | | *0* | *0* | *55,200* |

**Figure 15.11  continued**

*Basic Edition*

**Standard Estimate Report**
*Office-Warehouse*

Page 10
7/23/2002 2:41 PM

**Estimate Totals**

| | | | |
|---|---|---|---|
| Labor | 175,651 | 175,651 | 1,077.850 hrs |
| Material | 154,016 | 154,016 | |
| Subcontract | 596,183 | 596,183 | |
| Equipment | 37,803 | 37,803 | |
| Other | 11,830 | 11,830 | |
| | **975,483** | **975,483** | 975,483 |
| | | | |
| Small Tools | 5,270 | | 3.000 % C |
| Payroll Additive | 43,913 | | 25.000 % C |
| State Taxes | 18,482 | | 12.000 % C |
| Building Permit | 6,559 | | 6.000 $ / 1,000.000 T |
| 50% Bond | 7,925 | | 7.250 $ / 1,000.000 T |
| Insurance | 5,465 | | 5.000 $ / 1,000.000 T |
| | **87,614** | 1,063,097 | |
| | | | |
| Fee | 30,000 | | L |
| **Total** | | **1,093,097** | 45.002 /sf |

**Figure 15.11** continued

## SUMMARY

- This chapter considers the last stages in completing an estimate and submitting a bid. For this process to be concluded successfully, it has to be carefully planned and organized.
- Most contractors use a standard format of summary to capture all the main components of an estimate.
- A construction contract bid is compiled in three stages: the advance stage, the review stage and the closing stage.
- As much of the estimate as possible is finalized in the advance stage; also a plan for the delivery of the bid is formulated here.
- In addition, the contractor has to address the following items in the advance stage:
  1. Completion of bid forms
  2. Calculation and listing of unit prices
  3. Alternative prices (alternates)
  4. Separate prices
  5. Allowances
  6. Bid security—bid bonds and deposits
  7. Bid conditions
- A pre-bid review meeting will usually take place at least 2 workdays before the bid-closing day. This review is conducted for three main reasons:
  1. To ensure that there are no obvious omissions from the bid
  2. To consider ways of improving the competitiveness of the bid
  3. To consider an appropriate fee (bid markup) for the project
- The bid markup has two components: a contribution to cover company overheads and profit that is generally assessed as compensation for the risk taken by the contractor.
- A bid-closing can be a period of hectic activity in a contractor's office with a large number of competing subcontractors submitting bids for often more than 30 different trades.
- The bid-closing needs to be properly staffed and organized to effectively deal with all the fax, e-mail, and telephone bids received over a relatively short time period.
- One of the problems that can arise during the bid closing is the submission by sub-trades of conditional bids.
- Whereas lump-sum bids involve calculating a single final price and assembling a list of subtrades, with a unit-price bid, the contractor is also required to complete a "Schedule of Prices," that is, an itemized breakdown of the work of the project.
- A number of tasks follow the bid submission, including:
  1. Tabulation of bid results
  2. Gathering together all documents
  3. Returning bid drawings and specifications
  4. Storing bid documents
  5. Compiling estimate file
  6. Handling subcontractor inquiries
  7. Conducting the post-bid review
- Summarizing and preparing bids can be done manually or via computer.

## REVIEW QUESTIONS

1. At which point in the bid process are errors most likely to occur? Explain why.
2. What should happen to bid documents in the advance stage of a bid?
3. Why is the planning of the delivery of a bid an important part of the advance stage of the bidding process?
4. What is the usual function of unit prices when they are required to accompany a lump-sum bid?
5. Why do owners frequently call for alternative prices to be quoted with bids for construction work?
6. What are "separate prices" and what is their function in the bid process?
7. If a "cash allowance" for floor finishes was specified in the contract documents of a project, what action would the estimator who is preparing a bid on the project have to take?
8. What is the most common form of bid security for construction bids?
9. What is the purpose of bid security?
10. What are the objectives of, and who should attend, the pre-bid review?
11. Of all the prices that are contained in a contractor's bid, which items hold the most risk for the contractor?
12. Why do some contractors include "bid clarifications" with their bid even when they risk having their bid rejected by doing this?
13. Why is it often difficult for an estimator to prepare a cost-breakdown with the bid?
14. What is the major difference between a lump-sum bid and a unit-price bid?
15. What are "unbalanced" unit-price bids and why do contractors manipulate their bids in this way?
16. What is the purpose of a post-bid review?

# A

# EXTRACT FROM A TYPICAL SOILS REPORT

## Subsurface Conditions

The geological profile at the test hole locations for the depth investigated consists of the following strata in descending order of occurrence:

|  | Thickness |  |
|---|---|---|
| Topsoil | + 200 mm | (8") |
| Silt-Sand | 2.8 m to 4.3 m | (9'2" to 14'1") |
| Clay (till) | 0.3 m to 1.3 m | (1'0" to 4'2") |
| Bedrock (shale) | n.a. |  |

The topsoil at the site is approximately 200 mm (8") thick. No other surficial fill was observed at the ground surface; however, some fill can be expected over the gas pipeline that crosses the southwest corner of the site. The nature or degree of compaction of the trench backfill was not determined.

The uppermost soil at the site is a light brown fine silty sand. The sand is dry to damp with moisture contents generally in the 5–10% range. The upper approximately 1.5 m (5'0") is very silty and in some areas grades into a sandy silt. The results of grain size analyses of samples from the 0.75 m (2'6") depth are summarized on the following pages.

| Test Hole | Sand Content | Silt and Clay Content |
|---|---|---|
| 8 | 50.7% | 49.3% |
| 9 | 50.1% | 4.9% |
| 11 | 50.3% | 49.7% |
| 12 | 66.2% | 33.8% |

A moisture-density relationship for the upper silty sand, included as Plate 8*, indicates that the optimum moisture content is 13.5% and the maximum density is 1940 kg/m³. A CBR test on the same silty sand yielded a CBR value of 9 under soaked conditions.

Thin clay layers were encountered within the silt-sand in the upper 1.5 m (5'0"). The thickest of these clay layers was encountered at test hole 3, where a firm to stiff clay layer was present from 0.8 to 1.5 m (2'7" to 5'0") below the ground surface.

With increasing depth, the sand becomes coarser and cleaner. At depths of approximately 2.5 m to 3.0 m (8'0" to 10'0"), the sand becomes very gravelly and probably contains many cobble-size particles since drilling was very difficult and auger refusal occurred in four of the test holes. The sand stratum is medium dense as determined from the results of the SPT test, which yielded N values ranging from 13 to 32.

Beneath the gravelly layer is a brown, damp to moist, low to medium plastic clay till. Till is a glacial deposit and is characterized by a variety of grain sizes from cobbles to clay. The clay till is hard as determined by pocket penetrometer tests that yielded unconfined compressive strengths of >450 kPa.

Bedrock was encountered in test holes 1, 4, 5 and 6 at depths ranging from 3.3 m to 5.75 m (10'10" to 18'10"). The bedrock is typically a silty shale brown to rusty-brown in color and damp to moist. Unconfined compressive strength was >450 kPa as estimated with a pocket penetrometer.

No ground water was encountered in any of the test holes during drilling. The stand pipe in test hole 4 was dry approximately 24 hours after drilling. It should be noted that ground water levels are subject to wide fluctuations depending on many factors, especially the precipitation and hydrogeology of the area. The conditions at the Big River site are conducive to development of a perched water table at the surface of the bedrock due to the relatively high permeability of the overlying sand strata.

## Comments and Recommendations

Based on the results of the field investigation, the lab testing carried out, and our understanding of the proposed development, the following comments and recommendations are submitted.

### 1. Site Preparation

All topsoil should be stripped from areas of proposed buildings or parking. Fill required to bring the site up to design grade may consist of the natural silty sand or imported low plastic clay or gravel. The native soil is considerably dry of the optimum moisture content for compaction and, consequently, if the silty sand is used for general fill, water will need to be added during compaction. It should be noted that proper compaction of fine silty sand is dependent on proper moisture control. Fill placed for rough grading purposes should be placed in thin lifts and compacted to a minimum of 97% of the maximum Standard Proctor density.

The nature and degree of compaction of the trench backfill over the gas pipeline was undetermined. It is reasonable to assume that some settlement of the trench backfill will occur after development due to the increased soil moisture from lawn watering. A small portion of the access road will possibly be affected by the potential settlement of trench backfill. In this area, the trench backfill should be sub-cut to a depth of 600 mm below existing grade and the fill replaced with compacted pit run gravel.

---

*Plate 8, which is a graph showing moisture-density relationship, is not included here.

## 2. Foundation Systems

*Footings*

Shallow spread footings are the most feasible type of foundation system at this site. Footings based in native undisturbed silty sand may be designed on the basis of an allowable bearing pressure of 150 kPa. Exterior footings should be based at a minimum depth of 1.4 m (4'7") below final grade to ensure adequate frost protection. It should be noted that the native silty sand is very easily disturbed by construction equipment, particularly during wet weather conditions. It is imperative that the bearing surface for footings be protected from disturbance and accumulation of ponded water during construction. Also, if construction takes place during winter, precautions must be taken to ensure that there is no frost penetration beneath the footings.

*Piles*

End-bearing piles based in the hard clay till or bedrock may be considered as an alternative. However, construction difficulties can be expected during installation due to the presence of coarse gravel and cobbles. Casing may be required during installation to prevent sloughing of material into the pile hole. End-bearing piles based at least 6 m below existing grade may be designed on an allowable bearing pressure of 400 kPa. It is imperative that end-bearing piles have the bases inspected and cleaned of any loose or sloughed material prior to pouring of concrete.

No allowance for skin-friction should be made for piles designed as end-bearing.

## 3. Excavations

Excavations for footings and utility trenches will extend into the upper dry to damp silty sand to sandy silt material. In a dry state, the silt-sand will need to be cut back to angles of 1.5H:1V to ensure adequate stability. The silt-sand is also very susceptible to erosion and any soil washed into footing trenches must be removed prior to pouring foundation concrete. Ground water is not expected to be a problem during excavation.

## 4. Slab-on-grade Construction

The native silty sand sub-grade soil beneath the slab-on-grade should be scarified, moisture added and compacted to 97% of Standard Proctor maximum density. The slab should be underlain by a minimum 200 mm thickness of gravel compacted to 97% of its Standard Proctor maximum density. Gravel sizes larger than 50 mm should not be incorporated into under slab material to prevent stress concentrations.

## 5. Site Grading

Final grades should be such that surface water drains away from the building. Since the silt-sand is highly frost susceptible in the presence of free water and freezing temperatures, it is essential that good drainage on the site be maintained. Poorly drained areas that tend to collect surface water may experience considerable frost heave under freezing temperatures.

## 6. Pavement Design

For areas of light traffic, the following pavement section may be used:

    75 mm (3") — asphaltic concrete
    50 mm (2") — 20 mm (1") minus crushed gravel base
    300 mm (12") — 75 mm (3") minus pit run gravel sub-base
        — compacted sub-grade

If areas of concentrated heavy traffic can be defined, such as garbage bin/storage area, the following pavement section should be used:

125 mm (5") — asphaltic concrete
50 mm (2") — 20 mm (1") minus crushed gravel base
350 mm (14") — 75 mm (3") minus pit run gravel sub-base
— compacted sub-grade

Other alternative sections may also be considered. The sub-grade soil should be compacted to 98% of the Standard Proctor maximum density. As mentioned previously, the native silty sand is often difficult to compact due to its sensitivity to moisture content. At low moisture contents, the density may not reach the required level. At higher moisture contents, density can be achieved. However, deflection of the sub-grade is excessive. Proof-rolling of the sub-grade surface should be conducted after compaction to detect areas of soft or wet areas. Any areas so detected should be sub-cut a depth of 300 mm (12") and replaced with pit-run gravel.

If proper density or deflection properties cannot be easily achieved, then it is recommended that the native silt-sand be mixed with pit-run gravel to form the sub-grade surface.

It is recommended that samples of sub-base and base material proposed to be used by the paving contractor be submitted for grain size analyses and approval prior to parking lot construction.

## 7. Foundation Concrete

The soluble sulfate content of the soil ranged from 0.002% to 0.16% indicating the relative degree of sulfate attack on concrete will range from negligible to mild. In view of the well-drained nature of the native soil, it is recommended that Normal Portland Cement (Type I) be used in manufacture of all concrete in contact with the soil.

# B

# LIST OF ITEMS AND CSI-BASED CODES

These items with codes are based on the CSI specification coding system. They correspond to "Phases" in the Precision Estimating Database hierarchy and could serve as the core of a general contractor's estimating database.

| Code | Description | Takeoff Unit | Code | Description | Takeoff Unit |
|---|---|---|---|---|---|
| 1000.000 | GENERAL EXPENSES | | 2221.000 | Earthwk: Bulk Fill | CY |
| 1100.010 | Cash Allowances | Lsum | 2221.500 | Earthwk: Gravel Fill | CY |
| 1300.010 | Personnel: Supervision | Mo. | 2222.000 | Earthwk: Dispose Surplus | CY |
| 1301.010 | Personnel: Foreman | Mo. | 2225.000 | Backfill Trenches | CY |
| 1302.010 | Personnel: Project manager | Mo. | 2226.010 | Sand Bedding | CY |
| 1304.010 | Site Layout | No. | 2227.010 | Drain Gravel | CY |
| 1310.010 | First Aid | No. | 2230.000 | Backfill Pits | CY |
| 1400.010 | Travel & Accommodation | Mo. | 2235.000 | Backfill Basements | CY |
| 1510.010 | Utilities: Temporary | Mo. | 2315.021 | Earthwk: Excav Foot/Misc | CY |
| 1510.020 | Utilities: Final | Mo. | 2315.070 | Backfill: Foot Wall Misc | CY |
| 1520.030 | Temp: Site Office | Mo. | 2364.010 | Piling: Conc Fill Steel | lnft |
| 1530.010 | Site Access | No. | 2366.010 | Piling: Prestressed Conc | lnft |
| 1532.010 | Storage | Mo. | 2367.010 | Piling: Steel H Section | lnft |
| 1536.010 | Site Security | Mo. | 2368.010 | Piling: Steel Pipe | lnft |
| 1537.010 | Site Safety | Mo. | 2368.050 | Piling: Sheet Steel | sf |
| 1540.010 | Temp: Tools & Equipment | Mo. | 2369.010 | Piling: Wood | lnft |
| 1542.010 | Trucking | Mo. | 2370.010 | Piling: Wales & Shores | sf |
| 1550.010 | Dewatering | Mo. | 2380.010 | Piling: Caissons | lnft |
| 1600.010 | Photographs | No. | 2510.010 | Paving: Gravel Base | CY |
| 1630.010 | Testing | No. | 2511.010 | Paving: Sidewalks | sf |
| 1650.010 | Cleanup | Mo. | 2512.010 | Paving Concrete | CY |
| 1690.010 | Allowances | Lsum | 2513.010 | paving: Curb & Gutter | lnft |
| 1700.000 | DEMOLITION | | 2514.010 | Paving: Asphalt Curb | lnft |
| 1700.010 | Demolition Subtrades | Lsum | 2515.010 | Paving: Asphalt | SY |
| 1701.010 | Demolition | CY | 2721.010 | Drainage: Site Catch Basin | No. |
| 1720.030 | Layout | Lsum | 2726.010 | Drainage: Site Manholes | No. |
| 1740.010 | Cleanup | wk | 2730.010 | Drainage: Septic Tanks | No. |
| 2000.000 | SITEWORK | | 2831.010 | Improvmnts: Fencing | lnft |
| 2100.010 | Division 2 Subtrades | Lsum | 2860.010 | Site Equip: PlayingFields | Lsum |
| 2111.010 | Investigate: Soil Testing | Lsum | 2900.010 | Landscape: General | SY |
| 2141.010 | Dewater: General | Lsum | 2921.010 | Landscape: Replace Topsoil | CY |
| 2144.000 | Gravel @ Slab | CY | | CONCRETE | |
| 2201.000 | Earthwk: Remove Topsoil | CY | 3000.010 | Division 3 Subtrades | Lsum |
| 2202.010 | Earthwk: Grade Site | SY | 3110.100 | Forms: Footings | sf |
| 2205.000 | Earthwk: Excav Trench | CY | 3112.000 | Forms: Pile Caps | sf |
| 2210.000 | Earthwk: Excav Pits | CY | 3114.000 | Forms: Grade Beams | sf |
| 2215.000 | Earthwk: Excav Basements | CY | 3114.500 | Forms: Voids | lnft |
| 2220.000 | Earthwk: Bulk cut | CY | 3115.000 | Forms: Walls | sf |
| 2220.400 | Earthwk: Grade & Trim | sf | 3119.000 | Forms: Pilasters | sf |
| 2220.450 | Earthwk: Fine Grade | sf | 3122.010 | Forms: Columns | sf |

| Code | Description | Takeoff Unit | Code | Description | Takeoff Unit |
|---|---|---|---|---|---|
| 3123.000 | Forms: Pedestals | sf | 3950.000 | Misc: Anchor Bolts | No. |
| 3127.000 | Forms: Edgeform Slabs | sf | 3955.000 | Misc: Insulation | sf |
| 3129.010 | Forms: Curb & Gutter | sf | 3960.000 | Misc: Expansion Joint | lnft |
| 3134.010 | Forms: Soffit Susp Slab | sf | 3961.000 | Misc: Sawcut Concrete | lnft |
| 3135.000 | Forms: Edgeform Susp Slab | sf | 3965.000 | Misc: Bollards | No. |
| 3136.010 | Forms: Stairs | sf | 4000.000 | MASONRY | |
| 3138.010 | Forms: Beams | sf | 4000.010 | Division 4 Subtrades | Lsum |
| 3139.000 | Forms Curbs | sf | 4060.100 | Mortar: All Types | CY |
| 3140.010 | Forms: Grade Beams | sf | 4070.100 | Mortar: Grout Fill Conc | CY |
| 3146.010 | Forms: EdgeFm Metal Deck | sf | 4080.100 | Reinforce: Horizontal Wall | mlf |
| 3159.000 | Forms: Strip & Oil | sf | 4101.100 | Control Joints | lnft |
| 3160.000 | Screeds | sf | 4210.100 | Brick: All Types | M |
| 3161.010 | Forms: Notches & Chamfers | lnft | 4220.100 | Conc. Block: 12" | ea |
| 3206.010 | Rebar: Footings | lb | 4220.110 | Conc. Block: 8" | ea |
| 3209.010 | Rebar: Walls | lb | 4220.120 | Conc. Block: 6" | ea |
| 3210.010 | Rebar: Columns | lb | 4220.130 | Conc. Block: 4" | ea |
| 3214.000 | Rebar: Slab-On-Grade | sf | 4220.160 | Conc. Block: 12" Lintel | ea |
| 3300.910 | Misc: Set Grade Pins | lnft | 4220.170 | Conc. Block: 8" Lintel | ea |
| 3305.000 | Conc: Pile Caps | CY | 4220.180 | Conc. Block: 6" Lintel | ea |
| 3306.000 | Conc: Footings | CY | 4220.190 | Conc. Block: Opening Form | ea |
| 3307.000 | Conc: Walls | CY | 4220.196 | Misc: Temp Wall Bracing | ea |
| 3308.010 | Conc: Columns | CY | 4250.000 | Misc: Masonry Scaffold | sf |
| 3309.000 | Conc: Pedestals | CY | 4930.100 | Cleaning: Masonry | sf |
| 3310.000 | Conc: Slab-On-Grade | CY | 4930.200 | Cleaning: Rub Block | sf |
| 3310.500 | Conc: Sidewalks | CY | 5000.000 | METALS | |
| 3314.000 | Conc: Stairs | CY | 5000.010 | Division 5 Subtrades | Lsum |
| 3317.010 | Conc: Beams | CY | 5050.010 | Steel Testing | Lsum |
| 3318.000 | Conc: Grade Beams | CY | 5090.010 | Fastener: Col Anchor Bolt | ea |
| 3319.000 | Conc: Curbs | CY | 5110.010 | Fastener: Framing | ea |
| 3320.000 | Conc: Suspended Slabs | CY | 5500.010 | Misc: Lintels | ea |
| 3325.010 | Conc: Topping | CY | 6000.000 | WOOD & PLASTICS | |
| 3350.000 | Conc: Curing | sf | 6000.010 | Division 6 Subtrades | Lsum |
| 3380.000 | Finish: General | sf | 6090.010 | Fasteners: Frame Anchors | ea |
| 3450.010 | Precast: Structural | CY | 6091.010 | Fasteners: Nails | keg |
| 3451.010 | Precast: Architectural | sf | 6105.010 | Framing: Joists | mbf |
| 3470.010 | Tilt Up: Concrete | CY | 6110.010 | Framing: Plates | mbf |
| 3901.000 | Misc: Grout | cf | 6110.020 | Framing: Plates PT | mbf |
| | | | 6112.010 | Framing: studs 2 × 4 > 2 × 8 | mbf |
| | | | 6113.010 | Framing: Ceiling Joists | mbf |

| Code | Description | Takeoff Unit | Code | Description | Takeoff Unit |
|------|-------------|--------------|------|-------------|--------------|
| 6114.010 | Framing: Roof Rafters | mbf | 9124.010 | GWB Int Frame: Fur Matrl | lnft |
| 6115.010 | Framing: Bridging Sets | No. | 9125.010 | GWB Int Frame: Track | lnft |
| 6117.020 | Blocking: Misc. | mbf | 9126.010 | GWB Int Frame: Misc. | lnft |
| 7000.000 | THERMAL & MOISTURE PROT. | | 9127.010 | GWB: Hanging Labor | sf |
| | | | 9130.010 | GWB: Fasteners | ea |
| 7000.010 | Division 2 Subtrades | Lsum | 9131.010 | GWB: Boards & Sheathing | sf |
| 7100.000 | Vapor Barrier | sf | 9132.010 | GWB: Finish Mud/Tape | sf |
| 7110.010 | Damp-proofing: Cement | sf | 10000.000 | SPECIALTIES | |
| 7210.030 | Insulation: Sound Blankt | sf | 10000.010 | Division 10 Subtrades | Lsum |
| 7210.040 | Insulation: Board | sf | 10160.010 | Toilet Partition Metal | lnft |
| 7210.050 | Insulation: Loose Fill | sf | 10430.010 | Signs | No. |
| 8000.000 | DOORS & WINDOWS | | 10505.010 | Lockers | No. |
| 8000.010 | Division 8 Subtrades | Lsum | 10617.010 | Partitions: Movable Metal | sf |
| 8110.010 | Doors: Steel with Frames | No. | 10800.010 | Misc Toilet/Bath Accessories | No. |
| 8210.010 | Doors: Wood | No. | | | |
| 8510.010 | Windows: Steel | No. | 11000.000 | EQUIPMENT | |
| 8520.010 | Windows: Aluminum | No. | 12000.000 | FURNISHINGS | |
| 8610.010 | Windows: Wood | No. | 13000.000 | SPECIAL CONSTRUCTION | |
| 8710.010 | Hardware: Finishing | Lsum | 14000.000 | CONVEYING SYSTEMS | |
| 9000.000 | FINISHES | | 15000.000 | MECHANICAL | |
| 9000.010 | Division 9 Subtrades | Lsum | 15000.010 | Division 15 Subtrades | Lsum |
| 9120.010 | GWB Int Frame: Labor | lnft | 16000.000 | ELECTRICAL | |
| 9121.010 | GWB Int Frame: S Studs | ea | 16000.010 | Division 16 Subtrades | Lsum |

## A

**addenda**   A change to bid documents issued at the time bids are being prepared.

**add-on**   An estimate item (usually related to general expenses) calculated as a percentage of some part or the total of an estimate.

**assembly**   A component of a project that consists of a collection of takeoff items.

## B

**bank measure**   A volume of excavation or backfill calculated using the actual dimensions of the hole to be excavated or filled with no allowance for swell or compaction of the material.

**bar-chart**   A graph that depicts a project schedule that lists tasks, and by means of a time-line, shows when the tasks will occur and how long they will take.

**bid bond**   An indemnity from a surety that guarantees the contractor named is submitting a bid in good faith; if this contractor should fail to comply with the bid conditions, the surety will be liable to compensate the owner for damages suffered as a consequence.

**bid runner**   A person who hand delivers a bid for a construction project. This person is usually also required to enter the final details onto the bid documents before they are submitted.

**bond beam**   A reinforced concrete beam contained within U-shaped concrete blocks.

**brick-on-edge course**   A course of bricks laid so that the brick ends are visible on the face of a wall and the long dimension of the end is vertical rather than horizontal, which is normally the case.

**brick ties**   Sometimes referred to as "strap anchors" or "wall ties"; these are wire or sheet-metal devices inserted at regular intervals into horizontal masonry (brick or block) joints to attach one masonry wall to another or to attach a masonry wall to another part of the structure.

## C

**cash allowance**   A sum of money included in a bid as an allowance for a certain project expenditure; sometimes called a PC sum.

**compaction factor**   An amount added to the volume of backfill material to account for the additional material required to increase the density of the backfill material when it is in place.

**conceptual estimate**    An estimate prepared when the project is no more than an idea under consideration, when very little specific detail is known about the design of the work involved.

**construction manager**    The person or company charged with responsibility for the on-site activities of a construction project.

**contingency sum**    A sum of money included in a bid as an allowance for nonspecific project expenditures such as the cost of extra work in general.

**cost plan**    A summary of anticipated project expenditures prepared in the early (feasibility) stage of the project and expressed as a collection of budgets, one for each element or major component of the project.

**cost-plus contract**    A form of construction contract in which the main terms allow for the contractor to be paid all the costs incurred to complete the project plus an additional amount for profit.

**CSI MasterFormat**    A master list of numbers and descriptions for organizing information about construction requirements, products and activities in specifications.

**D**

**design-build**    A project delivery method whereby an organization undertakes to both design and construct a project for an owner.

**E**

**extra over**    The cost of providing an additional feature over and above the cost of the standard feature; for example, the extra cost of providing tinted glass rather than plain glass.

**F**

**fast tracking**    Organizing a construction project so that the work begins before the design is complete in order to reduce the overall duration of the project.

**fly forms**    A formwork system designed and built so that it can be moved as a unit, usually by crane, from use in one location to use in another location.

**form ties**    Metal wires or rods used to produce the desired amount of separation between the sheets of form material on each side of a wall formwork system, and also to resist the pressures that result from filling the form system with liquid concrete.

**G**

**gang forms**    See *fly forms*.

**general contractor**    A contractor hired, usually after a bid competition, to complete the work of a project. This contractor assumes responsibility for construction of the entire project although subcontractors may perform some or all of the work involved.

**I**

**instructions to bidders**    Usually found at the beginning of the bid documents, they describe the project in general terms and provide information such as the required time and place of the bid closing. They will usually also include a clause that calls for the bidder to leave the bid open for acceptance for a specified period of time.

**L**

**ladder reinforcement**    A masonry wall reinforcing system that is laid in horizontal joints. It comprises two wires that run the length of the wall and short wires that pass between the two long wires at intervals so that the complete assembly resembles a ladder.

**lump-sum contract**    A contract between an owner and a contractor whereby the contractor agrees to perform a defined scope of work for a stipulated sum. As a consequence, lump-sum contracts are applicable only when the work can be well defined in the contract documents.

**M**

**masonry ties**    See *brick ties*.

**Method of Measurement**    A uniform basis for measuring construction work that provides rules relating to how work is described and measured in the estimating process.

**N**

**net in place**    Quantities of work measured using the dimensions obtained from drawings with no adjustment for factors such as waste.

**P**

**PC sum**    Prime cost sum; see *cash allowance*.

**pilaster**    A column attached to a wall.

**preliminary estimate**    An estimate prepared in the early stages of a project when little detail is known about the specific design of the project.

**Q**

**query list**    A list of questions compiled by an estimator in the takeoff process. These lists are generally sent to the design consultants to obtain clarification in the form of *addenda* to bid documents.

**R**

**recap**    A summary listing of takeoff items and quantities that presents like items together in a prescribed order to facilitate ease of pricing.

**running bond**    A method of laying bricks in which the bricks are laid in courses with their long edge exposed and vertical joints staggered from one course to the next.

**S**

**slump**    A measure of how liquid a mix of concrete is. This characteristic is determined in a standard test carried out with a conical mould 12 inches (300 mm) high. In the test, the mould is first filled with concrete and then the mould is lifted free. The height of the cone of concrete formed is measured to determine the amount by which it is lower than 12 inches (the slump). Thus, a large slump signifies a more fluid mix of concrete.

**soldier course**    A course of bricks laid side-by-side on their ends so that the long dimension of the brick edge exposed is vertical rather than horizontal.

**superplasticizer**    A concrete additive that reduces the amount of water needed for a concrete mix while, at the same time, producing a more fluid concrete that flows more easily.

**swell factor**    The amount by which a volume of excavated material expands after it has been extracted from the ground in the excavation process.

**T**

**topsoil**    The top layer of vegetable soil on a site.

**truss reinforcement**    A masonry wall reinforcing system that is laid in horizontal joints. It comprises two wires that run the length of the wall and a third wire that zigzags between the two straight wires so that the complete assembly resembles a truss.

### U

**unit-price contract**   A contract between an owner and a contractor whereby the contractor agrees to perform a defined scope of work on the basis of so much per unit of work completed. For example, the contractor may offer a price per cubic yard for trench excavation, a price per cubic yard for 3000 psi concrete in footings and so on. Subsequently, the contractor is paid for the agreed quantity of each item completed multiplied by the unit price for that item.

### V

**value analysis**   A systematic and creative method of obtaining the desired function from a component of a construction project for the least expenditure. In the process, the value of a component to the owner or the end-user of a project is assessed and an effort is then made to obtain the desired component at the best price.

### W

**wall ties**   See *brick ties.*

**waterstops**   A feature introduced into a concrete joint to prevent the flow of water through that joint.

**weep holes**   Gaps left in a course of masonry wall usually by omitting mortar from some of the joints in that course to allow water to drain from the inside of the wall.

# D

# THE METRIC SYSTEM AND CONVERSIONS

## SI Units

These units belong to the International System of Units, which is abbreviated SI from the French **Le Système International d'Unités**. The system is constructed from seven base units:

### Base Units

| Quantity | Unit Name | Symbol |
|----------|-----------|--------|
| length | meter | m |
| mass | kilogram | kg |
| time | second | s |
| electric current | ampere | A |
| amount of substance | mole | mol |
| luminous intensity | candela | cd |

### Units Outside the SI That Are Acceptable for Use with the SI

| Unit Name | Symbol | Value in SI Units |
|-----------|--------|-------------------|
| minute (time) | min | 1 min = 60 s |
| hour | h | 1 h $\;$ = 60 min |
| day | d | 1 d $\;$ = 24 h |
| liter | L | 1 L $\;$ = $10^{-3}$ m$^3$ (or 1 m$^3$ = $10^3$ L) |
| metric ton | t | 1 t $\;$ = $10^3$ kg |
| hectare | ha | 1 ha $\;$ = $10^4$ m$^2$ |

**Prefixes**

| Multiplication Factor | Prefix Name | Prefix Symbol |
|---|---|---|
| $1\ 000\ 000\ 000 = 10^9$ | giga | G |
| $1\ 000\ 000 = 10^6$ | mega | M |
| $1\ 000 = 10^3$ | kilo | k |
| $100 = 10^2$ | hecto | h |
| $10 = 10^1$ | deka | da |
| $0.1 = 10^{-1}$ | deci | d |
| $0.01 = 10^{-2}$ | centi | c |
| $0.001 = 10^{-3}$ | milli | m |
| $0.000\ 001 = 10^{-6}$ | micro | μ |
| $0.000\ 000\ 001 = 10^{-9}$ | nano | n |

## Conversion Factors

### Length

| English Units | → | Metric Units |
|---|---|---|
| 1 inch | = | 25.4 mm |
| 1 inch | = | 0.0254 m |
| 1 foot | = | 0.305 m |
| 1 yard | = | 0.914 m |
| 1 mile | = | 1.609 km |

| Metric Units | → | English Units |
|---|---|---|
| 1 mm | = | 0.039 inches |
| 1 m | = | 39.37 inches |
| 1 m | = | 3.28 feet |
| 1 m | = | 1.094 yards |
| 1 km | = | 0.621 miles |

### Area

| English Units | → | Metric Units |
|---|---|---|
| 1 square foot | = | $0.093\ m^2$ |
| 1 square yard | = | $0.836\ m^2$ |
| 1 acre | = | 0.405 hectares |
| 1 square mile | = | $5.590\ km^2$ |

| Metric Units | → | English Units |
|---|---|---|
| $1\ m^2$ | = | 10.764 square feet |
| $1\ m^2$ | = | 1.196 square yards |
| 1 ha | = | 2.471 acres |
| $1\ km^2$ | = | 0.386 square miles |

## Volume

| English Units | → | Metric Units |
|---|---|---|
| 1 cubic foot | = | $0.028 \text{ m}^3$ |
| 1 cubic yard | = | $0.765 \text{ m}^3$ |
| 1 quart (liquid) | = | 0.946 liters |
| 1 gallon | = | 3.785 liters |

| Metric Units | → | English Units |
|---|---|---|
| $1 \text{ m}^3$ | = | 35.315 cubic feet |
| $1 \text{ m}^3$ | = | 1.308 cubic yards |
| 1 liter | = | 1.057 quarts |
| 1 liter | = | 0.264 gallons |

## Mass

| English Units | → | Metric Units |
|---|---|---|
| 1 ounce | = | 28.350 grams |
| 1 pound | = | 0.454 kg |
| 1 ton (2000 lb) | = | 907.185 kg |
| 1 ton (2000 lb) | = | 0.907 metric tons |

| English Units | → | Metric Units |
|---|---|---|
| 1 gram | = | 0.035 ounces |
| 1 kg | = | 2.205 pounds |
| 1 kg | = | 0.001 tons |
| 1 t | = | 1.102 tons |

# INDEX

**A**
Accommodation expenses, 327–328, *329*
Add-on, 38
Air barriers, 144
Air support structures, 331
Allowances, 356–357
Alternative prices, 354–356, *355*
Approximate quantities estimation, 9
Area, 35, *46*
Assembly, 42–43
    estimates, 13
Associations, construction, 21
Average annual investment, 197

**B**
Backfill
    measurement of, 59–62
    pricing, 209–233, *210, 213, 218–219, 224*
Bank measure, 57
Bar-chart schedule, 324, *325*
Basement excavations, 211
Bathroom accessories, 148
Bearing piles, 64
    classification of, 64
    poured-in-place concrete, 65
    precast concrete, 65
    steel, 65
    timber, 64–65
Bid bonds, 310, 357
Bid closing
    advance stage of, 350, 352
    allowances, 356–357
    alternative prices, 354–356, *355*
    bid breakdown, 364
    bid clarifications, 360

bid conditions, 357–358
bid form, 352
bid markup, 359-360
bid security, 357
bid summary sheet, 350, *351*
computer programs for, 364–365
naming subcontractors, 364
post-bid review, 371–373, *372*
pre-bid review, 358–359
separate prices, 356
staffing for, 361–362
subtrade conditions, 362–363
summary: house, 373–375, *374*
summary: office/warehouse, 375–388, *379–388*
telephones and fax machines for price submission, 362, *363*
unit prices, 352–354, *353–354*, 365–371, *366, 367, 369–370*
Bid depositories, 313
Bid runner, 361
Bids, 18, 20–27. *See also* Bid closing
    bid documents, 25, 27
    Bid Record, 25
    bid report, 25, *26*
    closed, 22
    conditional, 305
    decision to, 23–24
    errors in, 304
    information sources for, 18, 20
        construction news services, 22
    marketing to get, 22–23
    open, 20
    prequalified, 22
    scheduling for, 24–25
    shopping for, 313

Bids (*cont.*)
    subtrade, 311–313
    total add-ons, 341
    unbalanced, 371
    unsolicited, 308
Bid summary sheet, 350, *351*
Blocking furring, 145
Board foot (BF), 142
Board measure (BM), 142
Bond beams, 129
Bonding, 24, 309–310
Brick Institute of America (BIA), 127
Brick masonry, 127
    measurement of, 131
Brick-on-edge course, 127
*The Building Estimator's Reference Book* (Walker),
    127, 129, 135
Bulkhead forms, 106, 116

C
Cabinets, 146
Cant strips, 144
Carpentry. *See* Finish carpentry; Rough carpentry
Cash allowances, 148, 356–357
Catering facilities, 330
Centerline perimeter, calculating, 45, *46–52*
Chief estimator, 361
Clay soils, 214
Closed bid, 22
Coil ties, 246
Compaction factor, 57, 220–221
Computer estimating, 29, 32, 44–45, 176
Conceptual estimate, 2, 3–4
Concrete blocks, 127–129, *128*
    measurement of, 131
Concrete work, 99–124
    measurement of, 99–100
        finishes and miscellaneous work, 104–105
        formwork, 102–104
        notes, 100–101
    pricing
        formwork, 245–253
        materials, 241–243
        miscellaneous items, 254, 256, *256–257*
        placing, 236–241
        recap: house, 257–264, *258, 260, 263*
        recap: office/warehouse, 265–269, *265–266*
        reinforcing steel for, 253–254, *255*
        supplying, 236, 243–245
    productivity rates, 241, *242*
Consent of Surety, 357
Construction equipment pricing, 191–208
    company overhead costs, 199
    depreciation, 193–196
    equipment operator costs, 199
    examples, 200–203
    financing expenses, 197

    fuel and lubrication costs, 198
    maintenance and repair costs, 196–197
    renting versus purchasing equipment, 192–193
    spreadsheets for, 203, *204–206*
    taxes, insurance, storage costs, 197–198
Construction management contract, 6–7
*Construction Safety and Health Regulations* (OSHA),
    57–58
Contingency sum, 356
Contraction control joints, 116
Contractor. *See also* Subcontractor
    risks of, 176–177
Contracts, 8–9
    construction management, 6–7
    cost-plus, 9, 177
    lump-sum, 8, 177
    unit-price, 9, 38, 365–371, *366–367, 369–370*
Conversion factors
    masonry work, 129–130
Copings, 144
Cost planning, 4–5
Cost-plus contract, 9, 177
Cost reports, 181–184, *183*
Costs. *See also* General expenses
    allocation of, 193–196
    company overhead, 199
    equipment operator, 199
    equipment transportation, 211
    of financing, 197
    fuel and lubrication, 198
    maintenance and repair, 196–197
    spreadsheets for calculating, 203
    taxes, insurance, storage, 197–198
Counters, 146
CSI Masterformat, 13, 306, 311
Cupboards, 146
Curing slabs, 104
Cut and fill
    calculating with grid method, 59–60, *61*
    calculating with section method, 62, *63*
    equipment used for, 211

D
Depreciation, 193
    declining-balance, 194–195
    production or use, 195–196
    of rubber-tired equipment, 196
    straight-line, 193–194
Descriptions, take-off items, 40–41
Design-build project delivery system, 7–8
Detailed estimating
    stages of, 13–14
Dewatering, 337
Digitizers, 58
Dimensions, take-off items, 39–41
Doors, 146
Double handling (materials), 180

**E**

Efficiency of operations, 38
Electricity, 330
Electronic data processing (E.D.P.), 339
Employer's liability insurance, 340
English measurement system, 36
Equipment
    pricing
        cost reports for, 181–184, *183*
        strategy for, 184
    productivity of, 179–181
    rental rates, 179
    on site, 334–335
Equipment overhead rate, 198
Estimate—cost control cycle, 373
Estimates. *See also* Estimating process; Pricing
    assembly method, 13
    conceptual, 3–4
    and construction safety, 14–15
    and contract types, 8–9
    detailed, 13–14
    formulas for calculating, 45, *46–52*
    functions of, 2–3
    preliminary, 4–5
        approximate quantities method, 9
        price per unit area method, 12
        price per unit method, 12
        price per unit volume method, 12–13
Estimating process, 17–33. *See also* Bids
    bidding, 18, 20–27
    computer systems for, 29, 32, 44–45, 176
    flow chart of, 18, *19*
    query list, 27–28
    site visit, 29, *30–31*
    team approach to, 28
    technological advances in, 32
Excavation
    equipment and methods, 209–212, *210*
    measurement of, 59–62
    pricing, 209–212
        materials, 220–223
        productivity, 212–216
        recap: house, 223, *224*, 225–227
        recap: office/warehouse, 227–231, *228*
        work crews, 216–219
Expansion joints, 105
Expenses. *See* General expenses
Exterior finishes, 149
    pricing, 281, *283*
        recap: house, 293–298, *294–295*
    take-off, 166–173, *167–172*
Extra over, 127

**F**

Fascias, 144
Fast-tracking (project), 6–7
Fax machine, 28, 32, 310, 362

Feasible project
    definition of, 4
Fee markup, 360
Fiber reinforced concrete (FRC), 240
Financing expenses, 197, 339
Finish carpentry
    measuring, 145
        notes, 145–146
    pricing, 281, *282*
        recap: house, 289–293, *290–292*
    take-off, 165–166
Finish hardware, 148–149
Fire protection, temporary, 338
Firm price, 241
First aid, 327
Fly forms, 245, 246
F.O.B. (free on board), 185
Form hardware, 252
Formulas, 45, *46–48*, 48–49, *50–52*
Formwork
    measurement of, 102
        notes, 102–103, *103–104*
    pricing materials, 248–253, *249*, *250*
    productivity in, 245–247, *247*
Foundation concrete, 394
Foundation systems, 393
Frames, 146
Framing work, 144
Fringe benefits, 340

**G**

Gang forms, 245, 246
General contractors, 5
General expenses
    bid total add-ons, 341
    dewatering, 337
    General Expense Sheet, 321–323, *322*, *324*
    hoardings and temporary enclosures, 331, *332*
    labor add-ons, 339–341
    miscellaneous expenses, 337–339
    pricing notes: house, 343, *343*
    pricing notes: office/warehouse building, 344–347,
        *344–345*, *347*
    project schedule, 323–325, *325*
    safety and first aid, 326–327
    site access and storage space, 333–334
    site cleanup, 337
    site equipment, 334–335, *335*
    site personnel, 325, *326*
    site security, 334
    temporary heating, 331, 333, *333*
    temporary site offices, 328, 330
    temporary site services, 330
    travel and accommodations, 327–328, *329*
    trucking, 337
Glass and glazing
    scope of work, 317

Going rate, 274
Gravel supply price, calculating, 222–223
Grid method, excavation calculations by, 59–60, *61*

**H**
Hand excavation, 212
Headings, 40
Heating, temporary, 331, 333, *333*
Hoardings and temporary enclosures, 331, *332*

**I**
Inserts, 105
Insurance, 197–198, 340
Interior finishes, 149
  pricing, 281, *283*
    recap: house, 293–298, *296–297*
  take-off, 173

**J**
Joists, 144

**L**
Labor
  and management factors, 181, 216
  pricing
    with cost reports, 181–184, *183*
    strategy for, 184
    wage rates, 178
  productivity of, 179–181
Labor add-ons, 339–341
Ladder reinforcement, 126
Large projects, 42, 180–181, 215
Layout, 338
Length, 37
Lump-sum contract, 8, 177, 365

**M**
Maintenance and repair costs, 196–197
Manufactured beams, 144
Marketing, 22–23
Masonry work, 125–140
  measurement of, 126–127
    brick, 127, 131
    concrete blocks, 127–129, *128*, 131
    conversion factors, 129–130
    house take-off, 131–134, *132–134*
    notes, 130–131
    office/warehouse building take-off, 134–137, *135–136*
  pricing, 274–276, *275*
    recap: house, 284, *285*
    recap: office/warehouse, 298–300, *299*
  scope of, 314–315
  wind bracing, 338
Materials
  in concrete operations, 241, 243
  in excavation, 220–223
  moisture conditions of, 213–214

pricing, 184–186
  swell and compaction factors, 57, 220–221
Materials testing, 338
McGraw-Hill Dodge Reports, 22
Measurement
  accuracy of, 41–42
  of carpentry, 141–145, 145–146
  of concrete work, 99–105
  of excavation and backfill, 59–62
  of masonry work, 125–137
  of piling, 62, 64–66
  of sitework, 55–56
  units of, 36–37
Measurement and payment contract. *See* Unit price contract
Metals, miscellaneous, 147–148
Method of measurement, 36
Metric system, 36, 142–143, 403–405
Millwork, 145
  pricing, 281, *282*
Miscellaneous concrete items, 254, 256, *256–257*
Miscellaneous expenses, 337–339
Miscellaneous metals, 147–148
Municipal charges, 338–339

**N**
Net in place, 36, 37–39
News services, construction, 22
Nominated subcontractor, 308
Non-shrink grout, 105
Nonunion labor, 178, 328
Number, 37

**O**
Open bid, 20
OSHA (Occupational Safety and Health Act), 14, 57–58, 327
Overtime premium, 340
Owner-supplied materials, 186

**P**
Paneling, 146
Paper take-off, 44
Pavement design, 393–394
Payroll additive, 340–341
Payroll taxes, 339
Perimeter, 45
  of centerline, *46–52*
Photographs, 338
Pilasters, 102
Piling
  bearing, 64–65
    classification of, 64
    measurement of, 62, 64–66
  sheet, 65
    extraction of, 66
Pit and sump excavations, 212

Placing concrete
    amount of rebar in forms, 241
    method of, 236–240
        properties of concrete to be placed, 240
        size and shape of structures, 240–241
Positive action programs, 305
Post-bid review, 371–373, *372*
Poured-in-place concrete piles, 65
Pre-bid proposals, 310, *311*
Pre-bid review, 358–359
    bid clarifications, 360
    bid markup, 359–360
Precast concrete piles, 65
Preliminary estimate, 4–5
Prequalified bid, 22
Pre-Solicitation Notice, 20
Price per unit area estimation, 12
Price per unit estimation, 12
Price per unit volume estimation, 12–13
Pricing
    concrete work, 235–271
    construction equipment, 191–208
    excavation and backfill, 209–231
    exterior and interior finishes, 281, *283*, 293–298,
        *294–297*
    finish carpentry and millwork, 281, *282*, 289–293,
        *290–292*
    general expenses. *See* General expenses
    generally, 175–189
        contractor's risk, 176–177
        labor and equipment, 177–184, *183*
        materials, 184–186
        subcontractor's work, 186–187
    masonry work, 274–276, *275*, 284, *285*, 298–300,
        *299*
    rough carpentry, 276–281, *278*, 286–289,
        *287–288*
    sheet, *44*
    subcontractor's work, 303–309, *306*
Prime cost sums, 356
Productivity
    in concrete operations, 236–241, *242*
        formwork, 245–253, *247*
    in excavation, 212–217, *213*, *218–219*
    in exterior and interior finishes, 281, *283*
    in finish carpentry and millwork, 281, *282*
    of labor and equipment, 179–184
Profit, 360
Project
    schedule of, 323–325, *325*
    size of, 214–215
Project delivery systems, 5–8
Project estimator, 361
Project signs, 338
Property damage insurance, 340
Proposal, *See* Bids
Public liability insurance, 340

Q
Qualification statement, 22
Quantity sheet, *43*
Quantity take-off
    by assembly, 42–43
    carpentry/miscellaneous work example, 149,
        *150–162*, 163–166, *164–165*
    by computer, 44
    concrete work example, 105–116, *106–107*,
        *109–115*
    definition of, 35–36
    exterior/interior finishes example, 166–173,
        *167–172*
    masonry example, 131–134, *132–134*, 136–137,
        *135–136*
    measurement of, 36
        accuracy of, 41–42
    recapping items, 175–176
    risk of error in, 177
    sitework example, 80–88, *81–83*, *85–88*
    take-off rules, 39–41
Query list, 27–28

R
Rafters, 144
Rebar, 241
Recap breakdown, 44, 175–176
Reinforcing steel, 105
    installation productivity, 253–254, *255*
Rental equipment, 192–193
Rental rates, 179
Repetitive work, 180–181
Risk, 360
    of contractor, 176–177
Rough carpentry. *See also* Finish carpentry
    board measure, 142
    measuring, 141–142
        notes, 143–145
    pricing, 276–281, *278*
        recap: house, 286–289, *287–288*
    take-off
        floor system, 149, *150–162*
        roof system, 163–165, *164–165*
        wall system, 149, 163
Rough hardware, 145, 281–282
Rubber-tired equipment, 196
Running bond, 129

S
Safety
    costs of lack of, 14–15
    at excavation site, 57–58
    at project site, 326–327
Sales tax, 339
Saw cutting slabs, 105
Saw setup, 338
Schedule of Prices, 365, *366*

Scope of work, 313–314
  glass and glazing, 317
  masonry, 314–315
  windows, 315–316
Screeds, 104, 254
Section method, excavation calculations by, 62, *63*
Security systems, 334
Separate prices, 356
Sheathing, 144
She-bolt ties, 246
Sheet piling, 65–66
Shelving, 146
Shoring frames, 252
Side notes, 40
Sidings, 144
Site
  access to, 180, 214, 333–334
  cleanup of, 337
  cut and fill operations, 211
  equipment on, 334–335, *335*
  grading, 392
  personnel, cost of, 325, *326*
  safety and first aid requirements at, 326–327
  security of, *335*
  visit to, 29, *30–31*
Sitework
  measurement of, 55–56
  takeoff, 80–95
Slab construction joints, 116
Slab finishes, 104
Slab-on-grade construction, 393
Slump concrete mixes, 240
Small tools and consumables, 339–340
Snap ties, 246
Snow removal, 338
Social Security tax, 340
Soffits, 144
Software, 354–365
Soils report, 56–57
  excavations, 393
  foundation concrete, 394
  foundation systems, 393
  pavement design, 393–394
  site grading, 393
  site preparation, 392
  slab-on-grade construction, 393
  subsurface conditions, 391–392
Soldier course, 127
Sonic digitizer, 58
Southam Building Reports, 22
Space heaters, portable, 331
Specialties, 148
Spreadsheets, use of, 203
Staffing, for closing, 361–362
Stairs, 146
Stationery, 43–44, *45*

Steel piles, 65
Storage
  costs, 180, 197–198
  space, 333–334
Subcontractor
  analyzing subtrade bids, 311–313, *312*
  bid depositories, 313
  bonding, 309–310
  conditions imposed by, 362–363
  evaluating, 308–309
  list of subtrades, 305–306, *307*
  naming, 364
  pre-bid proposals, 310, *311*
  pricing work of, 186–187
    problems with, 304–305
  scope of work, 313–317
  unknown, 306, 308
Subfloors, 144
Subsurface conditions, 391–392
Subtrade estimating area, 25
Subtrade list, 305–306, *307*
Superplasticizer additives, 240
Surety bonds, 309–310
Swell factor, 57, 220–221

**T**
Tablet digitizer, 58
Take-off. *See* Quantity take-off
Taxes, 197–198, 339, 340
Team approach, 28
Technology, and estimating process, 32
Telephone bid, 362, *363*
Telephone service, 330
Temporary site
  enclosures, 331, *332*
  heating, 331, 333, *333*
  office accommodations, 328, 330
  services, 330
Tender. *See* Bids
Ties
  brick, 131
  form, 246
Timber piles, 64–65
Tower crane, 335–337, *336*
Trade definitions, 313
Traditional (design-bid-build) project delivery system,
  5–6
Traffic control, 338
Travel expenses, 327–328, *329*
Trench excavations, 211
Trim, 146
Trucking requirements, 221–222, 337
Trusses, 144
Truss joists, 144
Truss rafters, 144
Truss reinforcement, 130

**U**

Unbalanced bids, 371
Underlay, 144
Unemployment Compensation tax, 340
Uniform sales tax, 339
Unionized labor, 178, 328
Unit-price contract, 9, 38, 352–354, *353–354*
  bid form for, 368, *370*, 371
  recap format, 365, 367–368, *367–368*
  Schedule of Prices, 365, *366*
  summary, 368, *369*
  unbalanced bids, 371

**V**

Value-added tax, 339
Value analysis, 5
Vapor barriers, 144
Volume, 34, *47*
  of trench excavation, 45

**W**

Wage increases, 340
Wage rates, 178, 256, 281
Warranty, 338
Waste factors, 243
Water leakage tests, 338
Waterstops, 105
Weather conditions, 180, 214
Weep holes, 131
Weight, 37
Welded wire mesh, 104
Wind bracing masonry, 338
Windows, 146–147
  scope of work, 315–316
Worker's Compensation, 340

**Sage Timberline Office Estimating Basic, by Sage Software**

Save $250 on the purchase of Sage Timberline Office's advanced estimating software packages by Sage Software. (See reverse side for details.)

Sage Timberline Office's Estimating Basic software that comes with this book provides the essentials you need to create a construction estimate. You start with an easy-to-use spreadsheet-like user interface that is backed by a "price book" database, which you can easily customize to include your own unique formulas, productivity and conversions.

Pull-down menus give you instant access to everything you need to prepare an estimate, whether you use items from another estimate, do a takeoff, modify your pricing database, or print a report. Point-and-click, drag-and-drop, and other easy-to-use features are all a part of the estimating suite in Sage Timberline Office.

**How to install this software**

You must install the software using the Setup command. Please note: These instructions are based on the use of Microsoft® Windows XP Professional. Your installation may vary slightly depending on the system you are using.

1. If you use Microsoft Windows NT Workstation 4.0, Windows 2000 Professional, or Windows XP Professional, log on as the administrator (not as an operator with administration rights).
2. If you use anti-virus software, turn it off for the duration of the installation. Shut down other unnecessary programs or services for the duration of the installation.
3. Insert the CD-ROM into your CD drive.
4. From the Windows Start menu, select Run. In the Open box, type d:\Setup.exe (if d: is the letter assigned to your CD drive.)
5. In the Welcome window, click Next.
6. After you read the end user license agreement, click Accept.
7. To install the Pervasive Work Group Engine, click Yes. After installation, click Continue to restart your computer. Log in again, if necessary. Estimating Basic installation will resume.
8. Click Next to accept the default destination folder, or click Browse to select a different folder.
9. In the Select Components window, select Estimating--Basic and Sample Database-- Standard. Click Next. In the Select Program Folder window, accept the default choices and click Next.
10. In the Start Copying Files window, review the list of components and destination folders and click Next to begin the installation.
11. Click yes if you would like to view the release notice now. Click No to continue.
12. Restart your anti-virus software and any other programs that you closed in step 1.

For technical and installation support, call Thomson/Delmar Learning, 1-800-477-3692, Monday through Friday, 8:30 a.m. to 5:30 p.m. EST. You may also e-mail a request to help@delmar.com.

Now, you can take your estimating to the next level with Sage Timberline Office's advanced estimating solutions—and save $250. (See below for details).

**Estimating—Standard Edition**

With powerful takeoff, analysis and reporting tools, Estimating Standard offers everything you need to increase your estimating productivity. Features include: assembly takeoff and digitizer capabilities; additional spreadsheet sequences; price update methods and estimate reports; optional interfaces to CAD, scheduling and ODBC-compliant software. An extensive line of database solutions, including R.S. Means, is also available.

**Estimating—Extended Edition**

For estimators who need a little more, Sage Timberline Office offers Estimating Extended. As Sage Timberline Office's most comprehensive package of cutting-edge estimating tools, Estimating Extended builds on the power of Estimating Standard and offers even greater takeoff and analysis capabilities. Advanced features include enhanced assembly takeoff, crew capabilities, enhanced work breakdown structure (WBS) coding, enhanced reporting, unit conversion, and more.

To ensure your $250 discount toward your purchase of Estimating Standard or Estimating Extended, please complete this form and send this page to the address below. (Photocopies will not be accepted.) A Sage Software representative will contact you. Offer expires December 31, 2007.

Name: _____

Address: _____

City: _____  State: _____  ZIP: _____

Phone: _____

e-mail: _____

Send to:

Sage Software / Sage Timberline Office
Attn: Market Development/Dept. PEO
15195 NW Greenbrier Parkway
Beaverton, OR 97006

I would like to learn more about:

_____Estimating Standard      _____ Estimating Extended      _____ Other Estimating modules

For more information about purchasing Sage Timberline Office software products, you may also contact us at 1-800-628-6583, productinfo.timberline@sage.com, or www.sagetimberlineoffice.com.

Sage Timberline Office Estimating is part of a integrated suite of financial and operations software for professionals in construction and real estate.